ANALYSIS OF AIR POLLUTANTS
Peter O. Warner

ENVIRONMENTAL INDICES
Herbert Inhaber

URBAN COSTS OF CLIMATE MODIFICATION
Terry A. Ferrar, Editor

CHEMICAL CONTROL OF INSECT BEHAVIOR: THEORY AND APPLICATION
H. H. Shorey and John J. McKelvey, Jr., Editors

MERCURY CONTAMINATION: A HUMAN TRAGEDY
Patricia A. D'Itri and Frank M. D'Itri

POLLUTANTS AND HIGH RISK GROUPS
Edward J. Calabrese

METHODOLOGICAL APPROACHES TO DERIVING ENVIRONMENTAL AND OCCUPATIONAL HEALTH STANDARDS
Edward J. Calabrese

NUTRITION AND ENVIRONMENTAL HEALTH—Volume I: The Vitamins
Edward J. Calabrese

NUTRITION AND ENVIRONMENTAL HEALTH—Volume II: Minerals and Macronutrients
Edward J. Calabrese

SULFUR IN THE ENVIRONMENT, Parts I and II
Jerome O. Nriagu, Editor

COPPER IN THE ENVIRONMENT, Parts I and II
Jerome O. Nriagu, Editor

ZINC IN THE ENVIRONMENT, Parts I and II
Jerome O. Nriagu, Editor

CADMIUM IN THE ENVIRONMENT, Parts I and II
Jerome O. Nriagu, Editor

NICKEL IN THE ENVIRONMENT
Jerome O. Nriagu, Editor

ENERGY UTILIZATION AND ENVIRONMENTAL HEALTH
Richard A. Wadden, Editor

FOOD, CLIMATE AND MAN
Margaret R. Biswas and Asit K. Biswas, Editors

continued on back

Hutchins Library
of
Berea College

Berea, Kentucky

PLANT DISEASE CONTROL: RESISTANCE AND SUSCEPTIBILITY

PLANT DISEASE CONTROL

Resistance and Susceptibility

Edited by

RICHARD C. STAPLES

Boyce Thompson Institute
Cornell University

GARY H. TOENNIESSEN

The Rockefeller Foundation

A WILEY-INTERSCIENCE PUBLICATION
JOHN WILEY & SONS
New York • Chichester • Brisbane • Toronto

Copyright © 1981 by John Wiley & Sons, Inc.

All rights reserved. Published simultaneously in Canada.

Reproduction or translation of any part of this work beyond that permitted by Sections 107 or 108 of the 1976 United States Copyright Act without the permission of the copyright owner is unlawful. Requests for permission or further information should be addressed to the Permissions Department, John Wiley & Sons, Inc.

Library of Congress Cataloging in Publication Data:

Main entry under title:

Plant disease control.

(Environmental science and technology ISSN 0194-0287)
Papers drawn from a conference held at the Rockefeller Foundation's Bellagio Study and Conference Center, August 21–25, 1979.
"A Wiley-Interscience publication."
Includes index.
1. Plant diseases—Congresses. 2. Plants—Disease and pest resistance—Congresses. I. Staples, Richard C. II. Toenniessen, Gary H. III. Bellagio Study and Conference Center.

SB727.P5 632′.3 80-22923
ISBN 0-471-08196-5

Printed in the United States of America

10 9 8 7 6 5 4 3 2 1

632.3
P713

CONTRIBUTORS

Ivan W. Buddenhagen
International Institute for Tropical Agriculture
Oyo Road, PMB 5320
Ibadan, Nigeria

Richard M. Cooper
School of Biological Sciences
Claverton Down
University of Bath
Bath BA2 7AY, England

Elizabeth D. Earle
Department of Plant Breeding and Biometry
Cornell University
Ithaca, New York 14853

Vernon E. Gracen
Department of Plant Breeding and Biometry
Cornell University
Ithaca, New York 14853

Michèle C. Heath
Department of Botany
University of Toronto
Toronto, Ontario M5S 1A1, Canada

Noel T. Keen
Department of Plant Pathology
University of California
Riverside, California 92521

Joseph Kuć
Department of Plant Pathology
University of Kentucky
Lexington, Kentucky 40506

Alberto Matta
Instituto di Patologia Vegetale
Via Pietro Giuria, 15
10126 Torino, Italy

Harry Mussell
Boyce Thompson Institute for Plant Research
Tower Road, Cornell University
Ithaca, New York 14853

Hachiro Oku
Laboratory of Plant Pathology
College of Agriculture
Okayama University
Tsushima, Okayama 700, Japan

Seiji Ouchi
Laboratory of Plant Pathology
College of Agriculture
Okayama University
Tsushima, Okayama 700, Japan

Jack Paxton
Department of Plant Pathology
University of Illinois
Urbana, Illinois

George F. Pegg
Department of Agriculture and Horticulture
University of Reading, Earley Gate
Reading RG6 2AT, England

James E. Rahe
Department of Biological Sciences
Simon Fraser University
Burnaby, British Columbia V5A 1S6, Canada

Raoul A. Robinson
Trinity Hall
2 Balmoral Terrace
St. Helier, Jersey
C.I., United Kingdom

290278

Luis Sequeira
Department of Plant Pathology
University of Wisconsin
Madison, Wisconsin 53706

Bjørn Sölheim
Institute of Biology and Geology
University of Tromsö
P.O. Box 790
N-9001 Tromsö, Norway

Richard C. Staples
Boyce Thompson Institute for Plant Research
Tower Road, Cornell University
Ithaca, New York 14853

Gary H. Toenniessen
The Rockefeller Foundation
1133 Avenue of the Americas
New York, New York 10036

Edward J. Trione
Department of Botany and Plant Pathology
Oregon State University
Corvallis, Oregon 97331

Willard K. Wynn
Plant Pathology Department
University of Georgia
Athens, Georgia 30602

Olin C. Yoder
Department of Plant Pathology
Cornell University
Ithaca, New York 14853

To Elvin Charles Stakman (1885–1979) who devoted his life to increasing basic food crop production through advances in the sciences of plant protection.

SERIES PREFACE

Environmental Science and Technology

The Environmental Science and Technology Series of Monographs, Textbooks, and Advances is devoted to the study of the quality of the environment and to the technology of its conservation. Environmental science therefore relates to the chemical, physical, and biological changes in the environment through contamination or modification, to the physical nature and biological behavior of air, water, soil, food, and waste as they are affected by man's agricultural, industrial, and social activities, and to the application of science and technology to the control and improvement of environmental quality.

The deterioration of environmental quality, which began when man first collected into villages and utilized fire, has existed as a serious problem under the ever-increasing impacts of exponentially increasing population and of industrializing society. Environmental contamination of air, water, soil, and food has become a threat to the continued existence of many plant and animal communities of the ecosystem and may ultimately threaten the very survival of the human race.

It seems clear that if we are to preserve for future generations some semblance of the biological order of the world of the past and hope to improve on the deteriorating standards of urban public health, environmental science and technology must quickly come to play a dominant role in designing our social and industrial structure for tomorrow. Scientifically rigorous criteria of environmental quality must be developed. Based in part on these criteria, realistic standards must be established and our technological progress must be tailored to meet them. It is obvious that civilization will continue to require increasing amounts of fuel, transportation, industrial chemicals, fertilizers, pesticides, and countless other products; and that it will continue to produce waste products of all descriptions. What is urgently needed is a total systems approach to modern civilization through which the pooled talents of scientists and engineers, in cooperation with social scientists and the medical profession, can be focused on the development of order and equilibrium in the presently disparate segments of the human environment. Most of the skills and tools that are needed are already in existence. We surely have a right to hope a technology that has created such manifold environmental problems is also capable of solving them. It is our hope that this Series in Environmental Sciences and Technology will not only serve to make this challenge more explicit to the established professionals, but that it also will help to stimulate the student toward the career opportunities in this vital area.

Robert L. Metcalf
Werner Stumm

PREFACE

The total production and yields of the world's principal food crops have increased substantially over the past 50 years. Yet the predominant food situation in the developing countries is one of inadequate total supply, rising food imports, and chronic malnutrition. The International Food Policy Research Institute calculates that, if past production and consumption trends continue, by 1990 the developing countries will face a deficit in staple food crops of 145 million tons. Past accomplishments have been significant, but even greater increases in production are needed through the remainder of this century if adequate food for all people is to be provided.

The Rockefeller Foundation and the Boyce Thompson Institute for Plant Research are committed to supporting and conducting research that can provide the knowledge and technologies necessary for applied scientists and farmers to improve yields and to meet the world's future food needs. Preparation of this book and organization of the conference that led to it are representative of several cooperative efforts between these two institutions toward this objective. The substantive content of this volume, however, draws on the combined expertise of 22 scientists from eight countries. They represent a wide variety of disciplinary backgrounds, but all are concerned with the development of improved strategies for the management of the plant diseases that so often can cause or contribute to tragedies of hunger.

The United Nations Food and Agriculture Organization has estimated that field losses from pests (insects, diseases, weeds) average 35% for major food crops, with 12% being due to diseases. These cultivated plant crops are the principal source of the world's food supply, and improved pest management is clearly required if the necessary yield increases are to be attained. The contributors address this need by presenting the most current information on the basic physiological and biochemical relationships between plants and their pathogens and by identifying ways of using this information most effectively in developing promising new methods of disease control.

Constructive disagreement still exists between the basic and applied scientists concerning the applicability of certain research findings. The use of disease resistant plants has worked effectively as a plant protection measure. Numerous varieties with considerable resistance have been developed without recourse to an understanding of the mechanisms of that resistance. Basic scientists themselves often agree that the past record of applying the results of their research to plant protection has been disappointing. However, they think that recent findings concerning the plant-pathogen relationship could be used to expand and supplement the traditional plant breeding approach. For example, compounds toxic to many bacterial and fungal disease agents are produced by plants when challenged by pathogens. This induced response may be triggered by specific elicitor compounds in the pathogen, by certain synthetic chemicals, or by some microbial agents unrelated to the disease. Perhaps it would be possible to "immunize" plants against disease or to mimic the specific naturally occurring chemicals in order to produce effective and environmentally benign fungicides and bacteriocides. Similarly, if we understand the mechanisms by which plants arrest the growth of avirulent strains, it may be possible to assist the plant in developing a similar response to potentially virulent strains. Some plants on the other hand are able to tolerate heavy pathogen loads without a reduction in yield. Such a characteristic properly used could certainly contribute to greater food production.

The basic mechanisms and potential value of the foregoing and other aspects of the host-pathogen relationship were reviewed and discussed at the conference. The participants then rewrote their manuscripts to reflect the questions and comments raised by others. This book presents its chapters in three parts. The first on "Compatible Host-Pathogen Interactions" deals with processes that occur when pathogens successfully infect a plant. Included are reviews on surface structure, recognition phenomena, the role of toxins and other chemicals produced by pathogens that lead to symptom expression, and other factors that cause the plant to be or to become susceptible to infection.

The second part on "Host-Pathogen Incompatibility" reviews those attributes of the host that enable it to resist infection. The authors focus on facets of resistance that, they think, offer promise for control of plant disease but have not yet been exploited. Recognition phenomena again play an important role.

Chapters presented in the third part on "New Directions in Development of Plants Resistant to Disease," concentrate on how such basic information might contribute to practical control of plant diseases.

The last chapter presents our reflections on the closing session of the conference, with its lively and productive debate on future research directions. This broad-ranging discussion purposely included consideration of research opportunities not formally presented in the manuscripts.

We thank the Rockefeller Foundation for hosting the conference at its Study and Conference Center in Bellagio, Italy. The contribution of each author is greatly appreciated. We also thank Drs. John J. McKelvey, Jr., and

Carlton S. Koehler for assisting with the organization of the conference and the preparation of this book.

RICHARD C. STAPLES
GARY H. TOENNIESSEN

Richfield Hill Farm, New York
January 1981

ACKNOWLEDGMENTS

The conference from which the papers in this book were drawn was held at the Bellagio Study and Conference Center, Lake Como, Italy, on August 21 to 25, 1979. The Conference Center is operated by the Rockefeller Foundation.

Preparation of the book would not have been possible without the help of Greta Colovito, who handled the manuscript, and Kathryn W. Torgeson, who thoughtfully prepared the index.

R. C. S.
G. H. T.

CONTENTS

I

COMPATIBLE HOST–PATHOGEN INTERACTIONS

GENETIC ANALYSIS AS A TOOL FOR DETERMINING THE SIGNIFICANCE OF HOST-SPECIFIC TOXINS AND OTHER FACTORS IN DISEASE

Olin C. Yoder

It is important that concentrated efforts are directed toward basic research on the molecular mechanisms that control development of plant disease. Understanding molecular mechanisms should permit predictable, rather than empirically derived, solutions to disease problems, and allow for selection of resistance mechanisms that are stable and harmless to nonpests. The problem of safety becomes especially important if mechanisms of resistance are found to involve production of toxicants by plants. Such toxicants, at high concentrations, could result in the same kinds of environmental concerns that the use of synthetic pesticides have raised. Indeed, the potato cultivar Lenape is highly desirable for its resistance to insects, but was removed from production several years ago because its high content of toxic steroid glycoalkaloids made it potentially unsafe for human consumption (Miller, 1974). In the case of α-tomatine, a toxic alkaloid implicated in pest resistance of tomato, there is greater toxicity to a beneficial insect than to an insect pest, suggesting that plant-produced toxicants could interfere in biological pest control (Campbell and Duffey, 1979).

Stability of resistance also demands analysis. Why have some resistance genes provided protection for more than 40 years, whereas others have lasted only a short time? Is this analogous to the observation that some fungicides are effective continuously but other types are not? If we knew the mechanisms of action of various types of resistance genes as well as fungicides, perhaps the answers to these questions would be obvious.

In addition to practical benefits, studies of disease at the molecular level fulfill our obligation as biological scientists to explain the phenomena of plant disease in modern biological terms. Can the rapidly developing technology of recombinant DNA be used to dissect, analyze, and identify the elements of the diseased state that are biologically unique? It is now possible to clone both prokaryotic and eukaryotic DNA and to transform both prokaryotic and eukaryotic cells with cloned DNA (Case et al., 1979; Hinnen et al., 1978; Struhl et al., 1979). Thus, genetically isolated organisms can now be induced to exchange genetic information, and the promise of "genetic engineering" is gaining credibility. It is important that host–parasite systems be developed sufficiently to use the new technology to address such questions as the following: What is the difference between a plant pathogen and a saprophyte (Yoder and Scheffer, 1969)? What is a gene for pathogenicity (physically and metabolically), and how is it regulated? Have genes for pathogenicity evolved uniquely for that purpose, or do they also function in general metabolism? Can a gene for resistance control more than one biochemical mechanism? Are there mechanisms of pathogenic variation other than mutation, heterokaryosis, and sex? What are the primary protein products of genes for pathogenicity and how do they function?

HOST-SPECIFIC TOXINS AND GENETIC ANALYSIS

If it is important to understand molecular mechanisms of disease, what can be learned from studying host-specific toxins? There are 13 plant diseases in which host-specific toxins are now known to be involved (Gilchrist and Grogan, 1976; Kohmoto et al., 1979; Onesirosan et al., 1975; Scheffer and Yoder, 1972; Yoder, 1973). These toxins are compounds of low molecular weight and are produced by a variety of fungal plant pathogens. They are toxic specifically to plants that are susceptible to the respective toxin-producing fungi; resistant plants and nonhost plants are affected little or not at all. The toxins can cause the same visible and biochemical changes that the respective toxin-producing fungi cause (Scheffer, 1976). But the most convincing evidence that these toxins are involved in disease development has come from genetic analyses, which have included the manipulation of genetic variation in both the pathogens and their respective hosts (Scheffer, 1976; Scheffer and Yoder, 1972; Yoder and Gracen, 1975).

The most complete data have come from studies of *Helminthosporium maydis* race T (the cause of southern corn leaf blight) and *H. carbonum* race 1 (the cause of *Helminthosporium* leaf spot) on corn, and of *H. victoriae* (the cause of Victoria blight) on oats. In each of these cases, plants susceptible to the fungus and sensitive to the toxin have been crossed with plants resistant to the fungus and insensitive to the toxin. In each progeny there was, without exception, a correlation between susceptibility to the fungus and sensitivity to the toxin. Plants resistant to the respective toxins have been selected from populations of oats susceptible to *H. victoriae* and from populations of corn susceptible to *H. maydis* race T. In both cases the toxin-resistant mutants were also resistant to the fungus itself (Gengenbach et al., 1977; Scheffer, 1976).

Similar correlations were found between pathogenicity and toxin production in *H. victoriae* and *H. carbonum* race 1. For each pathogen all isolates from the field that produced toxin were also pathogenic. Mutants that lost the ability to produce toxin also lost pathogenicity at the same time. When toxin-producing and -nonproducing isolates were crossed and progeny tested, all pathogenic isolates produced toxin and all nonpathogenic isolates did not. Certain isolates of *H. victoriae* and *H. carbonum* are sexually compatible. When progeny from such crosses were tested, there were four classes: both of the parental types and 2 nonparental types. One parental type produced *H. victoriae*-toxin and was pathogenic only to oats, whereas the other produced *H. carbonum*-toxin and was pathogenic only to corn. One nonparental type produced both *H. victoriae*-toxin and *H. carbonum*-toxin and was pathogenic to both oats and corn, whereas the other produced neither toxin and was nonpathogenic to oats and corn. These correlations support the view that these two toxins are required for the pathogenicity of the fungi that produce them.

The situation with *H. maydis* race T was somewhat different. Nonpathogenic mutants are not yet available, but crosses have been made between race T, which produces toxin and race O, which does not produce toxin, but which is still pathogenic, although less virulent than race T. Progeny of such crosses were all parental types: either they produced toxin and were highly virulent race T, or they did not produce toxin and were weakly virulent race O. These data suggest that race T toxin is required for the high virulence of race T but that it is not required for pathogenicity. By comparison with results from *H. victoriae* and *H. carbonum*, it appears that host-specific toxins can play different roles in different diseases (Payne and Yoder, 1978; Yoder, 1976).

The genetic data for host-specific toxins give us confidence that they play significant roles in disease. The next step is to determine the molecular mechanisms by which they act. Biochemical and physiological analyses suggest that these toxins recognize susceptible cells by interacting with receptor substances probably located in one or more of the cellular membranes (Scheffer, 1976). Presumably the interaction between toxin and receptor sets in motion a sequence of biochemical events that permits growth of the pathogen and culminates in visible expression of disease symptoms; this constitutes the mechanism of disease.

What is the significance of understanding a mechanism by which disease occurs? Several points can be made in the case of the toxins.

1. The conceptual conclusion is that, in diseases involving host-specific toxins, the key molecular event between host and pathogen (i.e., interaction of toxin with receptor) results in susceptibility. This is sometimes

called the theory of induced susceptibility. Resistance occurs if this key molecular event does not happen. Thus it is not necessary to postulate a host response, or the accumulation of toxicants, to account for resistance. This concept is juxtaposed to that of induced resistance, which is supported by genetic studies of certain other diseases, in which the key molecular event appears to cause resistance (Ellingboe, 1976). Much additional work is needed to establish the generality of either of these proposed mechanisms.

2. Biochemical studies are simplified. The biochemical changes in diseased tissue can be studied in the absence of the pathogen, thus eliminating the complication of two interacting metabolic systems.
3. The ability of host-specific toxins to accurately identify susceptible plants has permitted development of highly efficient methods of selecting resistant plants in mixed populations (Scheffer, 1976). In fact, screening for resistance can be conducted in geographical regions in which the living pathogen is quarantined.
4. The specificity of these toxins at the cellular and subcellular levels has opened the way for genetic analysis and manipulation of genes for resistance *in vitro* (Earle et al., 1978; Gengenbach et al., 1977). This is not possible if the living pathogen must be used. Drs. Earle and Gracen describe the utility of toxins *in vitro* elsewhere in this volume.

The genetic arguments for host-specific toxins are among the best available in support of roles for specific molecules in plant disease development. But they are only a beginning in our understanding of the mechanism of the host–parasite interaction. For a more complete picture much more information is needed. Several of the host-specific toxins have not been subjected to genetic analysis because hosts and/or pathogens are not amenable to it. In no case have genes for toxin production been mapped, mutants with altered toxin structure have not been induced, the steps in toxin biosynthesis have not been elucidated by mutational analysis, the primary translation products of alleles for toxin production have not been identified and analyzed. Genes for toxin production have not been physically isolated, cloned, analyzed chemically, or used to transform nontoxin-producing strains. The extent to which host-specific toxins are involved in diseases in general is not known. In many diseases they have been sought but not found. Is this because a different type of mechanism is involved in those diseases, or is it because a toxin is involved but methodology is not sufficiently sensitive to detect the presence of toxin activity?

OTHER FACTORS AND GENETIC ANALYSIS

From a broader perspective, factors other than host-specific toxins also require rigorous genetic analysis for possible roles in disease. Examples of possible candidates include nonspecific toxins, enzymes, hormones, phytoalexins, glycoproteins, lectins, and nucleic acids. The roles of many of these proposed factors in disease have not been rigorously evaluated (Daly, 1976), even though basic research on the host–parasite interaction has concentrated on aspects of genetics and biochemistry in the host–parasite relationship. While genetic loci controlling virulence in the pathogen and/or susceptibility in the host have been identified for many diseases (Day, 1974), there have been few studies that have bridged the gap between genetic analysis and biochemical analysis, despite a large literature describing the biochemical and physiological changes that occur in infected plants (Wood, 1967). This is especially so for those gene products responsible for the biochemical changes that actually cause disease rather than result from it. However, there are cases, in addition to the host-specific toxins, where this has been done, and selected examples are listed to illustrate the approach.

1. Use of variation in a bacterial pathogen to evaluate indoleacetic acid (IAA) as a factor in virulence. Smidt and Kosuge (1978) induced a series of mutants in *Pseudomonas savastanoi,* which causes a gall disease of oleander. The mutants had altered abilities to produce IAA. Some produced no IAA and some produced more than wild type. Those that produced no IAA caused no galls, whereas those that overproduced IAA caused larger galls than wild type. All isolates reproduced at the same rate in host tissue. It appears that in this case IAA is responsible for production of the gall symptom, but that it has no role in controlling bacterial multiplication in the host.
2. Use of variation in a host plant to evaluate a proposed resistance factor. Zalewski and Sequeira (1975) found that various clones of a relative of potato, *Solanum phureja,* were resistant to *Pseudomonas solanacearum* and that they consistently contained high concentrations of an antibacterial factor that inhibited *P. solanacearum in vitro*. However, when a resistant clone was crossed with a susceptible clone and the progeny was tested for resistance and for inhibitor content, disease reaction segregated independently from inhibitor content, suggesting that the inhibitor was not involved in resistance.
3. Use of variation in a fungal pathogen to evaluate the significance of a post-infectional inhibitor in resistance. Tegtmeier and VanEtten (1979) have initiated a genetic analysis of the role of the phytoalexin pisatin in resistance of pea to *Fusarium solani* f. sp. *pisi*. Pisatin accumulates in both susceptible and resistant plants after infection. Thus, if pisatin is important in resistance, highly virulent isolates should be more tolerant of it than weakly virulent isolates. This correlation was found in a collection of field isolates. To further test the theory, progeny from a cross between a highly virulent tolerant isolate and a weakly virulent sensitive isolate were tested for virulence and sensitivity. Tolerant progeny ranged from highly to weakly virulent, but sensitive progeny were all weakly virulent.

This observation suggests that, to be highly virulent, the fungus must be tolerant of the environment presented by its host, which includes pisatin; this provides indirect evidence that pisatin has some role in resistance.

EXPERIMENTAL APPROACHES IN DISEASE PHYSIOLOGY

The foregoing examples illustrate that genetics can be a powerful analytical tool in studies of plant disease. Perhaps we should think formally about the rationale for selecting methods to analyze disease mechanisms. The simplicity of the experimental method is helpful: hold all variables in the system constant except for the one under study. Thus, if a molecule is pathologically significant, disease reaction should change if that molecule is specifically eliminated from the system, leaving everything else the same. Specific elimination is usually difficult to achieve, and a variety of approaches have been used. (1) Specific inhibitors are sometimes appropriate, but in a complex system like a host–parasite interaction, demonstration of a satisfactory level of specificity usually is not possible, except in cases where specific antibodies can be used (Maiti and Kolattukudy, 1979). (2) Biochemical disassembly and reconstruction with or without specific components give definitive information, but require relatively simple systems like viruses. (3) The genetic approach, using mutational analysis as described above, may be the most generally useful with complex host–parasite systems. Whether mutants are induced or naturally occurring variants, they provide more meaningful information if segregating progenies from a recombinational event can be studied. A variation on the mutant theme that has become available recently is the use of plasmids or other vectors to transfer specific genes from one cell type to another (Case et al., 1979; Hinnen et al., 1978; Struhl et al., 1979). But genetic analysis is limited to those host–parasite systems that are, to one degree or another, genetically defined. This means that both host and parasite should have a variety of markers under known genetic control, and requires that both host and parasite must permit recombinational analysis, either sexually or asexually. Mutants (either naturally occurring or induced) should be easily obtainable and a genetic map should be available. In addition, it is desirable that both host and parasite can be cultured and that the system can be manipulated efficiently.

Genetic definition alone does not lead to an understanding of biochemical mechanisms. For example, the genetics of *Ustilago maydis* (Holliday, 1974), *Venturia inaequalis* (Boone, 1971), and *Melampsora lini* (Flor, 1971) have received much attention, but this has not led to elucidation of the mechanism of pathogenicity in any case, even when mutants for pathogenicity or nonpathogenicity have been available. Analogy can be made with genetic dissection of metabolic processes, where metabolically active molecules (e.g., enzymes) are known before mutational analysis is done. Thus it is important that metabolites of host or pathogen be available for testing with the genetically defined system.

What experimental approach can be recommended for more rapid progress in the understanding of molecular mechanisms? If there was an obvious first choice, it would already have been generally adopted. It seems clear that in biological science in general and in plant host–parasite relationships in particular, the most convincing correlations have come from genetic analyses. But for purposes of discussion here, a number of other issues should also be considered. What should be done with systems that are not readily amenable to genetic analyses? Since our resources are limited, should we concentrate our efforts on one, or a few, model systems? How much emphasis should be placed on continued evaluation of known molecular candidates for roles in disease, as opposed to exploring unknown possibilities? If exploration is to be done, what system should be used? Since genetic analysis based on sexual recombination is either impossible or cumbersome with most plant pathogens, should more attempts be made to develop asexual recombinational systems? What are the possibilities for using the techniques of transformation and recombinant DNA in genetic analysis? For most of these questions there is a wide array of opinions.

My general recommendation is that (1) greater emphasis should be placed on genetic development of host–parasite systems for which molecular candidates are available to be analyzed, and (2) more attempts should be made to confirm published results independently in the area of disease physiology. Independent confirmation is one of the most rigorous tests of the validity of an observation; yet it is rarely done, primarily because there is little concentrated effort on any given experimental system. More people working on fewer diseases would help.

REFERENCES

Boone, D. M. (1971) Genetics of *Venturia inaequalis*. *Annu. Rev. Phytopathol.* **9**:297–318.

Campbell, B. C., and S. S. Duffey. (1979) Tomatine and parasitic wasps: Potential incompatibility of plant antibiosis with biological control. *Science* **205**:700–702.

Case, M. E., M. Schweizer, S. R. Kushner, and N. H. Giles. (1979) Efficient transformation of *Neurospora crassa* by utilizing hybrid plasmid DNA. *Proc. Natl. Acad. Sci. USA* **76**:5259–5263.

Daly, J. M. (1976) Some aspects of host–pathogen interactions. In *Encyclopaedia of Plant Physiology*, Vol. 4 (R. Heitefuss and P. H. Williams, eds.). Springer-Verlag, New York. Pp. 27–50.

Day, P. R. (1974) *Genetics of Host–Parasite Interaction*. W. H. Freeman Co., San Francisco, Calif. 238 pp.

Earle, E. D., V. E. Gracen, O. C. Yoder, and K. P. Gemmill. (1978) Cytoplasm-specific effects of *Helminthosporium maydis* race T toxin on survival of corn mesophyll protoplasts. *Plant Physiol.* **61**:420–424.

Ellingboe, A. H. (1976) Genetics of host–parasite interactions. In *Encyclopaedia of Plant Physiology*, Vol. 4 (R. Heitefuss and P. H. Williams, eds.). Springer-Verlag, New York. Pp. 761–778.

Flor, H. H. (1971) Current status of the gene-for-gene concept. *Annu. Rev. Phytopathol.* **9**:275–296.

Gengenbach, B. G., C. E. Green, and C. M. Donovan. (1977) Inheritance of selected pathotoxin resistance in maize plants regenerated from cell cultures. *Proc. Natl. Acad. Sci. USA* **74**:5113–5117.

Gilchrist, D. G., and R. G. Grogan. (1976) Production and nature of a host-specific toxin from *Alternaria alternata* f. sp. *lycopersici*. *Phytopathology* **66**:165–171.

Hinnen, A., J. P. Hicks, and G. R. Fink. (1978) Transformation of yeast. *Proc. Natl. Acad. Sci. USA* **75**:1929–1933.

Holliday, R. (1974) *Ustilago maydis*. In *Handbook of Genetics*, Vol. 1 (R. C. King, ed.). Plenum Press, New York. Pp. 575–595.

Kohmoto, K., R. P. Scheffer, and J. O. Whiteside. (1979) Host-selective toxins from *Alternaria citri*. *Phytopathology* **69**:667–671.

Maiti, I. B., and P. E. Kolattukudy. (1979) Prevention of fungal infection of plants by specific inhibition of cutinase. *Science* **205**:507–508.

Miller, J. (1974) Agriculture: FDA seeks to regulate genetic manipulation of food crops. *Science* **185**:240–242.

Onesirosan, P., C. T. Mabuni, R. D. Durbin, R. B. Morin, D. H. Rich, and D. C. Arny. (1975) Toxin production by *Cornyespora cassiicola*. *Physiol. Plant Pathol.* **5**:289–295.

Payne, G. A., and O. C. Yoder. (1978) Effect of the nuclear genome of corn on sensitivity to *Helminthosporium maydis* race T toxin and on susceptibility to *H. maydis* race T. *Phytopathology* **68**:331–337.

Scheffer, R. P. (1976) Host-specific toxins in relation to pathogenesis and disease resistance. In *Encyclopaedia of Plant Physiology*, Vol. 4 (R. Heitefuss and P. H. Williams, eds.). Springer-Verlag, New York. Pp. 247–269.

Scheffer, R. P., and O. C. Yoder. (1972) Host-specific toxins and selective toxicity. In *Phytotoxins in Plant Diseases* (R. K. S. Wood, A. Ballio, and A. Graniti, eds.). Academic Press, London. Pp. 251–272.

Smidt, M., and T. Kosuge. (1978) The role of indole-3-acetic acid accumulation by alpha methyl tryptophan-resistant mutants of *Pseudomonas savastanoi* in gall formation on oleanders. *Physiol. Plant Pathol.* **13**:203–214.

Struhl, K., D. T. Stinchcomb, S. Scherer, and R. W. Davis. (1979) High-frequency transformation of yeast: Autonomous replication of hybrid DNA molecules. *Proc. Natl. Acad. Sci. USA* **76**:1035–1039.

Tegtmeier, K. J., and H. D. VanEtten. (1979) Genetic analysis of sexuality, phytoalexin sensitivity and virulence of *Nectria haematococca* MP VI (*Fusarium solani*). *Am. Phytopathol. Soc., 71st Annu. Meet.*, Abstr. No. 338.

Wood, R. K. S. (1967) *Physiological Plant Pathology*. Blackwell Scientific Publications, Oxford. 570 pp.

Yoder, O. C. (1973) A selective toxin produced by *Phyllosticta maydis*. *Phytopathology* **63**:1361–1365.

Yoder, O. C. (1976) Evaluation of the role of *Helminthosporium maydis* race T toxin in Southern Corn Leaf Blight. In *Biochemistry and Cytology of Plant-Parasite Interaction* (K. Tomiyama, J. M. Daly, I. Uritani, H. Oku, and S. Ouchi, eds.). Elsevier, New York. Pp. 16–24.

Yoder, O. C., and V. E. Gracen. (1975) Segregation of pathogenicity types and host-specific toxin production in progenies of crosses between races T and O of *Helminthosporium maydis (Cochliobolus heterostrophus)*. *Phytopathology* **65**:273–276.

Yoder, O. C., and R. P. Scheffer. (1969) Role of toxin in early interactions of *Helminthosporium victoriae* with susceptible and resistant oat tissue. *Phytopathology* **59**:1954–1959.

Zalewski, J. C., and L. Sequeira. (1975) An antibacterial compound from *Solanum phureja* and its role in resistance to bacterial wilt. *Phytopathology* **65**:1336–1341.

THE ROLE OF NONSPECIFIC TOXINS AND HORMONE CHANGES IN DISEASE SEVERITY

George F. Pegg

The involvement of nonspecific toxins and growth substances in disease development is a wide field of research in host and pathogen physiology, with a voluminous literature. Nonspecific toxins of plant pathogens have been the subject of a number of reviews by Gäumann (1954), Wheeler and Luke (1963), Owens (1969), Luke and Gracen (1972), Patil (1974), Strobel (1974), Rudolph (1976), and Strobel (1977). The role of growth substances in diseased plants has been reviewed by Sequeira (1963, 1973), Pegg (1976a), and Daly and Knoche (1976).

A toxin in the broadest sense is a nonenzymic product of a microorganism that causes damage to the plant in low concentrations. Growth substances for the most part are not regarded as toxins. Their action on plants as microbial products causing metabolic changes and tissue overgrowths, however, are reminiscent of toxins, even though they usually are not associated with necrosis. Enzymes involved indirectly in cell death (Wood, 1972) or less certainly in wilting (Mussell, 1972, 1973) may be regarded as toxins but will not be considered.

Nonspecific toxins are phytotoxic to hosts and nonhosts of the pathogen, and their production is not necessarily related to the virulence of pathogenic isolates. The real argument surrounding nonspecific toxins, however, is their importance in symptom development or the degree to which they determine pathogenicity. Purified *host-specific toxins* may show a range of activity in hosts or test organisms other than the pathogen-susceptible ones, but by definition a precise correlation exists between pathogen virulence and toxin production. In this sense infectivity, pathogenicity, and toxin production are all related.

Most of the 150 characterized fungal and bacterial phytotoxins affect respiration, membrane permeability, and cell metabolism at different concen-

Table 1. Selected Nonspecific Pathogens and Phytotoxins and Their Effects on Plants

Major Effect	Pathogen	Toxin
Wilting	*Corynebacterium michiganense*	Glycopeptide
	C. insidiosum	Glycopeptide
	Ceratocystis ulmi	Glycopeptide
	Pseudomonas solanacearum	Polysaccharide
	Xanthomonas campestris	Polysaccharide
	Verticillium spp.	Protein lipopolysaccharide
	Fusarium oxysporum f. spp.	5-*n*-Butylpicolinic acid
	Fusicoccum amygdali	Carbotricyclic diterpene glucoside
Chlorosis	*Pseudomonas phaseolicola*	Phaseotoxin: *n*-phosphoglutamic acid
		Phaseolotoxin: tripeptide
	P. tabaci	Tabtoxin
		2-Serine tabtoxin
	Rhizobium japonicum	Enol-ether amino acid
	Alternaria alternata	Tenuazonic acid
Water soaking	*Pseudomonas lachrymans*	Lipomucopolysaccharide
Local necrosis	*Septoria nodorum*	8-Methyl isocoumarin
	Pseudomonas syringae	Syringomycin peptide

trations in a nonspecific way. They may also affect root and shoot growth of nonhost organisms, or spore germination when these are used as assays. More specifically, individual phytotoxins induce symptoms such as chlorosis or wilting by affecting chlorophyll synthesis or water conduction. While these correspond for the most part to the host syndrome, tissue necrosis and wilting may result inevitably from nonspecific chronic effects on membrane permeability and blocked metabolism, depending on the localized or systemic distribution of the toxin. Selected examples of these are shown in Table 1.

MODE OF ACTION OF SELECTED TOXINS

Chlorosis Toxins

Pseudomonas phaseolicola and Phaseotoxins

P. phaseolicola, the causal organism of halo blight of *Phaseolus*, produces a cultural toxin that can simulate the chlorotic lesion when applied exogenously to leaves. Patil et al. (1976) isolated a number of toxic components in culture filtrates, the major one of which was *n*-phosphoglutamic acid and was designated phaseotoxin A. Mitchell (1979) further characterized phaseolotoxin, a tripeptide of homoarginine, alanine, and ornithine to the N^5 of which is attached a phosphosulfamyl group.

Plants infected by *P. phaseolicola* or treated by toxin accumulate ornithine 100-fold (Rudolph and Stahmann, 1966) with a concomitant 50% loss of arginine in the chlorotic halo. Patil et al. (1976) showed that the enzyme ornithine carbomyl transferase (OCT) was inhibited by crude toxin preparations *in vitro*. It was postulated that OCT inhibition led to decreased protein synthesis through decreased arginine, which included enzymes of chlorophyll synthesis. Toxin-induced chlorosis can be inhibited by pretreatment with citrulline and reversed with L-arginine (Gnanamanickam and Patil, 1976). These authors showed that a nontoxin producing mutant of *P. phaseolicola* multiplied in susceptible tissue but did not inhibit OCT.

Of special interest was the finding that the pterocarpan phytoalexins, phaseollin, and phaseollinisoflavan inhibited bacterial growth completely or partially (Gnanamanickam and Patil, 1977a) and that phaseotoxin-treated bean leaves suppressed the hypersensitive response and phytoalexin accumulation, while permitting enhanced growth of the pathogen *in vivo* (Gnanamanickam and Patil, 1977b). Notwithstanding these results and those of Keen and Kennedy (1974) the role of phytoalexins in the resistance of plant cultivars to bacterial pathogens is still conjectural. High levels of phytoalexins still remain in toxin-treated plants and levels are naturally high in susceptible plants as a result of infection. Recent work by Mitchell (1979) has questioned the role of phaseotoxin A (cited as the toxin in most of the foregoing experiments), in favor of phaseolotoxin (*N*-phosphosulfamyl)ornithylalanylhomoarginine. Comparative studies showed that *N*-phosphoglutamic acid was readily hydrolyzed by host phosphatase and was less stable in weakly alkaline or acidic aqueous solution. Moreover, the tripeptide at 0.3 μm inhibited OCT by 50%. Mitchell observed that 44 nmol phaseotoxin A induced neither chlorosis nor ornithine accumulation, whereas 0.07 nmol phaseolotoxin induced chlorosis and 7 nmol, systemic symptoms.

The specific effect of *P. phaseolicola* antimetabolites on OCT leading to chlorosis and enhanced bacterial growth in toxin-treated tissue is indisputable. The restriction of bacterial growth in resistant cultivars, however, may be neither a concomitant of suppressed toxin production nor phytoalexin production.

Pseudomonas tabaci (Including *P. coronafaciens* and *P. garcae*) and Tabtoxin and 2-Serine Tabtoxin

Based on the symptoms of natural infection in tobacco and those induced by culture filtrates, the toxin is also known as the wildfire toxin. The definitive structure of tabtoxin was given by Stewart (1971) and Taylor et al. (1972). On hydrolysis, the toxin gives tabtoxinine (5-amino-2-aminomethyl-2-

hydroxyadipic acid) and threonine, or in the case of 2-serine tabtoxin, serine. The latter occurs in cultures of all species that produce tabtoxin. Wooley et al. (1955) considered incorrectly that the structure of the toxin was similar in part to methionine and suggested that the toxin acted as a competitive inhibitor of L-methionine in protein synthesis, based on its prevention of toxin and methionine sulphoximine (MSO) [a structural analog of L-methionine]-induced growth inhibition in *Chlorella*. Methionine does not inhibit chlorosis *in vivo*. Further support for the methionine-competitor concept was suggested by the work of Carlson (1973), who found MSO-resistant mutants of tobacco which were also resistant to *P. tabaci*. Since MSO is a potent inhibitor of glutamine synthetase, Sinden and Durbin (1968) proposed such a role for tabtoxin. Recent work, however, has shown that pure tabtoxin does not affect glutamine synthetase; neither does it affect reactions involving methionine.

Carlson (1973) proposed the novel idea that MSO could be used to screen protoplasts for resistance to *P. tabaci*. The relation between MSO and tabtoxin is at present not clear in view of the nonsimilarity in their structural formulas, and this and other aspects of tabtoxin remain to be resolved.

Wilt Toxins

Corynebacterium insidiosum and Glycopeptides

Ries and Strobel (1972a) isolated a large glycopeptide (5×10^6 daltons) from cultures of *C. insidiosum,* the causal agent of lucerne wilt. They found that 40% of the toxin consisted of fucosyl residues and 20% each of mannosyl and rhamnosyl residues. The remainder of the toxin consisted of an unidentified keto deoxy sugar acid (9%) and 2.5% amino acids. There were 77 peptide chains per mole of toxin. Ries and Strobel (1972b) isolated the toxin from *C. insidiosum*-infected plants. They suggested, as did Rai and Strobel (1969 a,b), working on the glycopeptide of *C. michiganense,* that physical occlusion of water transport was not the primary cause of wilting. Model polymers of similar molecular weight, administered to shoots in amounts similar to those of the toxin found *in vivo,* did not induce wilt. Van Alfen and Turner (1975), however, presented data to support toxin-induced wilting by preventing water movement across pit membranes.

Fusicoccum Amygdali and Fusicoccin

Fusicoccin and its derivatives, allofusicoccin and isofusicoccin, represent an extensively studied family of carbotricyclic diterpene glucosides. These metabolites produced by *Fusicoccum amygdali,* causing canker and vascular wilt of peach and almond (Graniti, 1966; Ballio et al., 1971), are unlike most other nonspecific toxins in possessing auxin and cytokinin-like properties (Ballio et al., 1971; Lado et al., 1973; Marré, 1979).

The major effect of the toxin *in vitro* and *in vivo* is an increased water permeability of infected or treated plants associated with an opening of stomata. Stomatal opening induced by light or fusicoccin is accompanied by an H^+/K^+ exchange and synthesis of malate in the guard cells. Turner (1973) demonstrated the effect of the toxin on proton extrusion, K^+ uptake, and the reduced pH of the bathing solution. Marré et al. (1974) and Marré (1979) have proposed a mechanism linking K^+ fluxes with proton extrusion and a negative transmembrane potential. Since toxin-induced stomatal opening can be reversed by 2,4-dinitrophenol in light and dark, Turner (1973) concluded that K^+ stimulation was driven by ATP from photosynthetic and oxidative phosphorylation. The relationship between stomatal opening and fusicoccin-induced water loss is not simple, however. Chain et al. (1971) showed that toxin-treated leaves whose stomata had been closed with abscissic acid (ABA) still lost water. These authors postulated an increased cuticular water loss from decreased resistance of the plasmalemma to water movement. In this sense, stomatal opening could be regarded as partly incidental to the wilting effect.

Fusicoccin induces large fresh weight increases in pea internodes and tissue elongation (Ballio et al., 1971) and water uptake in tomato, clover, and tobacco leaves (Lado et al., 1972). Dormancy in wheat and lettuce seeds can be reversed by fusicoccin (Lado et al., 1974). It is also effective in reversing ABA-induced inhibition of germination of radish and lettuce seeds. Although the toxin does not appear to inhibit root cell elongation at concentrations effective in stems as does indol-3-ylacetic acid (IAA), its growth-promoting effects and its role in breaking dormancy, in which it is more active than giberellic acid (GA_3) or the cytokinin benzyladenine (BA), place it unequivocally in the growth substance, as well as the toxin category. In view of this role, fusicoccin presents a valuable tool for the understanding of growth substance action where the toxin, unlike growth substances, may act on a *specific* physiological target, namely, the stimulation of ion transport.

Evidence for the *in Vivo* Presence of Toxins

Relatively few toxins have been isolated from diseased plants. Demonstration of the *in vivo* presence of a toxin does not provide *a priori* or unequivocal grounds for its involvement in symptom induction. Nevertheless, this association strongly favors such a conclusion. Conversely, failure to detect the presence of a-toxin in diseased tissue may result from (1) its low effective dose and concentration in diseased tissue; (2) the predisposing effect of other molecules not included in the test assay (Cronshaw and Pegg 1976); (3) toxin lability; and (4) binding of the toxin to a site of action that is destroyed during isolation.

Table 2 shows toxins that have been isolated from diseased host plants.

Table 2. Toxins Isolated from Diseased Hosts

Tobacco	*Alternaria alternata*, tenuazonic acid (Mikami et al., 1971)
Lucerne	*Corynebacterium insidiosium*, glycopeptide (Ries and Strobel, 1972a)
Potato	*C. sepedonicum*, glycopeptide (Strobel, 1970)
Rice	*Drechslera oryzae*, ophiobolins (tricyclic terpenoids) (Luke and Gracen, 1972)
Pea	*Fusarium* spp., napthazarine derivatives, marticin (Kern et al., 1972)
Soybean	*Pseudomonas glycinea*, peptide (Hoitink and Sinden, 1970)
Cabbage	*Xanthomonas campestris*, polysaccharide (Sutton and Williams, 1970)
Lemon	*Phoma tracheiphila*, glycopeptide (Nachmias et al., 1979)

Correlation Between Host and Toxin Sensitivity and Infectivity

Relatively few examples of close correlations between toxin sensitivity and infectivity are known. The best of these are phytonivein and *Fusarium oxysporum* f. sp. *niveum* (Hiroe and Nishimura, 1956), tenuazonic acid and *Alternaria alternata* (Mikami et al., 1971), lipoprotein polysaccharide and *Verticillium dahliae* (Keen et al., 1972), phaseotoxin, phaseolotoxin, and *Pseudomonas phaseolicola* (Gnanamanickam and Patil, 1976), mellein and *Septoria nodorum* (Ao, 1977).

The Use of Toxins in Plant Breeding Strategies

The major advantage in utilizing toxins as a substitute for conventional infectivity screens using pathogenic organisms lies with the host-specific toxins. Nevertheless, where nonspecific toxins can be shown to play a dominant if not exclusive role in pathogenicity, their use in selecting toxin-resistant lines may enable breeding programs to be expanded and perhaps shortened. Carlson (1973) proposed the elegant system whereby protoplasts could be incubated with the toxin at a suitable dilution and on a regenerative medium. Although novel in concept, this has not been attempted successfully with nonspecific toxins and is technically difficult to achieve. Advantage could be gained only from a system where protoplasts of differing genetic character could be obtained from a genetic chimera, for example, *Saccharum*. These, however, are not common in crop plants. Alternatively, as in tobacco, pollen haploid protoplasts from different lines could be prepared and screened either for haploid plant production or fused diploids.

The methodology, for potatoes at least, has been developed. Shepard and Totten (1977) and Shepard (1979) have successfully regenerated potato plants from mesophyll cell protoplasts. Matern et al. (1978) have described the reaction of such protoplasts to the *Corynebacterium sepedonicum* glycopeptide toxin. This work is currently being evaluated in several laboratories.

Advantage with a toxin, or highly specific culture filtrate-based resistant plant screen, lies in the scale of the program that could be attempted. Thus the use of seedlings (where juvenile resistance reflected mature plant resistance) or detached leaves of young plants would permit greater numbers of F_1 plants and subsequent generations to be screened. If toxin activity only partially represented pathogenicity, such a technique would permit a primary rapid selection. Only surviving seedlings, or resistant plants from which a leaf had been detached, would need to be inoculated with the pathogen.

The number of toxin–host combinations in which this has been achieved is limited. Straley et al. (1974) successfully screened lucerne shoots for *Corynebacterium insidiosum* wilt resistance, using a critical 0.05% concentration of the phytotoxin. Toxin and wilt resistance gave a correlation coefficient of 0.62. Ao (1977) similarly achieved a correlation coefficient of 0.97 between the resistance of nine wheat cultivars to *Septoria nodorum* and their shoot growth inhibition to the toxin mellein.

Mellein (ochrain) (8-hydroxy-3-methyl isocoumarin) represents the simplest of a series of isocoumarins produced variously by fungi, actinomycetes, and as a stress metabolite in carrots. *Marasmius ramealis*, *Aspergillus melleus*, and *A. ochraceous*, in addition to *Septoria nodorum*, produce mellein. The 6-methoxy derivative was shown by Condon and Kuć (1960, 1962) to be a phytoalexin induced in carrot by *Ceratocystis fimbriata* infection. Sondheimer (1957), however, had also established that methoxymellein formed in healthy carrot stored at 32°C for several weeks. Of particular interest are the substituted isocoumarins from *Sclerotinia sclerotiorum*—sclerotinin A (3-hydroxy-4,5-methyl-6-hydroxy-7-methylmellein), sclerotin B, and sclerin (Ballio, 1972)—all of which promote plant growth.

Mellein nonspecifically inhibits root and shoot growth in wheat and also in genera and species not associated with mellein-producing pathogens. The close correlation achieved by Ao (1977) for toxin sensitivity and disease severity using nine wheat cultivars is unusual since foliar lesions are only induced through punctures and in light. Test droplets of 150 ppm mellein were inactive on unpunctured leaves, or in darkness, or on nonhosts of the pathogen when puncture tested (Ao and Griffiths, unpublished data). The multiple role of the isocoumarin metabolites in general is worthy of further detailed investigations.

Carr (1970) reported the tolerance of lucerne populations to an unspecified toxin from *Verticillium albo-atrum*. Austin and Pegg (unpublished information) found that an isoline of tomato susceptible to *V. albo-atrum* was sensitive to a high molecular weight polysaccharide toxin only in the presence of ethylene.

A novel use for toxins was developed by Strobel and Rai (1968) employing the antigenic properties of the *Corynebacterium sepedonicum* glycopeptide. Using toxin antibodies, potato lines have been screened for ring-rot infection for seed certification.

The Involvement of Nonspecific Toxins in Resistant and Susceptible Plants

Schemes for the possible differential involvement, or non-involvement of toxins in resistant and susceptible plants are shown in Table 3. Relatively little work has been published on the role of the resistant plant in relation to toxin involvement and much of Table 3 is still speculative. Examples of specific toxins and their involvement are as follows:

1. *Breaking disease resistance by toxins.* Only phaseotoxin/phaseolotoxin from *P. phaseolicola* has been shown conclusively to act in this role (Rudolph, 1972).
2. *Correlation between host and toxin sensitivity.* Examples as cited previously.
3. *Differential toxin production on resistant and susceptible hosts.* Trione (1960) found that *Fusarium oxysporum lini* produced fusaric acid only on a wilt-susceptible cultivar. In this context Fuchs (1975) claimed that as-

Table 3. Possible Schemes Based on the Involvement, or Non-Involvement of Toxins in the Expression of Resistance or Susceptibility in Plants

Resistance	Susceptibility
1. Restriction or exclusion of the pathogen in, or from the host caused by (*a*) physical or chemical antifungal barriers or (*b*) unsuitable nutrient substrate for fungal growth	1. (*a*) Biomass of colonizing pathogen sufficient for toxin levels to exceed the tissue damage threshold or (*b*) toxin breaking of the resistance mechanism controlling hyphal growth
2. (*a*) Non-recognition of a toxin or (*b*) high toxin-tolerance threshold	2. (*a*) Toxin receptor site on cell membrane or (*b*) low toxin-tolerance threshold
3. Non-induction or suppression of toxin production	3. Possession of toxin inducers or precursors
4. Degradation of a toxin	4. No detoxification

cochitine biosynthesis in *Ascochyta fabae* cultures was inhibited in a feedback mechanism by pisatin formed in the resistant reaction.

4. *Toxin degradation*. Differential resistance to fusaric acid in two tomato cultivars was shown by Kluepfel (1957) to be related to different rates of detoxification in the two hosts. Fusaric acid has been shown variously to degrade to 3-*n*-butylpyridine by decarboxylation, pyridine, carboxylic acid, and *n*-methylfusaric acid amide, all of which are less toxic than fusaric acid. Resistance in *Gossypium herbarum* and *Pisum sativum* to fusarial pathogens has been interpreted similarly (Jost, 1965; Kern, 1972).

GROWTH HORMONE CHANGES IN DISEASED PLANTS

During the last 30 years the role of plant growth substances in diseases caused by fungi, bacteria, viruses, and nematodes has been the subject of much research. For the most part, work has concentrated on indolyl auxins following the almost simultaneous discovery in 1934 that both fungi and higher plants could synthesize IAA. The phenomenon of hyperauxiny in diseased tissue is now as well established as pathological respiration (see Sequeira, 1963, 1973; Pegg, 1976a,b; Daly and Knoche, 1976). The initial discovery of IAA, with its particular properties rather than gibberellins or cytokinins, conditioned much of the later thinking and interpretation of growth-substance involvement in plant physiology.

At the present time indolyl auxins, gibberellins, cytokinins, ethylene, and abscisic acid represent the known classes of growth substances, all of which are found in healthy as well as diseased plants (Figure 1). Gibberellic acid is of special interest in this context, since, until 1955, it was known only as a fungal toxic metabolite and occupied the same position that helminthosporal and fusicoccin do today.

Auxins and Other Compounds

Helminthosporal, the aldehyde of the sesquiterpenoid toxin from *Helminthosporium sativum*-inducing necrosis in oats and wheat, functions in rice and other test plants like a gibberellin, promoting growth and stimulating α-amylase synthesis in barley grain (Tamura et al., 1963; Hashimoto and Tamura, 1967). Fusicoccin and the sclerotinins (Figure 2) mentioned previously, simulate auxin, cytokinin, and some gibberellin effects. (Lado et al., 1973; Ballio, 1972; Marré, 1979). Similar growth-promoting effects have been observed for the host-specific toxins Victorin and *Helminthosporium* Race-T toxin (Evans, 1973, 1974). Hadicin-*n*-formylhydroxyaminoacetic acid (Figure 2) produced by cultures of *Penicillium frequentans* and *P. purpurescens* induces dwarfism in pea and bean in the absence of phytotoxicity.

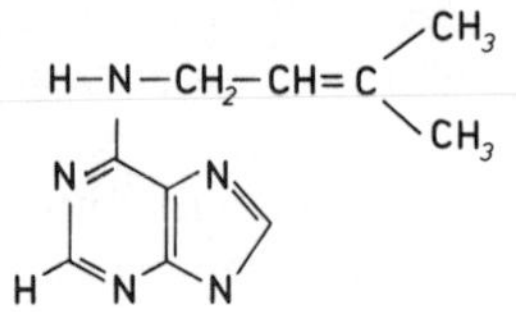

2- ^{2}-isopentenyl amino purine (2iP)

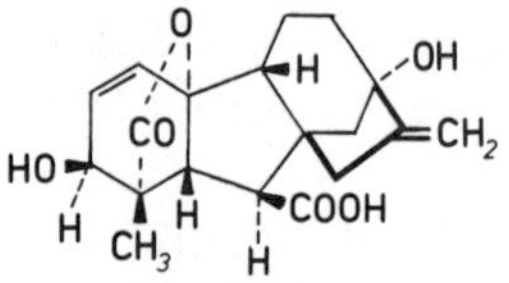

Gibberellic acid (GA_3)

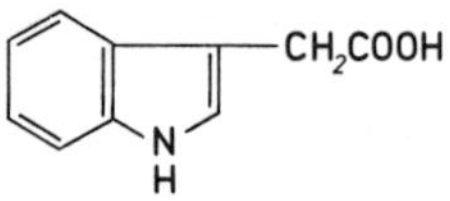

Indol-3-yl acetic acid (IAA).

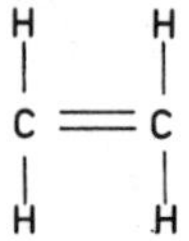

Ethylene (Ethene)

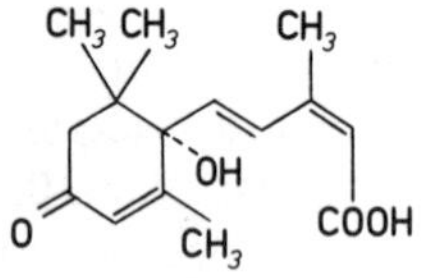

Abscisic acid (ABA)

Figure 1 Naturally-occurring plant growth regulators.

The basic questions regarding growth-substance involvement in plant disease which have been asked are as follows:

1. How are levels of hormones changed in diseased tissues?
2. Are they of host or pathogen origin?
3. What is their role in the diseased plant?

In general—with the exception of selected diseases displaying specific tissue overgrowth symptoms—it is felt that pathological increases in growth substances represent altered *host* activity with an insignificant contribution from the pathogen. This was demonstrated for IAA production in *Pseudomonas solanacearum*–wilt of tobacco by Sequeira and Williams (1964) using a discriminative, labeled precursor.

More recently Smidt and Kosuge (1978), working on gall formation in oleander caused by *Pseudomonas savastanoi,* have suggested that the ability of this pathogen to produce IAA is a prerequisite for pathogenicity, that is, gall formation. Mutants resistant to α-methyl tryptophan and α-tryptophan analog, which induced changed anthranilate synthetase activity, either failed to produce IAA and showed no tryptophan oxidative decarboxylase activity, or produced twice as much IAA as the wild types. *In vivo* growth of the wild-type and mutant strains showed no differences, but gall formation in artificially inoculated shoots was proportional to the levels of IAA produced in culture.

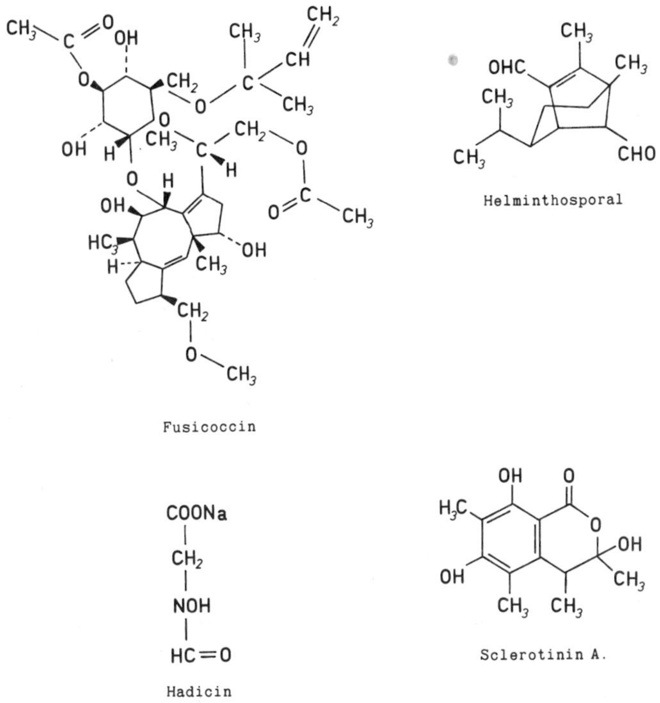

Fusicoccin

Helminthosporal

Hadicin

Sclerotinin A.

Figure 2 Non-specific toxins with growth-promoting properties.

From the knowledge of the complexity of plant ontogeny and morphology in different genera, the concept of a universal explanation for growth, abnormal and normal, involving all the five known classes of hormones is quite untenable, and yet much plant pathological research has attempted this in terms of *one or two substances*. No general interpretation of hormone involvement in diseased host metabolism should be attempted, but a clear distinction should be made between growth substances and diseases caused by nectotrophic and biotrophic pathogens.

Studies on gibberellins and cytokinins in disease have been fewer than with IAA, and are associated specifically with tissue overgrowths in which a cause and effect relationship seemed fairly obvious. Examples of these are fasciation and leafy gall in herbaceous plants caused by *Corynebacterium fascians*, club root of brassicas caused by *Plasmodiophora brassicae*, and crown gall caused by *Agrobacterium tumefasciens*. Thimann and Sachs (1966) first detected cytokinin activity in *C. fascians* culture filtrates, and Klambt et al. (1966) and Helgeson and Leonard (1966) isolated and characterized 2-isopentenyladenine. Reddy and Williams (1970) found three cytokinins in club root tissue and Dekhuijzen and Overeem (1971) identified zeatin from root galls. Wood et al. (1969, 1972) have isolated cytokinins from crown gall cells. One of these—3,7-dialkyl-2-alkythio-6-purinone—is a potent inhibitor of cAMP phosphodiesterase and an active inducer of cell division. It is also stimulated by kinetin.

The role of these compounds, however, may be multiple. Changes in host metabolism (of which the external symptoms represent an incidental feature) may be essential to the successful establishment of the pathogen. The relationship between nutrient flow and cytokinin activity in rusts and mildews and the "green-island effect" has been well-documented by Livne and Daly (1966), Kiraly et al. (1967), and Dekhuijzen and Staples (1968). It is a situation where the extended juvenility of the leaf acts as a nutrient sink to the site of the pathogen. Similarly, with gibberellins there is no evidence that the so-called "bakanae effect" in rice infected with *Gibberella fujikuroi* is in any way related to pathogenesis. Indeed the evidence suggests otherwise; however, Muromtsev and Globus (1976) have shown in a range of *Fusarium moniliforme* strains an inverse relationship between GA_3 and α-amylase production *in vitro*. All strains unable to produce GA_3 in culture possessed a constitutive enzyme α-amylase for the hydrolysis of starch, and, conversely, the strains capable of producing GA_3 had no α-amylase system. Here, then, could be a role for GA_3 in the diseased host either of host or fungal origin to provide a supply of hexose sugar for the pathogen.

The role of growth substance production by microorganisms has been considered from a different standpoint by Libbert and co-workers (Libbert et al., 1968; Libbert et al., 1969; Libbert and Risch, 1969; Libbert and Manteuffel, 1970; Libbert and Silhengst, 1970). These workers have presented convincing evidence that has been largely overlooked by physiological plant pathologists that saprophytic fungi add considerably to the apparent level of endogenous IAA in plants grown under nonsterile conditions. Few studies on hyperauxiny in virus, fungus, or bacterial diseases have been conducted in the total absence of saprophytic bateria. Thus, while the published results may not be invalidated by the findings of Libbert and co-workers, her results suggest that many of the absolute levels of endogenous auxins cited may be inaccurate. This, too, may apply to studies on cytokinins and gibberellins.

Ethylene

Ethylene, the simplest unsaturated carbon compound functions as a powerful growth regulator in most aspects of plant growth, development, senescence, and disease (see Pegg, 1976b; Lieberman, 1979). In *Verticillium albo-atrum* wilt of tomato, ethylene has been implicated as a phytotoxin (Weise and De Vay, 1970; Pegg and Cronshaw, 1976) as a toxin synergist (Cronshaw and Pegg, 1976) and as an inducer of resistance (Pegg, 1976c). Since ethylene is intimately involved in the metabolism of other growth substances, both in terms of stimulation and repression, and is produced in most tissue as a result of physical or physiological damage, it follows that it is involved in most, if not all, toxin-induced symptoms. Ethylene plays a central role in enzyme synthesis, some of which, notably peroxidase and phenol oxidases, are further involved in auxin and ethylene synthesis. Moreover, pectolytic enzymes and ethylene are both involved in reactions which lead to symptom development (see Pegg, 1979).

Only the toxin synergism of ethylene in vascular wilt pathogenesis will be considered here. Cronshaw and Pegg (1976) have shown in *Verticillium*-infected tomato plants that ethylene is produced as a peak immediately prior to, or coinciding with, symptom appearance. Recently Austin and Pegg (unpublished information) have shown that a number of components of *Verticillium albo-atrum* culture filtrates act as toxins at physiologically acceptable

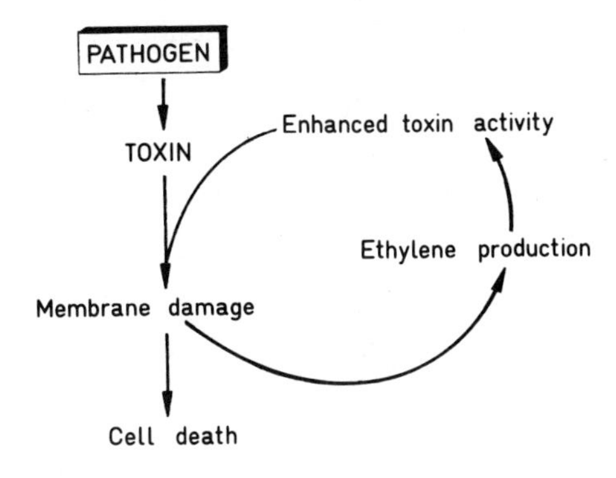

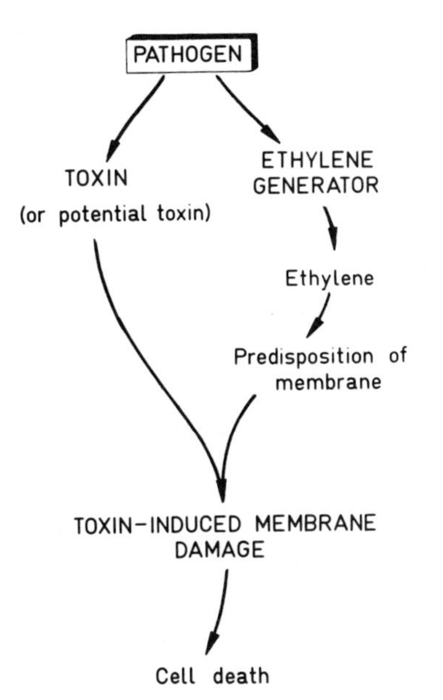

Figure 3 Two schemes for the synergistic involvement of ethylene in toxin-induced necrosis.

levels only when tested on shoots or leaves from ethylene-gassed tomato and cotton plants. Short-term exposure (12–48 h) of plants to ethylene, 1–5 ppm, had no deleterious effect on growth. Uptake of divalent cations Mg^{2+}, Ca^{2+}, and Mn^{2+} and the trivalent anion PO_4^{3-} at 5–10 mM concentrations in such tissue induced wilting, cell collapse, and necrosis comparable to natural infection. These ions at higher concentrations had no effect on ungassed control tissue. Similarly, monovalent cations and anions had no effect at subplasmolytic levels in control or ethylene-treated plants. Possible support for an ethylene–ion interaction in symptom development in wilt diseases was seen in naturally-infected plants where Mg^{2+}, Ca^{2+} and PO_4^{3-} concentrations increased prior to symptom appearance. Mg^{2+} levels increased up to 100% of healthy plants three days after symptom appearance.

The role of ethylene in pathogenesis is complex (see Abeles, 1973; Hislop et al., 1973; Pegg, 1976a). Pegg and Austin (unpublished information) have shown with a polysaccharide toxin from *Verticillium albo-atrum* that ethylene-treated isolines of tomato susceptible to *Verticillium* sustained damage, while gassed single gene-resistant plants remained symptomless. The action of ethylene in this predisposition of susceptible tissue is unknown. Ethylene lowers the resistance threshold of tissue to toxins and pectolytic enzymes, even when predisposition is not a prerequisite for toxicity. Thus where pathogenic metabolites are toxic *in vivo at very low concentrations* at cellular sites sensitized by ethylene, their concentration in the host plant may be below the level of detection.

In any host-pathogenic toxin combination where ethylene is generated, predisposition by ethylene (Figure 3) or other molecules, or enhanced host sensitivity to the toxin by ethylene remain a real possibility. The implications for such interactions with nonspecific toxins should be given consideration in any *in vitro* screening for disease resistance, using pure toxins or especially crude culture filtrates.

REFERENCES

Abeles, F. B. (1973) *Ethylene in Plant Biology*. Academic Press, New York. 302 pp.

Ao, H. C. (1977) Epidemiology of *Septoria* species on wheat. Ph.D. Thesis, University of Wales.

Ballio, A. (1972) Phytotoxins: An exercise in the chemistry of biologically active natural products. In *Phytotoxins in Plant Diseases* (R. K. S. Wood, A. Ballio, and A. Graniti, eds.). Academic Press, London. Pp. 71–90.

Ballio, A., F. Pocchiari, S. Russi, and V. Silano. (1971) Effects of fusicoccin and some related compounds on etiolated pea tissues. *Physiol. Plant Pathol.* **1**:95–105.

Carlson, P. S. (1973) Methionine sulphoximine-resistant mutants of tobacco. *Science* **180**:1366–1368.

Carr, A. J. H. (1970) Plant pathology. In *Jubilee Report, Welsh Plant Breeding Station, 1919*–1969. University College of Wales, Aberystwyth. Pp. 168–182.

Chain, E., P. G. Mantle, and B. V. Milborrow. (1971) Further investigations of the toxicity of fusicoccins. *Physiol. Plant Pathol.* **1**:495–515.

Condon, P., and J. Kuć. (1960) Isolation of a fungitoxic compound from carrot root tissue inoculated with *Ceratocystis fimbriata*. *Phytopathology* **50**:267–270.

Condon, P., and J. Kuć. (1962) Confirmation of the identity of a fungitoxic compound produced by carrot root tissue. *Phytopathology* **52**:182–183.

Cronshaw, D. K., and G. F. Pegg. (1976) Ethylene as a toxin synergist in *Verticillium* wilt of tomato. *Physiol. Plant Pathol.* **9**:33–44.

Daly, J. M., and H. W. Knoche. (1976) Hormonal involvement in metabolism of host–parasite interactions. In *Biochemical Aspects of Plant-Parasite Relationships* (J. Friend and D. R. Threlfall, eds.). Academic Press, London. Pp. 115–133.

Dekhuijzen, H. M., and J. C. Overeem. (1971) The role of cytokinins in clubroot formation. *Physiol. Plant Pathol.* **1**:151–162.

Dekhuijzen, H. M., and R. C. Staples. (1968) Mobilization factors in uredospores and bean leaves infected with bean rust fungus. *Contrib. Boyce Thompson Inst.* **24**:39–52.

Evans, M. L. (1973) Rapid stimulation of plant cell elongation by hormonal and non-hormonal factors. *BioScience* **23**:711–718.

Evans, M. L. (1974) Rapid responses to plant hormones. *Annu. Rev. Plant Physiol.* **25**:195–223.

Fuchs, A. (1975) Report at NATO Advanced Study Institute on *Specificity in Plant Diseases*. Porto Cante, Sardinia.

Gäumann, E. (1954) Toxins and plant diseases. *Endeavour* **13**:198–204.

Gnanamanickam, S. S., and S. S. Patil. (1976) Bacterial growth, toxin production and levels of ornithine carbamoyltransferase in resistant and susceptible cultivars of bean inoculated with *Pseudomonas phaseolicola*. *Phytopathology* **66**:290–295.

Gnanamanickam, S. S., and S. S. Patil. (1977a) Accumulation of anti-bacterial isoflavanoids in hypersensitively responding bean leaf tissue inoculated with *Pseudomonas phaseolicola*. *Physiol. Plant Pathol.* **10**:159–168.

Gnanamanickam, S. S., and S. S. Patil. (1977b) Phaseotoxin suppresses bacterially-induced hypersensitive reaction and phytoalexin synthesis in bean cultivars. *Physiol. Plant Pathol.* **10**:169–179.

Graniti, A. (1966) *Host-Parasite Relations in Plant Pathology* (Z. Király and G. Ubriszy, eds.). Research Institute for Plant Protection, Budapest. Pp. 211–217.

Hashimoto, T., and S. Tamura. (1967) Physiological activities of helminthosporol and helminthosporic acid. II. Effects of excised plant parts. *Plant Cell Physiol.* **8**:35–45.

Helgeson, J. P., and N. J. Leonard. (1966) Cytokinins: Identification of compounds isolated from *Corynebacterium fasciens*. *Proc. Natl. Acad. Sci. USA* **56**:60–63.

Hiroe, I., and S. Nishimura. (1956) Pathochemical studies on watermelon wilt. 1. On the wilt toxin phytonivein produced by the causal fungus. *Ann. Phytopathol. Soc. Jpn.* **20**:161–164.

Hislop, E. C., S. A. Archer, and G. V. Hoad. (1973) Ethylene production by healthy and *Sclerotinia fructigena*-infected apple peel. *Phytochemistry* **12**:1281–1286.

Hoitink, H. A. J., and S. L. Sinden. (1970) Partial purification of chlorosis-inducing toxins of *Pseudomonas phaseolicola* and *P. glycinea*. *Phytopathology* **60**:1236–1237.

Jost, J. P. (1965) Contribution à l'étude de la résistance toxicologique des végétaux à l'acide fusarique. *Phytopathol. Z.* **54**:338–378.

Keen, N. T., and B. W. Kennedy. (1974) Hydroxyphaseollin and related isoflavanoids in the hypersensitive resistant response of soybeans against *Pseudomonas glycinea*. *Physiol. Plant Pathol.* **4**:173–185.

Keen, N. T., M. Long, and D. C. Erwin. (1972) Possible involvement of a pathogen-produced protein–lipopolysaccharide complex in *Verticillium* wilt of cotton. *Physiol. Plant Pathol.* **2**:317–331.

Kern, H. (1972) Production and bioassay of phytotoxins. In *Phytotoxins in Plant Diseases* (R. K. S. Wood, A. Ballio, and A. Graniti, eds.). Academic Press, London. Pp. 49–71.

Kern, H., S. Naef-Roth, and F. Rufner. (1972) Der Einfluss der Ernahrung auf die Bildung von Napthazarin-derivaten durch *Fusarium martii* var. *pisi*. *Phytopathol. Z.* **74**:272–280.

Király, Z., M. El Hammady, and B. I. Pazsar. (1967) Increased cytokinin activity of rust-infected bean and broad bean leaves. *Phytopathology* **57**:93–94.

Klambt, D., G. Thies, and F. Skoog. (1966) Isolation of cytokinins from *Corynebacterium fasciens*. *Proc. Natl. Acad. Sci. USA* **56**:52–59.

Kluepfel, D. (1957) Über die Biosynthese und die Umwandlungen der Fusarinsäure in Tomatenplanzen. *Phytopathol. Z.* **29**:349–379.

Lado, P., A. Pennacchioni, F. Rasi-Caldogno, S. Russi, and V. Silano. (1972) Comparison between some effects of fusicoccin and indole-3-acetic acid on cell enlargement in various plant materials. *Physiol. Plant Pathol.* **2**:75–87.

Lado, P., R. Caldogno, A. Pennacchioni, and E. Marré. (1973) Mechanism of the growth-promoting action of fusicoccin. *Planta* **110**:311–320.

Lado, P., F. Rasi-Caldogno, and R. Colombo. (1974) Promoting effect of fusicoccin on seed germination. *Physiol. Plant.* **31**:149–152.

Libbert, E., and R. Manteuffel. (1970) Interactions between plants and epiphytic bacteria regarding their auxin metabolism. VII. *Physiol. Plant.* **23**: 93–98.

Libbert, E., and H. Risch. (1969) Interactions between plants and epiphytic bacteria regarding their auxin metabolism. V. *Physiol. Plant.* **22**:51–58.

Libbert, E., and P. Silhengst. (1970) Interactions between plants and epiphytic bacteria regarding their auxin metabolism. VIII. *Physiol. Plant.* **23**:480–487.

Libbert, E., S. Wichner, E. Duerst, W. Kaiser, R. Kunert, A. Manicki, R. Manteuffel, E. Riecke, and R. Schroder. (1968) Auxin content and auxin synthesis in sterile and non-sterile plants, with special regard to the influence of epiphytic bacteria. In *Biochemistry and Physiology of Plant Growth Substance* (F. Wightman and G. Setterfield, eds.). Runge Press, Ottawa, Canada. Pp. 213–230.

Libbert, E., W. Kaiser, and R. Kunert. (1969) Interactions between plants and epiphytic bacteria regarding their auxin metabolism. IV. *Physiol. Plant.* **22**:432–439.

Lieberman, M. (1979) Biosynthesis and action of ethylene. *Annu. Rev. Plant Physiol.* **30**:533–591.

Livne, A., and J. M. Daly. (1966) Translocation in healthy and rust-affected beans. *Phytopathology* **56**:170–175.

Luke, H. H., and V. E. Gracen, Jr. (1972) Phytopathogenic toxins. In *Microbial Toxins*, Vol. 8 (S. Kadis, A. Ciegler, and S. J. Ajl, eds.). Academic Press, New York. Pp. 131–137.

Marré, E. (1979) Fusicoccin: A tool in plant physiology. *Annu. Rev. Plant Physiol.* **30**:273–288.

Marré, E., R. Columbo, P. Lado, and F. Rasi-Caldogno. (1974) Correlation between proton extrusion and simulation of cell enlargement. Effects of fusicoccin and of cytokinins on leaf fragments and isolated cotyledons. *Plant Sci. Lett.* **2**:139–150.

Matern, U., G. Strobel, and J. Shepard. (1978) Reaction to phytotoxins in a potato population derived from mesophyll protoplasts. *Proc. Natl. Acad. Sci. USA* **75**:4935–4937.

Mikami, Y., Y. Nishijima, H. Iimura, A. Suzuki, and S. Tumara. (1971) Chemical studies on brown spot disease of tobacco plants. 1. Tenuazonic acid as a *vivo* toxin of *Alternaria longipes*. *Agric. Biol. Chem.* **35**:611–618.

Mitchell, R. E. (1979) Bean halo blight: Comparison of phaseolotoxin and *n*-phosphoglutamate. *Physiol. Plant Pathol.* **14**:119–128.

Muromtsev, G. S., and G. A. Globus. (1976) Adaptive significance of the ability to synthesize gibberellins for the phyto-pathogenic fungus. *Dokl. Akad. Nauk SSSR* **226**(1):204–206.

Mussell, H. W. (1972) Toxic proteins secreted by cotton isolates of *Verticillium albo-atrum*. In *Phytotoxins in Plant Diseases* (R. K. S. Wood, A. Ballio, and A. Graniti, eds.). Academic Press, London. Pp. 443–445.

Mussell, H. W. (1973) Endopolygalacturonase: Evidence for involvement in *Verticillium* wilt of cotton. *Phytopathology* **63**:62–70.

Nachmias, A., I. Barash, V. Buchner, Z. Solel, and G. A. Strobel. (1979) A phytotoxic glycopeptide from lemon leaves infected with *Phoma tracheiphila*. *Physiol. Plant Pathol.* **14**:135–141.

Owens, L. D. (1969) Toxins in plant disease: Structure and mode of action. *Science* **165**:18–25.

Patil, S. S. (1974) Toxins produced by phytopathogenic bacteria. *Annu. Rev. Phytopathol.* **12**:259–279.

Patil, S. S., P. Youngblood, P. Christiansen, and R. E. Moore. (1976) Phaseotoxin A, an antimetabolite from *Pseudomonas phaseolicola*. *Biochem. Biophys. Res. Commun.* **69**:1019–1027.

Pegg, G. F. (1976a) Endogenous auxins in healthy and diseased plants. In *Encyclopaedia of Plant Physiology*, Vol. 4 (R. Heitefuss and P. H. Williams, eds.). Springer-Verlag, New York. Pp. 560–616.

Pegg, G. F. (1976b) The involvement of ethylene in plant pathogenesis. In *Encyclopaedia of Plant Physiology*. Vol. 4 (R. Heitefuss and P. H. Williams, eds.). Springer-Verlag, New York. Pp. 582–591.

Pegg, G. F. (1976c) The response of ethylene-treated tomato plants to infection by *Verticillium albo-atrum*. *Physiol. Plant Pathol.* **9**:215–226.

Pegg, G. F. (1979) The biochemistry and physiology of pathogenesis. In *Fungal Wilt Diseases* (C. H. Beckman, A. A. Bell, and M. E. Mace, eds.). Academic Press, New York. Pp. 17–44.

Pegg, G. F., and D. K. Cronshaw. (1976) Ethylene production by tomato plants infected with *Verticillium albo-atrum*. *Physiol. Plant Pathol.* **8**:279–295.

Rai, P. V., and G. A. Strobel. (1969a). Phytotoxic glycopeptides produced by *Corynebacterium michiganense*. I. Methods of preparation, physical and chemical characterization. *Phytopathology* **59**:47–52.

Rai, P. V., and G. A. Strobel. (1969b) Phytotoxic glycopeptides produced by *Corynebacterium michiganense*. II. Biological properties. *Phytopathology* **59**:53–57.

Reddy, M. N., and P. H. Williams. (1970) Cytokinin activity in *Plasmodiophora brassicae*-infected cabbage tissue cultures. *Phytopathology* **60**:1463–1465.

Ries, S. M., and G. A. Strobel. (1972a) A phytotoxic glycopeptide from cultures of *Corynebacterium insidiosum*. *Plant Physiol.* **49**:676–684.

Ries, S. M., and G. A. Strobel. (1972b) Biological properties and pathological role of a phytotoxic glycopeptide from *Corynebacterium insidiosum*. *Physiol. Plant Pathol.* **2**:133–142.

Rudolph, K. (1972) The halo blight toxin of *Pseudomonas phaseolicola*: Influence on host–parasite relationships and counter effects of metabolites. In *Phytotoxins in Plant Diseases* (R. K. S. Wood, A. Ballio, and A. Graniti, eds.). Academic Press, London. Pp. 373–375.

Rudolph, K. (1976) Non-specific toxins. In *Encyclopaedia of Plant Physiology*, Vol. 4 (R. Heitefuss and P. H. Williams, eds.). Springer-Verlag, New York. Pp. 270–315.

Rudolph, K., and M. A. Stahmann. (1966) Accumulation of L-ornithine in halo blight infected bean plants induced by the toxin of the pathogen *Pseudomonas phaseolicola*. *Phytopathol. Z.* **57**:29–46.

Sequeira, L. (1963) Growth regulators in plant disease. *Annu. Rev. Phytopathol.* **1**:5–30.

Sequeira, L. (1973) Hormone metabolism in diseased plants. *Annu. Rev. Plant Physiol.* **24**:353–380.

Sequeira, L., and P. H. Williams. (1964) Synthesis of indoleacetic acid by *Pseudomonas solanacearum*. *Phytopathology* **54**:1240–1246.

Shepard, J. F. (1979) Mutant selection and plant regeneration from potato protoplasts. In *Emergent Techniques for the Genetic Improvement of Crops*. University of Minnesota Press, Minneapolis, Minn., in press.

Shepard, J. F., and R. E. Totten. (1977) Mesophyll cell protoplasts of potato: Isolation, proliferation, and plant regeneration. *Plant Physiol.* **60**:313–316.

Sinden, S. L., and R. D. Durbin. (1968) Glutamine synthetase inhibition: Possible mode of action of wildfire toxin from *Pseudomonas tabaci*. *Nature* **219**:379–380.

Smidt, M., and T. Kosuge. (1978) The role of indole-3-acetic acid accumulation by alpha methyltryptophan-resistant mutants of *Pseudomonas savastanoi* in gall formation on oleanders. *Physiol. Plant Pathol.* **13**:203–214.

Sondheimer, E. (1957) The isolation and identification of 3-methyl-6-methoxy-8-hydroxy-3,4-dihydroisocoumarin from carrots. *J. Am. Chem. Soc.* **79**:5036–5039.

Stewart, W. W. (1971) Isolation and proof structure of wildfire toxin. *Nature* **229**:174–178.

Straley, C. S., M. L. Straley, and G. A. Strobel. (1974) Rapid screening for bacterial wilt resistance in alfalfa with a phytotoxic glycopeptide from *Corynebacterium insidiosum*. *Phytopathology* **64**:194–196.

Strobel, G. A. (1970) A phytotoxic glycopeptide from potato plants infected with *Corynebacterium sepedonicum*. *J. Biol. Chem.* **245**:32–38.

Strobel, G. A. (1974) Phytotoxins produced by plant parasites. *Annu. Rev. Plant Physiol.* **25**:541–566.

Strobel, G. A. (1977) Bacterial phytotoxins. *Annu. Rev. Microbiol.* **31**:205–224.

Strobel, G. A., and P. V. Rai. (1968) A rapid serodiagnostic test for potato ring rot. *Plant Dis. Rep.* **52**:502–504.

Sutton, J. C., and P. H. Williams. (1970) Comparison of extracellular polysaccharide of *Xanthomonas campestris* from culture and from infected cabbage leaves. *Can. J. Bot.* **48**:645–651.

Tamura, S., A. Sakurai, A. Kainuma, and M. Takai. (1963) Isolation of helminthosporol as a natural plant growth regulator. *Agric. Biol. Chem.* **27**:738–739.

Taylor, P. A., H. K. Schnoes, and R. D. Durbin. (1972) Characterization of chlorosis-inducing toxins from a plant pathogenic *Pseudomonas* sp. *Biochim. Biophys. Acta* **286**:107–117.

Thimann, K. V., and T. Sachs. (1966) The role of cytokinins in the "fasciation" disease caused by *Corynebacterium fasciens*. *Am. J. Bot.* **53**:731–739.

Trione, E. J. (1960) Extracellular enzyme and toxin production by *Fusarium oxysporum* f. *lini*. *Phytopathology* **50**:480–482.

Turner, N. C. (1973) Action of fusicoccin on the potassium balance of guard cells of *Phaseolus vulgaris*. *Am. J. Bot.* **60**:717–725.

Van Alfen, N. K., and N. C. Turner. (1975) Changes in alfalfa stem conductance induced by *Corynebacterium insidiosum*. *Plant Physiol.* **55**:559–561.

Weise, M. V., and J. E. De Vay. (1970) Growth regulator changes in cotton associated with defoliation caused by *Verticillium albo-atrum*. *Plant Physiol.* **45**:304–309.

Wheeler, H., and H. H. Luke. (1963) Microbial toxins in plant disease. *Annu. Rev. Microbiol.* **17**:223–242.

Wood, H. N., A. C. Braun, H. Brandes, and H. Kende. (1969) Studies on the distribution and properties of a new class of cell division-promoting substances from higher plant species. *Proc. Natl. Acad. Sci. USA* **62**:349–356.

Wood, H. N., M. C. Lin, and A. C. Braun. (1972) The inhibition of plant and animal adenosine $3^1 4^1$-cyclic monophosphate, phosphodiesterase by a cell-division promoting substance from tissues of higher plant species. *Proc. Natl. Acad. Sci. USA* **69**:403–406.

Wood, R. K. S. (1972) The killing of cells by soft rot parasites. In *Phytotoxins in Plant Diseases* (R. K. S. Wood, A. Ballio, and A. Graniti, eds.). Academic Press, New York. Pp. 273–288.

Wooley, D. W., G. Schaffner, and A. C. Braun. (1955) Isolation of the phytopathogenic toxin of *Pseudomonas tabaci*, an antagonist of methionine. *J. Biol. Chem.* **197**:409–417.

SUSCEPTIBILITY AS A PROCESS INDUCED BY PATHOGENS

Seiji Ouchi and Hachiro Oku

In nature, disease is an exception rather than a rule. Only a limited number of microbes are pathogens of animals or plants. The majority of multicellular organisms defend themselves from potential pathogens by exerting physical and chemical mechanisms, which they acquire during an evolutionary process. Pathogens, as exceptional agents, must overcome these versatile barriers against infection establishment, hence must have been endowed with the abilities to suppress the operation of the static or dynamic defense mechanism by host cells or tissues.

In contrast to the enormous amount of information on disease resistance, practically nothing is known about the mechanism of susceptibility, except the role of host-specific toxins. It has long been known, however, that plants preliminarily infected by a virulent pathogen become susceptible to a weakly virulent pathogen or nonpathogen, and this phenomenon has been referred to as predisposition, induced susceptibility, acquired susceptibility (Yarwood, 1959, 1976) or induced accessibility (Ouchi et al. 1974a, b). This type of pathogen-induced susceptibility suggests that the active suppression of host defense mechanism is a prerequisite for the establishment of pseudosymbiotic association between the host cell and parasite, a basis of pathogenesis.

This treatise aims at summarising the current status of our understanding of the process of induced susceptibility.

ACCESSIBILITY INDUCTION

Viral Diseases

Pathogen–host interaction in viral infections is much more direct than in other diseases and could be characterized as a genomic interaction. A biochemical study of the function of split genomes such as satellite and multicomponent viruses provides a profound insight into the programming of viral infection.

When two closely related viruses were inoculated simultaneously or in a sequence with an interval, multiplication or symptom expression of one virus is inhibited by the other, and this phenomenon has been referred to as interference or cross-protection. There have been many hypotheses on the mechanism of the cross-protection (Matthews, 1970). In spite of the attractiveness of the exclusion hypothesis (Siegel, 1959), accumulated evidence supports the idea that two viruses enter and multiply in a single cell (Benda, 1956; Fujisawa et al., 1967; Otsuki and Takebe, 1976). Apparently there is competition between the two viruses for a single replication site.

Another aspect of interaction between two viruses is synergism. Since Salaman demonstrated in 1938 that potato virus X formed larger necrotic lesions in tobacco when inoculated together with tobacco mosaic virus (TMV) (Kassanis, 1963), much information has been accumulated about the synergistic multiplication of different viruses in a single plant host.

Atabekov et al. (1970) elucidated in tobacco–TMV system that a temperature-sensitive mutant was rescued when inoculated together with a wild-type strain. The synthesis of potato virus X in tobacco leaves is enhanced by the presence of potato virus Y or TMV in the same cell (Goodman and Ross, 1974). These findings together with other information suggest that the helper virus may complement some factors essential for the multiplication of the aberrant or otherwise weakly virulent virus. Of particular interest are those infections in which nonpathogenic virus multiplied to an appreciable extent in association with a virulent virus. In barley leaves inoculated with barley stripe mosaic virus or brome mosaic virus (BMV), TMV multiplication, otherwise subliminal, was markedly enhanced and became systemic (Hamilton and Dodds, 1970; Dodds and Hamilton, 1972, 1974; Hamilton and Nichols, 1977). Hamilton and Nichols (1977) interpreted these data to indicate that the helper virus perhaps affected the growth and development of the host plant to predispose it to systemic infection by TMV. However, the fact that alfalfa mosaic, cowpea chlorotic mottle, southern bean mosaic, tobacco necrosis and turnip yellow mosaic viruses did not cause systemic infection in BMV-infected barley leaves seems to make another interpretation possible; that is, the replication system encoded on the helper virus genome might have complemented the need for an enhanced multiplication of TMV. It could not have complemented those of other viruses, perhaps because the level of their restriction differed from that of the helper virus.

Furusawa and Okuno (1979) inoculated barley protoplasts with TMV-RNA with or without BMV-RNA and found that only a few cells (1–2%) fluoresced with anti-TMV antibody when inoculated with TMV–RNA alone. In contrast, 32% and 31% of protoplasts fluoresced with anti-BMV and anti-TMV antibodies, respectively, in the batch inoculated with the both viral RNAs. This is a clear indication that BMV in fact functioned as a helper in the translation and replication of TMV–RNA in barley leaves. The role of helper virus in replication and symptom expression should be studied with an emphasis on the involvement of host genome in these processes.

Bacterial Diseases

The virulence of pathogenic bacteria in general is related to their ability to multiply in the host tissue to the level that is sufficient for the expression of characteristic symptoms. Inoculation of a plant with mixed inoculum of virulent and avirulent or nonpathogenic bacteria usually results in less disease (Leben, 1965; Sequeira, 1979). In some cases, however, disease incited by a mixed inoculum becomes more severe than those caused by singular inoculation.

The rate of wilting and the extent of stem rot of carnation was higher when the plants were inoculated with a mixture of *Pseudomonas caryophylli* and *Corynebacterium* sp. (Brathwaite and Dickey, 1970). Synergism was also reported in the halo blight of bean. The presence of *Achromobacter* sp. in the inoculum enhanced not only the number of lesions, but also the multiplication of *P. phaseolicola* in the primary leaves (Maino et al., 1974). However, the population of *Achromobacter* sp. did not increase in the mixed inoculation. Leaves of rice plant inoculated with a mixture (1:1 ratio) of high-virulent and low-virulent isolates of *Xanthomonas oryzae* gave a high infection score regardless of the methods of inoculation (Reddy and Kauffman, 1974). By employing streptomycin-resistant mutants Fujii (1976) clearly demonstrated that avirulent isolates multiplied in rice leaves when inoculated together with a virulent isolate.

A similar population increase of the avirulent strain was found in apple shoot inoculated with mixtures of virulent and avirulent strain of *Erwinia amylovora*. Avirulent strains of *E. amylovora* and *E. herbicola* protect susceptible apple shoot from infection by a virulent strain (Goodman, 1967). When a streptomycin-resistant avirulent strain was mixed with a virulent strain and used to inoculate the petioles of young shoots of susceptible Jonathan apples, the avirulent strain multiplied at the same rate as that of virulent strain. This induced accessibility was not effective with some other strains of *E. amylovora* and nonpathogenic bacteria to apple shoot. The results thus suggest that nutritional supply by the virulent bacteria is not the major mechanism of the induced accessibility (Ouchi et al., unpublished data).

Fungal Diseases

As exemplified by the classical example of predisposition, increased susceptibility of bunt-infected wheat to yellow rust, induced susceptibility has been reported in many plant-fungus combinations, especially obligately parasitic diseases (Yarwood, 1959). In some cases predisposition was absolute, while others were relative. The description given was phenomenal, but the possible mechanism was discussed in few cases. Since there were difficulties in analyzing this type of induced susceptibility in necrotrophic diseases, most work has been confined to biotrophic diseases. The description here will be restricted to those works which aimed at disclosing the mechanism of induced susceptibility.

Hiura (1962) at first succeeded in hybridizing some races of different *formae speciales* of *Erysiphe graminis,* and noted that in some crosses white mycelial mats were formed without producing a cleistothecium. This would be the first observation that suggests a possible participation of induced susceptibility in the powdery mildew interaction. Moseman and Greeley (1964) and Moseman et al. (1965) then demonstrated that both the barley and wheat leaves could each be predisposed to nonpathogenic *forma specialis* by preliminary inoculation with a compatible race. Tsuchiya and Hirata (1973) extended this to other powdery mildews and found that on mildewed barley leaves 45 of 51 powdery mildew fungi (49 from dicotyledonous plants) established infection near the colonies of the barley fungus, and 30 produced conidia on these predisposed leaves. This type of predisposition was considered as a physiological process essential for pathogenesis, and the concept of accessibility was established (Ouchi et al., 1974a,b).

Preliminary inoculation with a compatible race of *E. graminis hordei* induced in barley leaves an accessibility not only to incompatible races of that fungus, but also to the wheat fungus (*E. graminis tritici*) and the melon powdery mildew fungus (*Sphaerotheca fuliginea*). On the other hand, barley leaves inoculated with an incompatible race were found, which become resistant, inaccessible in our term, to an originally compatible race of the barley fungus. For the completion of cellular conditioning toward accessibility, 15 to 18 hr was required, while conditioning toward inaccessibility required only 6 hr of incompatible interaction. Both the induced accessibility and inaccessibility were localized near the site of primary interaction, at least in the early phase of infection (Ouchi et al., 1979; Ouchi et al., 1976). Recent work in our laboratory disclosed that when the abaxial surface of barley leaves was inoculated with a compatible race, accessibility in the adaxial epidermis was found established by 72 hr and became maximum by 5 days after the primary inoculation. This accessible state lasted at least for 6 days. On the contrary, inaccessibility was established in the opposite epidermis by 18 hr after inoculation with an incompatible race (Ouchi et al., unpublished data).

Of particular interest was the irreversible nature of the cellular conditioning. Triple inoculation experiments indicated that once cells were conditioned toward the accessible state, even the melon fungus was not recognized as incompatible. The leaves that had been induced to become inaccessible then challenged with a compatible race responded with more extensive yellowing than those challenged by an incompatible race. This suggested that the cells that had been conditioned to an inaccessible state recognized an originally compatible race as incompatible. Irreversibility of recognition was also reported in crown rust of oat. An expression of resistance by oat leaves against incompatible race of *Puccinia coronata avenae* decreased when leaves were challenged by a compatible race within 8 hr of incompatible interaction. However, no appreciable effect was observed when the compatible race was inoculated 12 hr after the incompatible inducer. It was demonstrated that leaves that had been conditioned toward resistance recognized the subsequently inoculated compatible race as incompatible, leading to amplification of the resistance reaction (Tani et al., 1975). Furthermore, preliminary inoculation with a compatible race induced in oat leaves an accessibility to an incompatible race. This induced accessibility was effective against nonpathogens such as *Uromyces alopecuri* or *Puccinia graminis tritici,* but not to *P. coronata festucae* (Tani, 1979). Since resistance expression involved an activation of RNA (mRNA) and protein syntheses as a process sequel to the recognition and determination (Tani et al., 1976; Tani and Yamamoto, 1979), the compatible race must have suppressed either the recognition or the subsequent conditioning process by an as yet unknown mechanism.

CHEMICAL BASIS FOR INDUCED SUSCEPTIBILITY

Although the molecular mechanism of the induced susceptibility is beyond the subject of this treatise, an *a priori* assumption was that pathogens must produce molecules reversibly or irreversibly to suppress or deteriorate the intrinsic ability of host tissues to recognize and reject invading microbes. The most basic attribute of all pathogens is to impede the versatile defense mechanisms of the host plants. Some pathogens may produce an extracellular polysaccharide or capsule and passively escape from the host-cell defense reaction by means of the negation of the recognition process. Some will produce toxins or enzymes to impede the inherent resistance reaction by directly or indirectly impairing membrane function or other cellular activities. Still another group of pathogens will produce nontoxic compounds such as suppressors or hypersensitivity-inhibiting factors to interfere with most probably the recognition process or subsequent cellular conditioning toward resistance. The following are some examples indicating the presence of these types of pathogen products.

Toxins

Some pathogens produce toxins, either host-specific or nonspecific, to conquer the host defense reaction. The so-called host-specific toxins are, in most cases, characterized as pathotoxin and treatment of a toxin-sensitive host with these toxins usually renders the tissues susceptible to an avirulent mutant or nonpathogen. *Helminthosporium victoriae* toxin predisposed toxin-sensitive cultivars to avirulent *H. victoriae* and nonpathogenic *H. carbonum* when it was added to the conidial suspension of these fungi. HV-toxin also assisted the colonization of *N. crassa,* although the growth pattern was different (Yoder and Scheffer, 1969).

A similar phenomenon was reported for the pear-*Alternaria kikuchiana* system. Avirulent isolates became capable of establishing infection of pear when AK-toxin was added to a conidial suspension (Otani et al., 1975). Some saprophytic isolates of *A. alternata* behaved similarly in the presence of toxin, avirulent isolates contributing to lesion formation when inoculated together with the virulent conidia. Since preliminary inoculation with a nonpathogen reduced the number of lesions by subsequently inoculated virulent isolates, pear leaves were potentially capable of resisting these nonpathogens. AK-toxin then is a chemical entity which suppresses ability of tissues to reject nonpathogens, probably partially impairing membrane function as evidenced by electrolyte leakage in the very early stage of infection (Nishimura et al., 1976; Otani, 1979).

Nishimura et al. (1979) proposed that the pathogenicity of these *Alternaria* species known to produce host-specific toxins consisted of a potential for aggressiveness (ability to penetrate), common to all the fungi belonging to *A. alternata,* and a specific virulence exhibited by production of the respective host-specific toxin. It has recently been unequivocally demonstrated that while HMT-toxin is not required for the establishment of a compatible relationship, nevertheless it contributes to the virulence of *H. maydis* race T (Yoder, 1976; Payne and Yoder, 1978).

Phaseotoxin was also found to inhibit the hypersensitive reaction and phytoalexin accumulation in bean tissues inoculated with *P. phaseolicola* (Patil and Gnanamanicham, 1976). Thus the role of toxins seems to be versatile; some will function as a pathogenicity factor and others as a virulence factor.

Hypersensitivity-Inhibiting Factor and Suppressors

Hypersensitive cell death is one of the most characteristic resistance responses of plant tissues against pathogens. Potato tissues were not exceptional and responded with a typical hypersensitivity to incompatible races of *Phytophthora infestans.* The preliminary inoculation of potato tissue with a compatible race, however, inhibited the hypersensitive response against an incompatible race (Tomiyama, 1966). This phenomenon has been recently explained as a process involving interaction between molecular constituents of the host tissue and the fungus. Zoospores of compatible and incompatible races contained components that induced hypersensitive cell death and phytoalexin accumulation (Doke et al., 1976; Kuć et al., 1976). This component became bound to membrane fractions of potato tissue homogenate

(Doke et al., 1976). The race-cultivar specificity seems to reside in the inhibition of the binding of these components of the host and fungus, by a component produced by a compatible race. The hypersensitivity-inhibiting factor (or suppressor) isolated from zoospores or germination fluids of a compatible race inhibited the binding, but components from an incompatible race did not (Doke, 1975). This race specific factor was found to be a glucan with 1–3 and 1–6 linkages (Doke et al., 1977; Kuć et al., 1979).

A similar type of pathogenicity factor was found in *Mycosphaerella* blight of pea plants. It is well known that pea plants produce pisatin in response to infection and other type of stress. Pisatin inhibits establishment of infection by pisatin-resistant *Erysiphe pisi* even at a very low concentration, provided the leaf tissue had been treated in the early phase of infection (Oku et al., 1979). *Mycosphaerella pinodes,* another pathogen of pea plants, was found to produce a high molecular weight elicitor in the pycnospore germination fluid as well as a low-molecular weight suppressor of pisatin production. The elicitor appears to be a glucan of approximate molecular weight of 7×10^4 daltons. The suppressor seems to contain an oligopeptide as an active group that is comprised of glycine, aspartic acid, and serine at the ratio of 1:1:2 (Oku et al., 1980). Application of this peptide to pea leaves suppressed pisatin induction in response to the nonpathogen, *Erysiphe graminis* (Shiraishi et al., 1978). Presence of this peptide in a conidial suspension rendered pea leaves accessible to nonpathogens such as *Stemphylium sarcinaeforme* (target spot of clover), *Mycosphaerella ligulicola* (ray blight of chrysanthemum), *Alternaria kikuchiana* (black spot of pears), *Ascochyta punctata* (*Ascochyta* leaf spot of vetches). The peptide did not cause any observable changes in leaf tissues, hence could hardly be considered a toxin. Recent work in our laboratory also indicated the presence of a low molecular weight inducer of accessibility in the water extracts of barley leaves that had been infected by a compatible race (Ouchi et al., unpublished data).

No information is yet available in regard to the function of suppressors as a virulence factor. Pathogens may be producing these impediments to negate the dynamic defense reaction of host tissues during pathogenesis. Or they may function only in the initial stage of establishment of the pseudosymbiotic association between the host and parasite, and the subsequent host-cell metabolism may give rise to a class of compounds, which in turn suppress the operational defense of plant tissues.

In view of meager information on the mode of action of these pathogen products, the model based on metabolic alterations in compatible interactions (Daly, 1976) is certainly a milestone toward the molecular basis of induced susceptibility.

CONCLUDING REMARKS

Almost a century has passed since the concept of predisposition was established. The concept has changed since then, but unfortunately has been considered as one contradicting that of pathogenesis. The concept of accessibility treated in this chapter blends, in some respects, these ideas of predisposition and pathogenesis; that is, pathogens predispose host cells or tissues by producing factor(s), toxic or nontoxic, as the most basic mechanism of pathogenesis.

The factor(s) involved in conditioning the primary host cell toward accessibility could be named "aggressin" or "impedin," both of which have been used in medical science to mean a nontoxic bacterial factor capable of inhibiting any defense mechanism of multicellular organisms against infection. As the term "aggressin" literally gives the impression that it actually damages host cells, Glynn (1972) prefers to use the term "impedin," originally used by R. Torikata in 1917 to define a bacterial factor inhibiting an antigen–antibody reaction. These terms are both defined as bacterial factors, but origin may not be essential for the use of term. The term aggressin has already been used by Chou for the pollen factor that induces aggressive legion by *Botrytis cinerea* (Warren, 1972). In support of the idea of Glynn, we propose the use of the term impedin for the factor that is produced by a pathogen and actively or passively suppresses the defense reaction of host cells or tissues, enabling the producer or other potential pathogens to establish infection. As is summarized in this treatise, the term is no longer of conceptual significance but is of practical importance in the interpretation of pathogenesis. Susceptibility is indeed a process induced by pathogens.

At the moment this type of work may not contribute to the formulation of plant disease management concepts. However, we should make more effort to elucidate the biochemical basis of disease, not only from the aspect of the resistance mechanism, but also from the aspect of genuine and noble capability of pathogens to impede the operation of host defense mechanisms. This is the prerequisite for the theoretical approach that the blockage of the elaborate means of pathogens to suppress the host defense is the most promising strategy that we could develop. Certainly the task is in the future.

ACKNOWLEDGMENT

This work was supported in part by a grant in aid from the Ministry of Education, Science and Culture, Japan (No. 336005). Financial support from the Rockefeller Foundation is also acknowledged.

REFERENCES

Atabekov, J. G., N. D. Schaskolskaya, T. I. Atabekova, and G. A. Sacharovskaya. (1970) Reproduction of temperature sensitive strain of TMV under restrictive conditions in the presence of temperature-resistant helper strain. *Virology* **41**:397–407.

Benda, G. T. A. (1956) Infection of *Nicotiana glutinosa* L. following injection of two strains of tobacco mosaic virus into a single cell. *Virology* **2**:820–827.

Brathwaite, C. W. D., and R. S. Dickey. (1970) Synergism between *Pseudomonas caryophylli* and a species of *Corynebacterium*. *Phytopathology* **60**:1046–1051.

Daly, J. M. (1976) Induced susceptibility or induced resistance as the basis of host–parasite specificity. In *Biochemistry and Cytology of Plant–Parasite Interaction* (K. Tomiyama, J. M. Daly, I. Uritani, H. Oku, and S. Ouchi, eds.). Elsevier, Amsterdam. Pp. 144–156.

Dodds, J. A., and R. I. Hamilton. (1972) The influence of barley stripe mosaic virus on the replication of tobacco mosaic virus in *Hordeum vulgare* L. *Virology* **50**:404–411.

Dodds, J. A., and R. I. Hamilton. (1974) Masking of the RNA genome of tobacco mosaic virus by the protein of barley stripe mosaic virus in doubly infected barley. *Virology* **59**:418–426.

Doke, N. (1975) Prevention of the hypersensitive reaction of potato cells to infection with an incompatible race of *Phytophthora infestans* by constituents of zoospores. *Physiol. Plant Pathol.* **7**:1–7.

Doke, N., K. Tomiyama, H. S. Lee, N. Nishimura, and N. Matsumoto. (1976) Mode of physiological response of potato tissue induced by constituents of *Phytophthora infestans* in relation to host–parasite specificity. In *Biochemistry and Cytology of Plant–Parasite Interaction* (K. Tomiyama, J. M. Daly, I. Uritani, H. Oku, and S. Ouchi, eds.). Elsevier, Amsterdam. Pp. 157–167.

Doke, N., N. A. Garas, and J. Kuć. (1977) Partial characterization and mode of action of the hypersensitivity-inhibiting factor isolated from *Phytophthora infestans*. *Proc. Am. Phytopathol. Soc.* **4**:165 (Abstr.).

Fujii, H. (1976) Variation in the virulence of *Xanthomonas oryzae*. 3. Interaction between avirulent mutants and virulent wild type isolates. *Ann. Phytopathol. Soc. Jpn* **42**:526–532.

Fujisawa, I., T. Hayashi, and C. Matsui. (1967) Electron microscopy of mixed infections with two plant viruses. I. Intracellular interaction between tobacco mosaic virus and tobacco etch virus. *Virology* **33**:70–76.

Furusawa, I., and T. Okuno. (1979) Could BMV be the helper of TMV? *Proc. 1979 Annual Meeting of the Phytopathological Society of Japan*, V-54 (In Japanese).

Glynn, A. A. (1972) Bacterial factors inhibiting host defense mechanisms. In *Microbial Pathogenicity in Man and Animals* (H. Smith and J. H. Pearce, eds.). Cambridge University Press, Cambridge. Pp. 75–112.

Goodman, R. N. (1967) Prevention of apple stem tissue against *Erwinia amylovora* infection by avirulent strains and three other bacterial species. *Phytopathology* **57**:22–24.

Goodman, R. M., and A. F. Ross. (1974) Enhancement of potato virus X synthesis in doubly infected tobacco occurs in doubly infected cells. *Virology* **58**:16–24.

Hamilton, R. I., and J. A. Dodds. (1970) Infection of barley by tobacco mosaic virus in single and mixed infection. *Virology* **42**:266–268.

Hamilton, R. I., and C. Nichols. (1977) The influence of bromegrass mosaic virus on the replication of tobacco mosaic virus in *Hordeum vulgare*. *Phytopathology* **67**:484–489.

Hiura, U. (1962) Hybridization between varieties of *Erysiphe graminis*. *Phytopathology* **52**:664–666.

Kassanis, B. (1963) Interactions of viruses in plants. *Adv. Virus Res.* **10**:219–255.

Kuć, J., W. Currier, J. Elliston, and J. McIntyre. (1976) Determinants of plant disease resistance and susceptibility: A perspective based on three plant–parasite interactions. In *Biochemistry and Cytology of Plant–Parasite Interaction* (K. Tomiyama, J. M. Daly, I. Uritani, H. Oku, and S. Ouchi, eds.). Elsevier, Amsterdam. Pp. 168–180.

Kuć, J., J. Henfling, N. Garas, and N. Doke. (1979) Control of terpenoid metabolism in the potato–*Phytophthora infestans* interaction. *J. Food Prot.* **42**:508–511.

Leben, C. (1965) Epiphytic microorganisms in relation to plant diseases. *Annu. Rev. Phytopathol.* **3**:209–230.

Maino, A. L., M. N. Schroth, and V. B. Vitanza. (1974) Synergy between *Achromobacter* sp. and *Pseudomonas phaseolicola* resulting in increased disease. *Phytopathology* **64**:277–283.

Matthews, R. E. F. (1970) *Plant Virology*. Academic Press, New York.

Moseman, J. G., and L. W. Greeley. (1964) Predisposition of wheat by *Erysiphe graminis tritici* to infection with *Erysiphe graminis hordei*. *Phytopathology* **54**:618.

Moseman, J. G., A. L. Scharen, and L. W. Greeley. (1965) Production of *Erysiphe graminis* f. sp. *tritici* on barley and *Erysiphe graminis* f. sp. *hordei* on wheat. *Phytopathology* **55**:92–96.

Nishimura, S., K. Kohmoto, H. Otani, H. Fukami, and T. Ueno. (1976) The involvement of host-specific toxins in the early step of infection by *Alternaria kikuchiana* and *A. mali*. In *Biochemistry and Cytology of Plant–Parasite Interaction* (K. Tomiyama, J. M. Daly, I. Uritani, H. Oku, and S. Ouchi, eds.). Elsevier, Amsterdam. Pp. 94–101.

Nishimura, S., K. Kohmoto, and H. Otani. (1979) The role of host-specific toxins in saprophytic pathogens. In *Recognition and Specificity in Plant–Parasite Interactions* (J. M. Daly and I. Uritani, eds.). Japan Scientific Societies Press, Tokyo. Pp. 133–146.

Oku, H., T. Shiraishi, and S. Ouchi. (1979) The role of phytoalexin in host–parasite specificity. In *Recognition and Specificity in Plant–Host Parasite Interactions* (J. M. Daly and I. Uritani, eds.). Japan Scientific Societies Press, Tokyo. Pp. 317–333.

Oku, H., T. Shiraishi, S. Ouchi, M. Ishiura, and R. Matsueda. (1980) A new determinant of pathogenicity in plant disease. *Naturwissenschaften* **67**:310.

Otani, H. (1979) Regulation of infection response by a host-specific toxin. In *Proceedings of Symposium on Regulation Mechanisms in Plant Infections*. Phytopathological Society of Japan, July 14–16, Nagano, Japan. Pp. 105–112. (Written in Japanese.)

Otani, H., S. Nishimura, K. Kohmoto, K. Yano, and T. Seno. (1975) Nature of specific susceptibility to *Alternaria kikuchiana* in Nijisseiki cultivar among Japanese Pears. V. Role of host-specific toxin in early step of infection. *Ann. Phytopathol. Soc. Jpn* **41**:467–476.

Otsuki, Y., and I. Takebe. (1976) Double infection of isolated tobacco leaf protoplasts by two strains of tobacco mosaic virus. In *Biochemistry and Cytology of Plant-Parasite Interaction* (K. Tomiyama, J. M. Daly, I. Uritani, H. Oku, and S. Ouchi, eds.). Elsevier, Amsterdam. Pp. 213–222.

Ouchi, S., H. Oku, C. Hibino, and I. Akiyama. (1974a) Induction of accessibility and resistance in leaves of barley by some races of *Erysiphe graminis*. *Phytopathol. Z.* **79**:24–34.

Ouchi, S., H. Oku, C. Hibino, and I. Akiyama. (1974b) Induction of accessibility to a nonpathogen by preliminary inoculation with a pathogen. *Phytopathol. Z.* **79**:142–154.

Ouchi, S., H. Oku, and C. Hibino. (1976) Some characteristics of induced susceptibility and resistance demonstrated in powdery mildew of barley. In *Biochemistry and Cytology of Plant–Parasite Interaction* (K. Tomiyama, J. M. Daly, I. Uritani, H. Oku, and S. Ouchi, eds.). Elsevier, Amsterdam. Pp. 181–194.

Ouchi, S., C. Hibino, H. Oku, M. Fujiwara, and H. Nakabayashi. (1979) The induction of resistance or susceptibility. In *Recognition and Specificity in Plant–Parasite Interactions* (J. M. Daly and I. Uritani, eds.). Japan Scientific Societies Press, Tokyo. Pp. 49–65.

Patil, S. S., and S. S. Gnanamanicham. (1976) Suppression of bacterially induced hypersensitive reaction and phytoalexin accumulation in bean by phaseotoxin. *Nature* **259**:486–487.

Payne, G. A., and O. C. Yoder. (1978) Effect of the nuclear genome of corn on sensitivity to *Helminthosporium maydis* race T toxin and susceptibility to *H. Maydis* race T. *Phytopathology* **68**:331–337.

Reddy, A. P. K., and H. E. Kauffman. (1974) Population studies of mixed inoculum of *Xanthomonas oryzae* in susceptible and resistant varieties. *Ann. Phytopathol. Soc. Jpn* **40**:93–97.

Sequeira, L. (1979) The acquisition of systemic resistance by prior inoculation. In *Recognition and Specificity in Plant–Parasite Interactions* (J. M. Daly and I. Uritani, eds.). Japan Scientific Societies Press, Tokyo. Pp. 231–251.

Shiraishi, T., H. Oku, M. Yamashita, and S. Ouchi. (1978) Elicitor and suppressor of pisatin induction in spore germination fluid of pea pathogen, *Mycosphaerella pinodes*. *Ann. Phytopathol. Soc. Jpn* **44**:659–665.

Siegel, A. (1959) Mutual exclusion of strains of tobacco mosaic virus. *Virology* **2**:820–827.

Tani, T. (1979) Personal communication, Faculty of Agriculture, Kagawa University, Miki-cho, Kagawa, Japan.

Tani, T., and H. Yamamoto. (1979) RNA and protein synthesis and enzyme changes during infection. In *Recognition and Specificity in Plant–Parasite Interactions* (J. M. Daly and I. Uritani, eds.). Japan Scientific Societies Press, Tokyo. Pp. 273–287.

Tani, T., S. Ouchi, T. Onoe, and N. Naito. (1975) Irreversible recognition demonstrated in the hypersensitive response of oat leaves against crown rust fungus. *Phytopathology* **65**:1190–1193.

Tani, T., H. Yamamoto, T. Onoe, and N. Naito. (1976) Primary recognition and subsequent expression of resistance in oat leaves hypersensitively responding to crown rust fungus. In *Biochemistry and Cytology of Plant–Parasite Interaction*

(K. Tomiyama, J. M. Daly, I. Uritani, H. Oku, and S. Ouchi, eds.). Elsevier, Amsterdam. Pp. 124–135.

Tomiyama, K. (1966) Double infection by an incompatible race of *Phytophthora infestans* of a potato plant cell which has previously been infected by a compatible race. *Ann. Phytopathol. Soc. Jpn* **32**:181–185.

Tsuchiya, K., and K. Hirata. (1973) Growth of various powdery mildew fungi on the barley leaves infected preliminarily with the barley powdery mildew fungus. *Ann. Phytopathol. Soc. Jpn* **39**:396–403.

Warren, R. C. (1972) Attempts to define and mimic the effects of pollen on the development of lesions caused by *Phoma betae* inoculated onto sugarbeet. *Ann. Appl. Biol.* **71**:193–200.

Yarwood, C. E. (1959) Predisposition. In *Plant Pathology,* Vol. 1, *The Diseased Plants* (J. G. Horsfall and A. E. Dimond, eds.) Academic Press, New York. Pp. 521–562.

Yarwood, C. E. (1976) Modification of the host response—Predisposition. In *Encyclopedia of Plant Physiology,* Vol. 4 (R. Heitefuss and P. H. Williams, eds.). Springer-Verlag, New York. Pp. 703–718.

Yoder, O. (1976) Evaluation of the role of *Helminthosporium maydis* race T toxin in southern corn leaf blight. In *Biochemistry and Cytology of Plant–Parasite Interaction* (K. Tomiayama, J. M. Daly, I. Uritani, H. Oku, and S. Ouchi, eds.). Elsevier, Amsterdam. Pp. 16–24.

Yoder, O. C., and R. P. Scheffer. (1969) Role of toxin in early interactions of *Helminthosporium victoriae* with susceptible and resistant oat tissue. *Phytopathology* **59**:1954–1959.

TROPISMS OF FUNGI IN HOST RECOGNITION

Willard K. Wynn and Richard C. Staples

Tropisms are positive or negative responses to a source of stimulation by an organism or by one of its parts. These include orientations by turning, changing, or moving and may also involve other processes such as affinity and differentiation. We feel that the diverse responses occurring during host recognition can all be classified as tropisms.

Our purpose in this chapter is to reevaluate the information available on tropisms of plant pathogens to assess the possibility of exploiting tropic responses in disease control. As we examined the literature it became obvious that an attempt to utilize constructively the relevant information on all types of pathogens would be too diffuse. Therefore we decided to restrict our discussion to tropisms of those fungi that respond to specific stimuli and that are intimately associated with their host plants and not attempt to include taxes of fungal zoospores, bacteria, and nematodes. Several aspects of fungal tropisms that are not covered here are discussed in another review (Staples and Macko, 1980).

OVERALL ROLE OF TROPISMS IN PLANT PATHOLOGY

Tropisms are a critical part of the preinfection phase of disease development. Any method of disease management that functions during this initial period of fungus–host relationships must involve tropic stimuli and responses, either directly or indirectly. For fungal pathogens in which spores are the inoculum, preinfection development can be stopped at four stages: spore germination, germ-tube growth, preparation for penetration, and penetration. Germ-tube growth and preparation for penetration are the stages most profoundly influenced by tropisms, and most of the following discussion will be focused on them.

Preparation for penetration commonly involves appressorium formation. Some fungi, however, do not produce distinct appressoria, since their germ-tube tips do not swell before penetration pegs are formed, and others produce large lobed structures called infection cushions. To simplify things we would like to use the term *appressorium* in the broad sense, as defined by Emmett and Parbery (1975): a germ-tube or hyphal tip, without regard to morphology, which has the ability to initiate host penetration.

Several types of tropisms play a role in host recognition by fungi: chemotropism, thigmotropism (contact), hydrotropism (moisture), and phototropism (light). Since only the first two of these have been shown to be important, the evidence presented in this chapter is based on chemical and contact stimuli. Chemotropism includes both negative and positive responses to external sources of chemicals, including gases, as well as to compounds produced and secreted by the spores or germ tubes of the fungus itself (Fulton, 1906; Graves, 1916). The response of a fungus to its own metabolic products is called autotropism (Jaffe, 1966; Robinson et al., 1968; Stadler, 1952). Contact tropism is response to a surface, the mechanism of which is not understood. Since contact between two biological surfaces must involve a chemical interaction of some sort, it could be argued that contact tropism is simply a type of chemical response. But until we learn about contact phenomena, it seems logical to define chemotropism as response to small, soluble (or volatile) molecules and contact tropism as response to large, insoluble molecules, including polymers in the cuticle. It is generally accepted that a response is induced by contact if it can occur on an inert surface.

To analyze the observations that have been made on several specific tropisms, we have separated plant pathogenic fungi into two groups based on their patterns of tropic responses: the uredospore stage of the rust fungi, in which penetration occurs through stomates (Table 1), and those fungi that penetrate directly through the cuticle (Table 3).

RUST FUNGI

Preinfection initiated by uredospores of most rust fungi is a remarkable sequence of events, which has been perfected through coevolution on a wide range of host plants. When a uredospore germinates on the cuticle of a leaf or stem, its germ tube locates a stomate over which it forms an appressorium. An infection peg grows through the pore and enlarges into a vesicle within the substomatal cavity; then an infection hypha (often more than one) grows to the surface of a mesophyll cell where it forms a haustorial mother cell, which in turn produces the final penetration peg that develops into the intracellular haustorium (Littlefield and Heath, 1979). Since the haustorial mother cell initiates cellular penetration, it can be considered the functional

Table 1. Summary of Tropisms of Stomate-Entering Rust Fungi During Preinfection

Tropism	Specific Response	Primary Stimulus	Species of *Puccinia*, *Uromyces*, and *Melampsora* for Which Tropisms Have Been Shown[a]
1. Germ-tube adherence	Growth in close contact with plant surface	Contact	*P. graminis tritici* (15, 23); *P. recondita* (5, 23); *P. coronata* (10, 21); *P. hordei* (23); *P. sorghi* (23); *U. phaseoli typica* (22)
2. Germ-tube orientation	Directional growth across surface ridges or lines	Contact	*P. graminis tritici* (14, 15); *P. recondita* (8); *P. striiformis* (8); *P. coronata* (8, 9, 10); *P. sorghi* (8); *P. antirrhini* (18); *U. phaseoli typica* (22)
3. Appressorium formation[b]	Location only over stomate	(a) Contact	*P. sorghi* (23); *U. phaseoli typica* (22)
		(b) Chemical	*P. graminis tritici* (11, 12, 17)
4. Directional emergence of peg from appressorium	Growth downward into stomate	Contact/chemical[c]	*P. graminis tritici* (1); *P. recondita* (2, 4, 6); *U. phaseoli typica* (19); *U. phaseoli vignae* (16)
5. Haustorial mother-cell adherence	Close contact with mesophyll cell wall	Contact	*P. graminis tritici* (1); *P. recondita* (2, 7); *P. coronata* (9); *U. phaseoli typica* (20); *U. phaseoli vignae* (13); *M. lini* (3)

[a]References: (1) Allen, 1923b; (2) Allen, 1926; (3) Bracker and Littlefield, 1973; (4) Caldwell and Stone, 1936; (5) Dickinson, 1949a; (6) Dickinson, 1949b; (7) Dickinson, 1949c; (8) Dickinson, 1969; (9) Dickinson, 1971; (10) Dickinson, 1977; (11) Grambow and Riedel, 1977; (12) Grambow and Grambow, 1978; (13) Heath, 1974; (14) Johnson, 1934; (15) Lewis and Day, 1972; (16) Littlefield and Heath, 1979; (17) Macko et al., 1978; (18) Maheshwari and Hildebrandt, 1967; (19) Mendgen, 1973; (20) Mendgen, 1978; (21) Onoe et al., 1972; (22) Wynn, 1976; (23) Wynn, 1979.

[b]See Table 2 for the various factors that have been shown to induce appressorium formation.

[c]Insufficient evidence to determine whether the stimulus is contact or chemical.

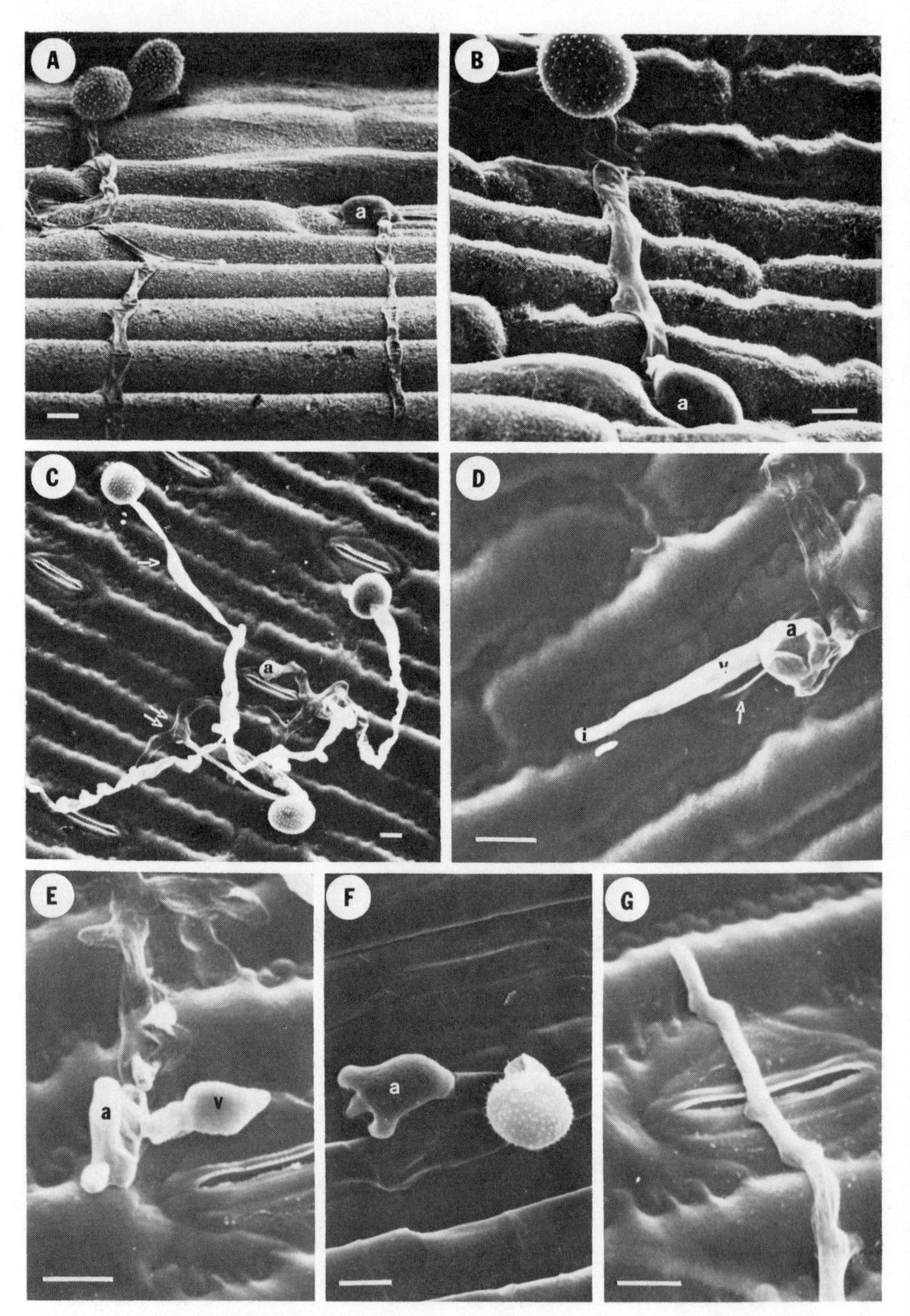

Figure 1 Scanning electron micrographs of normal preinfection development after 6 hours germination (*A, B*) and tropic mistakes after 24 hours germination (*C–G*) of two rust fungi on host and nonhost leaves. Each scale line = 10 μm; a = appressorium; i = infection hypha; v = vesicle. (*A*) Normal directional growth and appressorium formation over stomate by *Puccinia*

equivalent of the appressorium of direct-penetrating fungi such as powdery mildews (Bracker and Littlefield, 1973).

Exceptions to the typical pattern of stomatal penetration occur, notably with some of the rusts of tropical dicotyledonous plants. Uredospores of two species produce germ tubes and appressoria that penetrate the cuticle directly (Hunt, 1968).

We feel that five tropisms, the first three of which have been well documented, are involved in the preinfection sequence of stomate-entering rust fungi (Table 1). If any of these tropisms fail (Figure 1*C–G*), infection can be successfully prevented.

Germ-Tube Adherence

The close adherence of germ tubes to cuticular surfaces has been observed repeatedly with various rust fungi (Allen, 1923a, 1926; Lewis and Day, 1972; Onoe et al., 1972; Wynn, 1976; also Figure 1*A,B*) and clearly involves contact tropism. Dickinson (1949a) was the first to demonstrate that the nature of the surface on which spores are germinating determines whether or not the germ tubes grow in contact with the surface. He germinated *Puccinia recondita* uredospores on membranes prepared by different treatments under controlled humidity conditions and observed that the germ tubes would grow either up into the air or along the surface, depending on the type of membrane. He concluded that growth on the surface is controlled primarily by positive contact tropism; however, if humidity gradients are present, positive hydrotropism might play a secondary role. He interpreted germ-tube growth into the air as a negative thigmotropic response, but this does not explain why germ tubes grow away from certain surfaces. It is logical to assume that some type of repulsion exists between germ tubes and surfaces to which they do not adhere. Additional experiments by Dickinson (1970, 1971, 1972) with several nitrocellulose and plastic membranes leave little doubt that germ tubes either "like" or "dislike" different surfaces and that adherence or nonadherence is determined by surface characteristics. The general response of some of the rusts of gramineous plants is apparently that

---

recondita on wheat. (*B*) Normal directional growth and appressorium formation over stomate (obscured) by *P. sorghi* on corn. (*C*) Failure of germ-tube adherence and directional growth by *P. sorghi* on waxless corn. The germ tubes that appear white (single arrow) grew up in the air from the spores and then later fell down on the surface where they appear transparent (double arrows); the appressorium did not form over the stomate. (*D*) Failure of directional infection peg emergence from appressorium (over stomate, arrow) by *P. sorghi* on waxless corn. Vesicle and infection hypha grew on the surface. (*E*) Failure of appressorium of *P. recondita* to form over a stomate on waxless corn. As in (*D*) the vesicle grew on the surface. (*F*) Aberrant appressorium of *P. sorghi* over stomate (obscured) on waxless barley. Infection peg and vesicle did not form. (*G*) Failure of germ tube of *P. recondita* to form an appressorium as it contacted a stomate on waxless corn.

they adhere tightly to hydrophobic surfaces and grow away from hydrophilic ones.

Recently this phenomenon has been demonstrated with uredospores germinating on leaves of host and nonhost plants (Wynn, 1979). When germ-tube growth of *P. graminis tritici, P. recondita, P. hordei,* and *P. sorghi* was compared on waxy and waxless leaves, the differences in behavior were striking. On leaves with normal epicuticular wax to which these fungi are adapted, germ tubes adhered closely to the surface, grew directionally across the epidermal cells, and formed appressoria over stomates as is typical for rusts of gramineous plants (Figure 1*A*,*B*). On leaves without epicuticular wax the germ tubes initially grew up into the air; then as they elongated they fell down on the leaf surface but usually failed to adhere tightly (Figure 1*C*). The result was confusion: the germ tubes lost their ability to grow directionally, to locate and recognize stomates, and to form appressoria. We do not know the mechanisms behind the opposite responses on hydrophobic (waxy) and hydrophilic (waxless) leaves. Although small moisture differences within the surface microenvironments could possibly lead to a hydrotropic response, it seems more likely that direct reactions between the germ tubes and the surfaces are involved.

Germ-Tube Orientation

Directional growth of germ tubes has also been well documented as a contact tropism. It was first reported by Johnson (1934) for *P. graminis tritici* and has since been demonstrated for several other rusts (Dickinson, 1969, 1970; Maheshwari and Hildebrandt, 1967; Wynn, 1976). When germ tubes adhere to surfaces of plants or to artificial membranes, they grow at right angles to parallel ridges or lines on these surfaces (Figure 1*A*,*B*). Dickinson (1969, 1970, 1971, 1972, 1977) has meticulously analyzed the stimuli, interactions, and responses that make up directional growth by using model membranes in which the inducing surface features were experimentally defined. These studies have contributed more than any others to our understanding of contact tropism in the rusts. Although his results are too involved for us to discuss adequately here, the major concepts can be summarized as follows: (1) Growth patterns (amount of branching, zigzagging, turning) of adhering germ tubes are entirely controlled by the topography of the surface; and (2) directional growth is induced by contact of germ tube tips with a repetitive series of straight structural features whose size and frequency are critical in order to be effective. On plants these features are usually considered to be cuticular lines that are part of the ridges over the epidermal cells. Lewis and Day (1972) hypothesized that germ tubes of *P. graminis tritici* on wheat leaves are guided by lattice lines formed by the epicuticular wax projections. The specific features that induce directional growth by different rust fungi must vary with the hosts, since plants differ considerably in their surface topographies. However, the final result is the same: the germ tubes "feel" their way toward stomates with a surprising accuracy. On gramineous leaves locating stomates is fairly simple, since germ tubes are directed to grow perpendicularly to staggered rows of stomates; therefore a germ tube will eventually hit a stomate if it grows far enough (Johnson, 1934). On dicot leaves such as bean, the epidermal cells with their cuticular ridges form loose concentric circles around the stomates; as a germ tube is directed across the ridges, it usually finds a stomate after one or two attempts (Wynn, 1976).

Appressorium Formation

The most critical tropism that a rust fungus experiences during its maneuvers to establish itself as a pathogen is the formation of its appressorium specifically over a stomate (Table 1; Figure 1*A*). On host as well as many nonhost plants the efficiency of this response is amazing. Frequently 50–90% of the germ tubes on a leaf produce appressoria, 99% of which may be over stomates (Heath, 1977; Wynn, 1976). However, if a germ tube does not perceive or misreads a stimulus, it can make two types of mistakes—it forms an appressorium over interstomatal areas and thus cannot penetrate (Figure 1*E*) or it continues to grow across a stomate without recognizing it (Figure 1*G*). Both of these tropic failures can contribute to preinfection resistance (Heath, 1974, 1977).

Since this is a complex tropism, it has been studied more intensively than any other phase of preinfection. Several cytological and biochemical events are now clear. When a germ-tube tip contacts a stomate, germ-tube elongation and mitochondrial DNA synthesis stop (Staples, 1974). A round of nuclear DNA synthesis, replicated by a newly activated DNA α-polymerase (Staples, 1979), is necessary for appressorium induction. The first nuclear division then accompanies, but is not a prerequisite for, appressorium development (Staples et al., 1975). The four nuclei present in the mature appressorium migrate into the differentiating vesicle where the second nuclear division occurs; often one or two additional divisions take place as the infection hypha develops from the vesicle (Heath and Heath, 1978; Maheshwari et al., 1967b; Staples et al., 1975).

Table 2 summarizes much of the present knowledge on induction of appressoria. Available evidence is convincing that the entire infection structure (appressorium, infection peg, vesicle, and infection hypha) is triggered by a single stimulus. Appressoria that were induced by any one of several different stimuli, including heat shock (Dunkle and Allen, 1971), acrolein (Macko et al., 1978), host-plant fractions (Grambow and Grambow, 1978), various membranes (Dickinson, 1972; Maheshwari et al., 1967a), and leaf replicas (Wynn, 1976), all normally developed vesicles and infection hyphae without additional stimuli (Figure 1*D*).

Table 2. Summary of Stimuli and Conditions During Germination That Induce Infection Structure Formation by Uredospore Germ Tubes of Rust Fungi

	Induction of Infection Structures by:[a]							
Species of Rust Fungi (*Puccinia* and *Uromyces*)	Certain Nonhost Plants	Isolated Cuticles	Supplemented Artificial Membranes[b]	Inert Artificial Surfaces[c]	Inert Leaf Replicas[d]	Heat Shock	Volatile Fraction (Acrolein) from Uredospores	Volatile and Phenolic Fractions from Host
P. graminis tritici	+ (10)	+ (14)	+ (1)	+ (16)	0 (18)	+ (5)	+ (13)	+ (7,8)
P. recondita	+ (18)		+ (1)	+ (2)				
P. coronata	+ (15)		+ (3)	+ (4)				
P. sorghi	+ (9)	+ (14)	+ (14)	+ (18)	+ (18)	0 (12)	0 (12)	
P. helianthi	+ (9)	+ (14)	+ (14)			0 (12)	0 (12)	
P. antirrhini		+ (14)	+ (14)			0 (12)	0 (12)	
P. arachidis			+ (6)			0 (12)	0 (12)	
U. phaseoli typica	+ (17)	+ (14)	+ (17)	+ (17)	+ (17)	0 (12)	0 (12)	
U. phaseoli vignae	+ (9)		+ (11)					

[a]Results reported for consistent appressorium formation, usually with vesicles and infection hyphae. Symbols: (+), satisfactory evidence for infection structure induction; (0), failure to obtain infection structures; (blank), no data available. References: (1) Dickinson, 1949b; (2) Dickinson, 1970; (3) Dickinson, 1971; (4) Dickinson, 1972; (5) Dunkle and Allen, 1971; (6) Foudin and Macko, 1974; (7) Grambow and Riedel, 1977; (8) Grambow and Grambow, 1978; (9) Heath, 1977; (10) Leath and Rowell, 1966; (11) Littlefield and Heath, 1979; (12) Macko et al., 1976; (13) Macko et al., 1978; (14) Maheshwari et al., 1967a; (15) Romig and Caldwell, 1964; (16) Rowell and Olien, 1957; (17) Wynn, 1976; (18) Wynn, 1979.

[b]Various collodion membranes prepared with the addition of paraffin oil, wax, protein, or isolated plant cell walls.

[c]Appressoria formed in association with specific topographical features on plastic membranes and films without added chemicals or other materials.

[d]Appressoria formed over reproductions of stomates on plastic replicas.

Three generalizations about induction are apparent or can be safely inferred (Table 2): (1) Nonhost plants effectively induce appressoria; the more distantly related the plants are to the hosts, the poorer their induction (Heath, 1974, 1977; Wynn, 1976). (2) Isolated cuticles, provided they are not damaged during preparation, can be as effective as whole leaves (Maheshwari et al., 1967a). (3) Heterogeneous artificial membranes can be prepared that will induce appressoria; however, no one type of membrane is effective for all the different rusts, and it is sometimes difficult to relate membrane properties to specific stimuli.

The germ tubes of some rust fungi form infection structures in response to a contact stimulus. Dickinson (1970, 1971, 1972) has demonstrated this clearly with *P. recondita* and *P. coronata* on vinyl plastic and nitrocellulose membranes that were prepared with defined topographies. By simply altering the size and frequency of surface lines and granules he was able to control appressorium induction, and he concluded that the minimum size of the stimulus is 1.2 nm high and 120 nm wide. Similarly *P. graminis tritici* (Rowell and Olien, 1957), *P. sorghi*, and *Uromyces phaseoli typica* (Wynn, 1976, 1979) formed appressoria specifically over scratches on polyethylene film. The size of these "furrows" (the grooves together with the associated ridges) was apparently critical for induction. *U. phaseoli typica* formed appressoria over "stomates" of polystyrene bean leaf replicas with the same frequency as it did on leaves; thus the topography of the stomate, apparently the stomatal lips, provides the contact stimulus in nature (Wynn, 1976). Host leaf replica experiments with other rusts have given mixed results. The response of *P. sorghi* is similar to that of bean rust; however, attempts to induce *P. graminis tritici* appressoria on replicas of wheat leaves have failed repeatedly (Wynn, 1979).

P. graminis tritici is unique in that infection structures can be induced by specific chemicals alone or by heat shock (Table 2). Allen (1957) was the first to recover a fraction from uredospores of wheat stem rust, which stimulated germ-tube differentiation. Macko et al. (1978) purified the active compound and identified it as acrolein (2-propenal), an aliphatic aldehyde which requires only 0.4 nmole to induce infection structure formation by spores germinated in closed vessels. Because of its volatility, acrolein could possibly be a chemical intermediate in differentiation by heat shock (Macko et al., 1978).

Grambow (1977; Grambow and Riedel, 1977; Grambow and Grambow, 1978) has found that phenols extracted from the epicuticular wax and cell walls of wheat leaves induced infection structures in *P. graminis tritici* when applied together with volatile fractions from the leaves. Both components were necessary for activity. He postulated that the intercellular volatiles activate phenols present on the leaf surface and on the walls of the substomatal cavities and that these activated phenols are the inducers that the germ tubes contact during preinfection. Support for the activation theory comes from the fact that certain phenolic compounds (e.g., nordihydroguajaretic acid) were capable of inducing differentiation in the absence of a volatile component (Grambow, 1978).

In view of the preceding discussion, what can we conclude about the stimulus for inducing appressoria specifically over stomates? The simplest explanation is that rust fungi fall into two groups: those that respond only to stomatal contact and those that can respond to chemicals (Table 1).

We feel that all of the species except *P. graminis tritici* listed in Table 2 belong in the first group. This conclusion is based on the evidence that the infection structures of four common species (*P. recondita, P. coronata, P. sorghi,* and *U. phaseoli*) are clearly induced by germ-tube contact with specific surface features. It seems logical to us, then, that the capacity to translate a topographical stimulus into a trigger for appressorium formation is a widespread characteristic among the rust fungi and that the majority of the rusts must use this simple but effective mechanism in their uncanny ability to recognize and respond to stomates.

P. graminis tritici is an apparent anomaly, since it has a different pattern of response, including a remarkable sensitivity to chemical stimuli; therefore we place it alone in the second group. It is difficult to rationalize the various data available on wheat-stem rust to provide a satisfactory explanation of stomate recognition by this fungus except to say that infection structure induction appears to be more complex than in other rusts [Table 2; see Staples and Macko (1980) for further discussion].

Although infection structure formation is a single tropic response, it can be interrupted after appressorium formation on susceptible, resistant, or immune plants. A commonly observed example of this is the phenomenon that has been called stomatal exclusion or functional resistance. Figure 1*F* shows an appressorium that failed to produce an infection peg and vesicle; the aberrant, lobed shape has been described by Romig and Caldwell (1964). Allen (1923a,b) observed that on one variety of wheat *P. graminis tritici* formed normal appressoria but did not penetrate, and she concluded that the infection pegs could not get through the stomatal openings. Hart (1929) proposed, without making infection structure counts, that certain varieties of wheat were resistant to stem rust because their closed stomates prevented penetration. However, this mechanism has been disproved, since it is now clear that rust fungi can penetrate closed stomates; in fact, stomates close in response to appressorium formation (Caldwell and Stone, 1936; Yirgou and Caldwell, 1968). Similar failures of appressoria to produce infection pegs and vesicles have been observed for *P. graminis tritici* on a nonhost (Leath and Rowell, 1966), *P. recondita* on wheat (Romig and Caldwell, 1964), *P. coronata* and *P. graminis avenae* on oats (Kochman and Brown, 1975), and *U. phaseoli vignae* on host and nonhosts (Heath, 1974, 1977). Naito et al. (1971) showed that *P. coronata* germinating on epidermal strips can be inhibited at different stages of infection structure formation by applying various antimetabolites.

Another example of the interruption in infection structure development is the failure of *P. graminis tritici* to penetrate stomates of wheat leaves in the dark (Sharp et al., 1958). Yirgou and Caldwell (1968) demonstrated that vesicle formation is inhibited by high CO_2 and stimulated by low CO_2 concentrations independently of light. It seems clear that light-induced penetration is mediated by the photosynthetic removal of CO_2 in the substomatal spaces. Although the mechanism of the CO_2 effect is not known, it appears to act directly on the fungus as well as indirectly through the host. In contrast to *P. graminis tritici,* another rust of wheat, *P. recondita,* does not require light for penetration and is not affected by CO_2 concentrations.

A reasonable conclusion from the preceding observations is that stomatal exclusion, failure to penetrate stomates of resistant and immune plants, and the CO_2 effect with wheat-stem rust are all the same basic reaction: an inhibition of vesicle formation that disrupts the programmed infection structure sequence. With the exception of CO_2, the specific sources of inhibition are unknown, probably vary with different rusts, and may include compounds present in substomatal cavities, surface features of guard cells, and environmental conditions. Obviously penetration failures occur frequently enough to make them potentially useful in preventing rust infections.

Directional Peg Emergence

A response which is distinct from stimulation and inhibition of the infection peg is its direction of growth downward into the stomate. Since the peg always emerges from the floor of the appressorium instead of from the side or top during normal preinfection development, directional emergence cannot occur by chance (references in Table 1). Either a contact or chemical stimulus could be involved, and scant evidence is available for any mechanism. When an appressorium is induced on a collodion membrane, its wall is thinner where it contacts the surface (Littlefield and Heath, 1979). Also appressoria on membranes have been observed to form pegs which penetrate and then produce vesicles beneath the membranes (Dickinson, 1949b). Ultrastructural cross-sections of appressoria over host stomates show the intimate contact and conformation of the appressorium floor to the guard-cell surface (Mendgen, 1973). Based on these observations a possible explanation for directional peg emergence is that the outer appressorial wall is thinnest at the point of contact with the projecting guard-cell lips, which outline the stomatal pore. Since the infection peg develops from the inner appressorial wall (Littlefield and Heath, 1979), it breaks through the weak area in the outer wall where it grows directly into the stomate, forcing its way between the closed guard cells (Mendgen, 1973). The importance of appressorium contact with the stomate is illustrated in Figure 1*D* by the failure of the peg to grow downward, apparently the result of poor appressorium adherence. Romig and Caldwell (1964) reported that this type of tropic mistake is partly responsible for the failure of *P. recondita* to infect wheat sheaths and peduncles.

Haustorial Mother-Cell Adherence

The haustorial apparatus comprises the haustorial mother cell, penetration peg, and haustorium. Its development is well documented in several rusts—the tip of an infection hypha contacts an exposed mesophyll cell, stops elongating, and differentiates to form the haustorial mother cell, which adheres tightly to the host cell wall. The appressed wall of the haustorial mother cell thickens, and the peg emerges to penetrate the host wall and form the haustorium (references in Table 1; also Littlefield, 1972; Littlefield and Heath, 1979). It is obvious from the available information that close contact between fungus and host is an essential part of haustorium development and that adherence of the haustorial mother cell, like that of the germ tube to the cuticle, is a contact response (Table 1).

We would like to carry this concept a step further by proposing that contact by the infection hypha does more than simply provide the physical attachment necessary for penetration—that it actually induces the haustorial apparatus. A reasonable argument can be made for an induction, analogous to that of the infection structure, in response to the proper stimulus on the surface of the mesophyll cell walls. Evidence can be summarized as follows:

1. Haustorial mother-cell formation is not usually a part of infection structure development. When infection structures of *U. phaseoli typica* (Staples et al., 1975), *P. graminis tritici* (Allen, 1957; Dunkle and Allen, 1971; Grambow and Riedel, 1977), *P. antirrhini,* and *P. helianthi* (Maheshwari et al., 1967a) were produced under artificial conditions, haustorial mother cells did not develop. Although exceptions to this have been reported for differentiation of *U. phaseoli typica* (Maheshwari et al., 1967b) and *U. phaseoli vignae* (Heath, 1977) on collodion membranes, haustorial mother-cell formation seems to require a separate stimulus.
2. Infection hyphae do not respond to the same stimuli that germ tubes do. When bean rust (Wynn, 1976) and gramineous rusts (Dickinson, 1969, 1977) were induced to form infection structures on leaf surfaces, replicas, or membranes, the infection hyphae did not grow directionally, even though adjacent germ tubes did (Figure 1*D,E*).
3. Dickinson (1971, 1972) has induced haustoria on defined artificial membranes; the size and frequency of the surface features that were effective for haustorium formation were different from those necessary for induction of infection structures.
4. Development of the haustorial apparatus can be halted after an infection hypha contacts a mesophyll cell (Leath and Rowell, 1966); however, this does not mean that induction is absent. Such an interruption, which ap-

pears similar to the halt in infection structure development sometimes occurring after appressorium formation (see the section on Appressorium Formation above), involves several possible interactions and inhibitions (Heath, 1974; Littlefield and Heath, 1979) that may not be related to the induction process.

5. Mendgen (1978) has obtained evidence that adherence of *U. phaseoli typica* to host tissue is specific, since preparations of germ-tube walls bound only to cells of those plants in which the fungus is capable of forming haustoria. Although infection hyphae, because they differ from germ tubes, should have been used in this study to make it definitive, the results are encouraging that a contact recognition may occur at the haustorial induction stage and that adherence may be a determining factor.

DIRECT-PENETRATING FUNGI

The diversity of the fungi that enter their hosts by direct penetration makes it difficult to generalize about their preinfection process, even though it is considerably less sophisticated than that of the rusts. However, the following pattern emerges from observations made on the six most frequently studied genera. A germ-tube or hyphal tip contacts the plant surface, adheres, and grows until it forms an appressorium, which is commonly associated with a topographical feature of the cuticle such as the junction between two epidermal cells. A penetration peg then enters the plant by perforating the cuticle and underlying cell wall. Five tropisms, some of which are similar to those of the rusts, appear to be involved in directing these activities (Table 3).

Directional Germ-Tube Emergence

When spores of *Rhizopus, Geotrichum,* or *Botrytis* were germinated on artificial media they showed positive or negative autotropism, or perhaps responses to oxygen and carbon dioxide, by the direction in which the germ tubes emerged from the spores (Jaffe, 1966; Stadler, 1952; Robinson, 1973). This tropic germination was highly sensitive to spore density and other environmental factors.

At least one fungal pathogen germinates tropically on its host. On onion leaves and on inert leaf replicas prepared with fingernail lacquer, germ tubes of *B. squamosa* emerged from the side of the spores facing the epidermal cell junctions and then continued to grow toward these junctions where they terminated to form appressoria (Clark and Lorbeer, 1976). This is good evidence that all three tropic responses (directional emergence, directional growth, and appressorium formation over cell junctions) were induced by contact stimuli. The chemical environment interfered with these stimuli,

Table 3. Summary of Tropisms of Several Direct-Penetrating Fungi During Preinfection

Tropism	Specific Response	Primary Stimulus	Genera of Fungi for Which the Most Evidence for Tropisms is Available[a]
1. Directional emergence of germ tube from spore	Growth toward epidermal cell wall junctions	Contact/chemical[b]	*Botrytis* (3)
2. Germ-tube or hypha adherence	Growth in close contact with plant surface	Contact	*Erysiphe* (6); *Colletotrichum* (4, 12); *Botrytis* (2); *Sclerotinia* (1); *Helminthosporium* (8); *Rhizoctonia* (9, 10)
3. Germ-tube or hypha orientation	Directional growth toward or along epidermal cell wall junctions	Contact	*Botrytis* (3); *Rhizoctonia* (10)
4. Appressorium formation	Location commonly over or near epidermal cell wall junctions[c]	(a) Contact[d]	*Erysiphe* (7, 16, 21); *Colletotrichum* (4, 5, 13, 14, 19); *Botrytis* (2, 3); *Sclerotinia* (17); *Helminthosporium* (8, 11, 18, 19, 20)
		(b) Chemical	*Rhizoctonia* (9, 10)
5. Directional emergence of peg from appressorium	Growth downward into cuticle	Contact/chemical	*Erysiphe* (7); *Colletotrichum* (5, 15)

[a]References: (1) Boyle, 1921; (2) Brown and Harvey, 1927; (3) Clark and Lorbeer, 1976; (4) Dey, 1919; (5) Dey, 1933; (6) Dickinson, 1949a; (7) Dickinson, 1949c; (8) Endo and Amacher, 1964; (9) Flentje, 1957; (10) Flentje et al., 1963; (11) Hau and Rush, 1979; (12) Lapp and Skoropad, 1978a; (13) Lapp and Skoropad, 1978b; (14) Mercer et al., 1971; (15) Politis and Wheeler, 1973; (16) Preece et al., 1967; (17) Purdy, 1958; (18) Wheeler, 1977; (19) Wood, 1967; (20) Wynn, 1979; (21) Yang and Ellingboe, 1972.

[b]Insufficient evidence to determine whether the stimulus is contact or chemical.

[c]Fungi which most consistently form appressoria at epidermal cell junctions are *Erysiphe polygoni*, species of *Colletotrichum*, *Botrytis*, and *Helminthosporium*, together with several other genera not listed in this table (16, 19).

[d]Although induced by contact, appressorium formation can often be modified by the presence of nutrients on the surface.

since directional germ-tube emergence was suppressed in the presence of nutrient solution and enhanced with leaf diffusates.

At first it may seem difficult to visualize the control of tropic germination by contact, unless we accept the possibility that compounds secreted by spores can have highly localized effects and that these effects can be modified by the surface through absorption or inactivation. Robinson et al. (1968) showed that directional germ-tube emergence, presumably an autotropic response, in *Rhizopus*, *Mucor*, *Trichoderma*, and *Botrytis*, was influenced by the nature of the surface (agar as compared with cellophane) or by the addition of extraneous materials (powdered cellulose, charcoal) to a surface. If such a mechanism does occur, it is conceivable that inhibitors and stimulators of spore germination (Macko et al., 1976) could act as tropic chemicals.

Germ-Tube or Hypha Adherence

Adherence to a host surface, as with the rust fungi, is an inherent part of preinfection and a contact response (references in Table 3; also Corner, 1935; Staub et al., 1974). A characteristic of many direct-penetrating fungi is that only the germ tube or hyphal tips, in contrast with the entire germ tubes of the rusts, seem to adhere tightly to the surface (compare Figure 1*A*,*B* with Figure 2*A*,*B*). This may help explain why germ-tube orientation (see section on Germ-Tube or Hypha Orientation below) is not as pronounced in the direct-penetrating fungi as in the rusts.

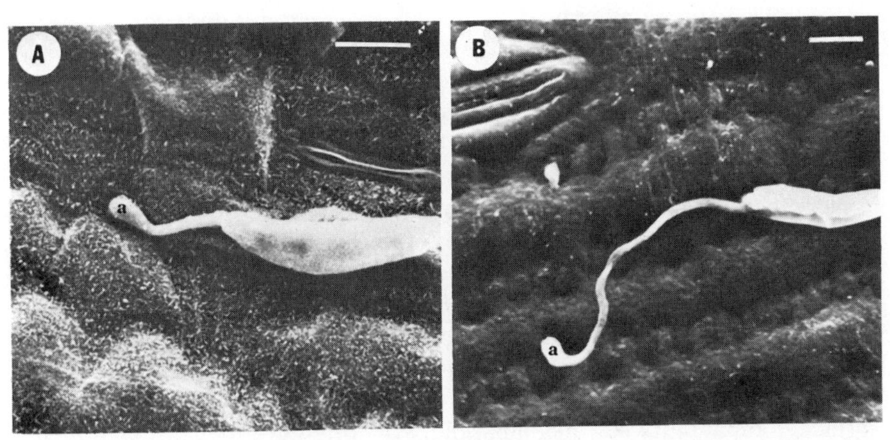

Figure 2 Scanning electron micrographs of preinfection development of *Helminthosporium maydis* after 6 hours germination. Each scale line = 10 μm; a = appressorium. (*A*) Conidium and germ tube with appressorium formed over junction between two epidermal cells on corn leaf. (*B*) Conidium and germ tube with appressorium formed over cell junction on polyethylene replica of corn leaf.

Dickinson (1949a) showed that, along with *P. recondita* (see section on Germ-Tube Adherence above), germ tubes of *Erysiphe, Peronospora, Cladosporium,* and *Verticillium* either grow in contact with membranes or up in the air, depending on the nature of the surface. Since each of these fungi responded differently to various types of membranes, their germ tubes can apparently discriminate against certain surfaces. Four other genera were found to grow in contact with all the membrane types. In relation to preinfection, these observations suggest that germ-tube or hypha adherence depends on properties of the fungus as well as those of the plant surface.

The remarkable adhesion to a suitable surface has repeatedly been attributed to the "mucilaginous substance" produced by germ tubes of several fungi (Blackman and Welsford, 1916; Boyle, 1921; Dey, 1919; Flentje, 1957; Marks et al., 1965; Paus and Raa, 1973). However, the only definitive study of this glue was that by Lapp and Skoropad (1978a), who tested different chemicals for their ability to remove *Colletotrichum* appressoria attached to glass coverslips. They concluded that the adhesive material is a hemicellulose.

Germ-Tube or Hypha Orientation

Besides the directional growth of *B. squamosa* toward epidermal cell junctions on leaves and replicas discussed in the section on Directional Germ-Tube Emergence above, the only clear-cut oriented growth of direct-penetrating fungi is that of *Rhizoctonia solani* hyphae, which characteristically follow the valleys of the cell-wall junctions on a host surface (Dodman and Flentje, 1970). Hyphal-tip adherence is necessary for directional growth; and these two responses, which are usually observed together, occur on epidermal strips and isolated cuticles in the absence of plant exudates (Kerr and Flentje, 1957; Flentje et al., 1963). By inoculating several different plants with six *R. solani* isolates, Flentje (1957) found that attachment–directional growth occurred in compatible combinations but frequently not in incompatible ones and that isolates distinguished between root and stem surfaces. De Silva and Wood (1964) showed that behavior of two species of *Rhizoctonia* on epidermal strips was specific for plants that were susceptible or resistant to attack by the fungus. It seems safe to conclude that hyphal attachment–directional growth is a contact tropism induced by an interaction between the host surface and the fungal isolate. The fact that these results with *Rhizoctonia* lead independently to the same conclusion about surface compatibility and incompatibility as different types of experiments with other fungi on artificial membranes (see section on Germ-Tube or Hypha Adherence above) emphasizes that surface recognition may play a greater role in specificity than we have assumed in the past.

Appressorium Formation

Emmett and Parbery (1975) have comprehensively reviewed appressorial development, and the present discussion is limited to those tropic aspects summarized in Table 3.

Although some fungi (e.g., *E. graminis*) form appressoria at random over the cuticular surface, the location of appressoria over or near the anticlinal walls of epidermal cell junctions (Figure 2*A*,*B*), including those between guard and subsidiary cells, is so common that it can be taken as a generalization. At least 22 species in 14 genera have been documented to form appressoria frequently at cell junctions: the four genera listed in Table 3; six genera described by Wood (1967, page 79); *Drechslera* (Hunt, 1976); *Puccinia psidii* (a direct-penetrating rust fungus; Hunt, 1968); *Peronospora* (Preece et al., 1967); and *Cladosporium* (Paus and Raa, 1973). A few fungi produce appressoria at other specific locations, such as over bulliform cells or at the base of trichomes (Hau and Rush, 1979; Wood, 1967).

The tight adhesion of appressoria is more obvious than that of germ tubes and hyphae (see section on Germ-Tube or Hypha Adherence above). In fact, only appressorium adherence is observed with fungi such as *Colletotrichum*, which produce very short germ tubes (Lapp and Skoropad, 1978a; Marks et al., 1965; Politis and Wheeler, 1973).

Most of the direct-penetrating fungi form appressoria in response to a surface contact stimulus (Table 3). This has been demonstrated most convincingly with species of *Collectotrichum*, in which appressoria are large and readily induced (Dey, 1919, 1933; Lapp and Skoropad, 1978b). The sequence of nuclear and biochemical events leading to appressorium formation has been described by Staples et al. (1976).

Three lines of evidence have been obtained for appressorium induction by contact. (1) *Leaf replica experiments*: *C. graminicola* appressoria are located at epidermal cell junctions but not exactly over the anticlinal walls. Lapp and Skoropad (1978b) measured the distances from appressoria to the nearest anticlinal wall centers for spores germinated on barley leaves and three types of inert leaf replicas; the measurements were comparable on all four surfaces. *B. cinerea* and *B. squamosa* formed appressoria equally well (70–80% of total appressoria) over cell junctions on onion leaves and replicas (Clark and Lorbeer, 1976). Similarly, *Helminthosporium oryzae* (Hau and Rush, 1979) and *H. maydis* (Wynn, 1979; Figure 2*A*,*B*) formed appressoria as readily over junctions on plastic leaf replicas as on host leaves. (2) *Germination on inert surfaces*: Appressoria are induced on hard surfaces (glass, hard membranes, plastic) in the absence of added chemicals; soft surfaces (agar, soft membranes) are ineffective. Results of this type have been obtained with *Erysiphe* (Dickinson, 1949c, 1979), *Colletotrichum* (Dey, 1919, 1933; Staples et al., 1976; Van Burgh, 1950), *Botrytis* (Brown and Harvey, 1927), *Sclerotinia* (Abawi et al., 1975; Boyle, 1921; Purdy, 1958),

Helminthosporium (Endo and Amacher, 1964), and *Phyllachora*, in which appressoria were abnormal if they were produced without contact (Parbery, 1963). (3) *Cuticular surfaces*: Yang and Ellingboe (1972) showed by using isolated cuticles and wax mutants that appressoria of *E. graminis* were induced by germ-tube contact with the epicuticular wax on cuticles of wheat and barley. Appressorium formation by *Alternaria porri* on isolated cuticles and inert surfaces was increased by the presence of wax (Akai et al., 1967).

Of the fungi whose tropisms have been studied extensively, *R. solani* is typical of those which depend on chemical stimuli for appressorium formation (Dodman and Flentje, 1970; Flentje, 1957; Flentje et al., 1963). The evidence with host exudates suggests that chemicals play a significant role in the induction of appressoria (infection cushions). We feel that the contact-induced hypha attachment and directional growth (see section on Germ-Tube or Hypha Orientation above) is a separate tropism from appressorium formation that occurs as a result of attachment but which may depend on different stimuli.

A phenomenon that is distinct from chemical induction of appressoria is the modification of contact induction by chemicals present on the surface during germination. This effect was particularly striking with species of *Colletotrichum* and *Gloeosporium* in which added nutrients suppressed appressorium formation and stimulated germ-tube elongation (Hasselbring, 1906; Mercer et al., 1971; Van Burgh, 1950). Conversely, appressorium formation by *S. sclerotiorum* required nutrients in addition to contact (Abawi et al., 1975; Purdy, 1958). Appressoria of *H. sorokinianum* were either increased or decreased, depending on the type of nutrients (Endo and Amacher, 1964).

With a number of fungi, as with the rusts, penetration peg formation does not seem to require a separate stimulus: *E. graminis* (Dickinson, 1949c, 1979), *C. lindemuthianum* (Dey, 1919), *B. cinerea* (Brown and Harvey, 1927), *S. sclerotiorum* (Boyle, 1921), and *H. sorokinianum* (Endo and Amacher, 1964). However, in some species of *Colletotrichum* and *Gloeosporium* the thick-walled appressoria may remain latent for long periods after maturing and thus act as survival structures; penetration pegs can then be induced by several treatments, including nutrient availability, high humidity, and ethylene (Brown, 1975; Dey, 1933; Hasselbring, 1906). The mechanism of appressorium dormancy is not understood, but it could involve an inhibition of peg formation in a manner analogous to that by CO_2 in wheat rust (see the section on Appressorium Formation, pp. 51–55).

Directional Peg Emergence

The evidence that downward emergence of the peg from an appressorium is controlled by the location of a weak point in the appressorium floor, as hypothesized for the rusts (see the section on Directional Peg Emergence, pp.

55–56) is more convincing for direct-penetrating fungi. The penetration peg is always formed at the center of the area where the appressorium adheres to the cuticle or to an inert surface (Boyle, 1921; Dey, 1919; Dickinson, 1949c, 1979; Hasselbring, 1906; Marks et al., 1965). Abnormal emergence from the sides of the appressorium occurred with *C. gloeosporioides* when peg formation was induced by nutrients (Dey, 1933). A well-defined pore through which the peg emerges has been observed repeatedly by light microscopy (Boyle, 1921; Hasselbring, 1906; Marks et al., 1965).

Politis and Wheeler (1973) described the ultrastructure of pore formation in *C. graminicola*. A new inner wall developed at the bottom of the appressorium only; then a small area of the outer wall disappeared in the center of the floor and provided the hole through which the penetration peg, formed from the inner wall, emerged. Although the stimulus inducing the activity at the appressorium floor could be due to chemicals exuded through the cuticle, the fact that directional emergence occurs on inert surfaces supports contact induction.

CONCLUSIONS

It seems clear that surface contact is the major stimulus for tropisms controlling preinfection development. From the five tropisms identified for each group of fungi, eight different specific responses have been documented (Tables 1 and 3): (1) germ-tube emergence toward cell junctions, (2) germ-tube adherence, (3) growth across surface ridges, (4) growth along cell junctions, (5) appressorium formation over stomates, (6) appressorium formation over cell junctions, (7) peg emergence downward, and (8) haustorial mother cell adherence. Evidence is persuasive that all of these except directional emergence of germ tubes, appressorium formation by *P. graminis tritici* and *R. solani*, and directional emergence of pegs are induced by contact. Thus contact tropisms are more prevalent than generally accepted in the past, although their importance was recognized by some (e.g., Wood, 1967) on the basis of considerably less evidence than is available today.

Close adherence is a critical prerequisite for proper response to contact stimuli (Dickinson, 1960). Directional germ-tube growth, appressorium formation (when induced by contact), and haustorium formation in rusts are entirely dependent on adhesion of fungi to suitable plant surfaces. Although it could be argued that adherence is simply a part of other tropisms, we feel that because of its importance it warrants classification as a separate response. Not only can it occur independently in the absence of other tropisms, but it can fail completely on certain surfaces. Also spores can actively adhere to cuticles before they germinate (Paus and Raa, 1973; Rapilly and Foucault, 1976).

Although our understanding of tropic stimuli is just beginning to emerge, some of the more obvious features of plants that control preinfection can be identified (see the sections on Rust Fungi and Direct-Penetrating Fungi above): (1) ridges and lines formed by cuticles and periclinal walls; (2) fine texture and nature of cuticles; (3) epicuticular wax; (4) epidermal-cell junctions at anticlinal walls; (5) stomatal- and guard-cell structure; and (6) chemicals in the cuticle, secreted through the cuticle, or present in the substomatal cavity. The diversity of these features within the plant kingdom (Martin and Juniper, 1970) emphasizes their potential role in host recognition.

Tropisms, as precise and orderly as they are in leading to successful infection, sometimes fail. Failures provide the basis for exploiting tropisms in disease control, since they represent mistakes made by pathogens in nature. The basic reasons for tropic failures are the lack of, or an incorrect, stimulus from a plant; faulty perception of a stimulus by a fungus; and environmental interference with a stimulus or a response. Plant surface features that pathogens cannot recognize properly are responsible for many mistakes. These features are components of horizontal resistance because the resulting mistakes interfere with tropisms and subsequently reduce disease (Russell, 1978).

Although fully capitalizing on tropic failures for agricultural production will require techniques not presently available, we are convinced that preinfection resistance plays a significant role in overall disease management. An increased awareness of the struggles that fungi experience on plant surfaces should promote an appreciation of this role. It is important to realize that manipulation of tropic stimuli will not usually provide complete resistance because it can only reduce the number of infections and has no effect on subsequent development. Tropic mistakes rarely stop all members of a pathogen population, and a few frustrated individuals will manage to penetrate and cause disease if a plant is otherwise susceptible. The heavier the inoculum load, the greater will be the chances of serious disease. Perhaps the most effective way to utilize preinfection resistance is to combine it with postinfection resistance in a multicomponent system that reduces both incidence and severity of disease (Russell, 1978).

ACKNOWLEDGMENT

We thank V. A. Wilmot for the scanning electron microscopy which was done specifically for this review.

REFERENCES

Abawi, G. S., F. J. Polach, and W. T. Molin. (1975) Infection of bean by ascospores of *Whetzelinia sclerotiorum*. *Phytopathology* **65**:673–678.

Akai, S., M. Fukutomi, N. Ishida, and H. Kunoh. (1967) An anatomical approach to the mechanism of fungal infections in plants. In *The Dynamic Role of Molecular Constituents in Plant Parasite Interaction* (C. J. Mirocha and I. Uritani, eds.). American Phytopathological Society, St. Paul, Minn. Pp. 1–20.

Allen, P. J. (1957) Properties of a volatile fraction from uredospores of *Puccinia graminis* var. *tritici* affecting their germination and development. I. Biological activity. *Plant Physiol.* **32**:385–389.

Allen, R. F. (1923a) A cytological study of infection of Baart and Kanred wheats by *Puccinia graminis tritici*. *J. Agric. Res.* **23**:131–152.

Allen, R. F. (1923b) Cytological studies of infection of Baart, Kanred, and Mindum wheats by *Puccinia graminis tritici* forms III and XIX. *J. Agric. Res.* **26**:571–604.

Allen, R. F. (1926) A cytological study of *Puccinia triticina* physiologic form 11 on Little Club wheat. *J. Agric. Res.* **33**:201–222.

Blackman, V. H., and E. J. Welsford. (1916) Studies in the physiology of parasitism. II. Infection by *Botrytis cinerea*. *Ann. Bot.* **30**:389–398.

Boyle, C. (1921) Studies in the physiology of parasitism. VI. Infection by *Sclerotinia libertiana*. *Ann. Bot.* **35**:337–347.

Bracker, C. E., and L. J. Littlefield. (1973) Structural concepts of host–pathogen interfaces. In *Fungal Pathogenicity and the Plant's Response* (R. J. W. Byrde and C. V. Cutting, eds.). Academic Press, New York. Pp. 159–318.

Brown, G. E. (1975) Factors affecting postharvest development of *Colletotrichum gloeosporioides* in citrus fruits. *Phytopathology* **65**:404–409.

Brown, W., and C. C. Harvey. (1927) Studies in the physiology of parasitism. X. On the entrance of parasitic fungi into the host plant. *Ann. Bot.* **41**:643–662.

Caldwell, R. M., and G. M. Stone. (1936) Relation of stomatal function of wheat to invasion and infection by leaf rust (*Puccinia triticina*). *J. Agric Res.* **52**:917–932.

Clark, C. A., and J. W. Lorbeer. (1976) Comparative histopathology of *Botrytis squamosa* and *B. cinerea* on onion leaves. *Phytopathology* **66**:1279–1289.

Corner, E. J. H. (1935) Observations on resistance to powdery mildews. *New Phytol.* **34**:180–200.

de Silva, R. L., and R. K. S. Wood. (1964) Infection of plants by *Corticium solani* and *C. praticola*—Effect of plant exudates. *Trans. Br. Mycol. Soc.* **47**:15–24.

Dey, P. K. (1919) Studies in the physiology of parasitism. V. Infection by *Colletotrichum lindemuthianum*. *Ann. Bot.* **33**:305–312.

Dey, P. K. (1933) Studies in the physiology of the appressorium of *Colletotrichum gloeosporioides*. *Ann. Bot.* **47**:305–312.

Dickinson, S. (1949a) Studies in the physiology of obligate parasitism. I. The stimuli determining the direction of growth of the germ-tubes of rust and mildew spores. *Ann. Bot., N.S.* **13**:89–104.

Dickinson, S. (1949b) Studies in the physiology of obligate parasitism. II. The behavior of the germ-tubes of certain rusts in contact with various membranes. *Ann. Bot., N.S.* **13**:219–236.

Dickinson, S. (1949c) Studies in the physiology of obligate parasitism. IV. The formation on membranes of haustoria by rust hyphae and powdery mildew germ-tubes. *Ann. Bot., N.S.* **13**:345–353.

Dickinson, S. (1960) The mechanical ability to breach the host barriers. In *Plant Pathology*, Vol. 2. *The Pathogen* (J. G. Horsfall and A. E. Dimond, eds.). Academic Press, New York. Pp. 203–232.

Dickinson, S. (1969) Studies in the physiology of obligate parasitism. VI. Directed growth. *Phytopathol. Z.* **66**:38–49.

Dickinson, S. (1970) Studies in the physiology of obligate parasitism. VII. The effect of a curved thigmotropic stimulus. *Phytopathol. Z.* **69**:115–124.

Dickinson, S. (1971) Studies in the physiology of obligate parasitism. VIII. An analysis of fungal responses to thigmotropic stimuli. *Phytopathol. Z.* **70**:62–70.

Dickinson, S. (1972) Studies in the physiology of obligate parasitism. IX. The measurement of thigmotropic stimulus. *Phytopathol. Z.* **73**:347–358.

Dickinson, S. (1977) Studies in the physiology of obligate parasitism. X. Induction of responses to a thigmotropic stimulus. *Phytopathol. Z.* **89**:97–115.

Dickinson, S. (1979) Growth of *Erysiphe graminis* on artificial membranes. *Physiol. Plant Pathol.* **15**:219–221.

Dodman, R. L., and N. T. Flentje. (1970) The mechanism and physiology of plant penetration by *Rhizoctonia solani*. In *Rhizoctonia Solani: Biology and Pathology* (J. R. Parmeter, ed.). University of California Press, Berkeley, Calif. Pp. 149–160.

Dunkle, L. D., and P. J. Allen. (1971) Infection structure differentiation by wheat stem rust uredospores in suspension. *Phytopathology* **61**:649–652.

Emmett, R. W., and D. G. Parbery. (1975) Appressoria. *Annu. Rev. Phytopathol.* **13**:147–167.

Endo, R. M., and R. H. Amacher. (1964) Influence of guttation fluid on infection structures of *Helminthosporium sorokineanum*. *Phytopathology* **54**:1327–1334.

Flentje, N. T. (1957) Studies on *Pellicularia filamentosa* (Pat.) Rogers. III. Host penetration and resistance, and strain specialization. *Trans. Br. Mycol. Soc.* **40**:322–336.

Flentje, N. T., R. L. Dodman, and A. Kerr. (1963) The mechanism of host penetration by *Thanatephorus cucumeris*. *Aust. J. Biol. Sci.* **16**:784–799.

Foudin, A. S., and V. Macko. (1974) Identification of the self-inhibitor and some germination characteristics of peanut rust uredospores. *Phytopathology* **64**:990–993.

Fulton, H. R. (1906) Chemotropism of fungi. *Bot. Gaz.* **41**:81–108.

Grambow, H. J. (1977) The influence of volatile leaf constituents on the *in vitro* differentiation and growth of *Puccinia graminis* f. sp. *tritici*. *Z. Pflanzenphysiol.* **85**:361–372.

Grambow, H. J. (1978) The effect of nordihydroguajaretic acid on the development of the wheat rust fungus. *Z. Pflanzenphysiol.* **88**:369–372.

Grambow, H. J., and G. E. Grambow. (1978) The involvement of epicuticular and cell wall phenols of the host plant in the *in vitro* development of *Puccinia graminis* f. sp. *tritici*. *Z. Pflanzenphysiol.* **90**:1–9.

Grambow, H. J., and S. Riedel. (1977) The effect of morphogenically active factors from host and nonhost plants on the *in vitro* differentiation of infection structures of *Puccinia graminis* f. sp. *tritici*. *Physiol. Plant Pathol.* **11**:213–224.

Graves, A. H. (1916) Chemotropism in *Rhizopus nigricans*. *Bot. Gaz.* **62**:337–369.

Hart, H. (1929) Relation of stomatal behavior to stem-rust resistance in wheat. *J. Agric. Res.* **39**:929–948.

Hasselbring, H. (1906) The appressoria of the anthracnoses. *Bot. Gaz.* **42**:135–142.

Hau, F. C., and M. C. Rush. (1979) Leaf surface interactions between *Cochliobolus miyabeanus* and susceptible and resistant rice cultivars. (Abstract) *Phytopathology* **69**:527.

Heath, M. C. (1974) Light and electron microscope studies of the interactions of host and non-host plants with cowpea rust—*Uromyces phaseoli* var. *vignae*. *Physiol. Plant Pathol.* **4**:403–414.

Heath, M. C. (1977) A comparative study of non-host interactions with rust fungi. *Physiol. Plant Pathol.* **10**:73–88.

Heath, M. C., and I. B. Heath. (1978) Structural studies of the development of infection structures of cowpea rust, *Uromyces phaseoli* var. *vignae*. I. Nucleoli and nuclei. *Can. J. Bot.* **56**:648–661.

Hunt, P. (1968) Cuticular penetration by germinating uredospores. *Trans. Br. Mycol. Soc.* **51**:103–112.

Hunt, P. (1976) Personal communication. Department of Botany, University of the West Indies, Kingston, Jamaica.

Jaffe, L. F. (1966) On autotropism in *Botrytis*: Measurement technique and control by CO_2. *Plant Physiol.* **41**:303–306.

Johnson, T. (1934) A tropic response in germ tubes of urediospores of *Puccinia graminis tritici*. *Phytopathology* **24**:80–82.

Kerr, A., and N. T. Flentje. (1957) Host infection in *Pellicularia filamentosa* controlled by chemical stimuli. *Nature* **179**:204–205.

Kochman, J. K., and J. F. Brown. (1975) Development of the stem and crown rust fungi on leaves, sheaths, and peduncles of oats. *Phytopathology* **65**:1404–1408.

Lapp, M. S., and W. P. Skoropad. (1978a) Nature of adhesive material of *Colletotrichum graminicola* appressoria. *Trans. Br. Mycol. Soc.* **70**:221–223.

Lapp, M. S., and W. P. Skoropad. (1978b) Location of appressoria of *Colletotrichum graminicola* on natural and artificial barley leaf surfaces. *Trans. Br. Mycol. Soc.* **70**:225–228.

Leath, K. T., and J. B. Rowell. (1966) Histological study of the resistance of *Zea mays* to *Puccinia graminis*. *Phytopathology* **56**:1305–1309.

Lewis, B. G., and J. R. Day. (1972) Behaviour of uredospore germ-tubes of *Puccinia graminis tritici* in relation to the fine structure of wheat leaf surfaces. *Trans. Br. Mycol. Soc.* **58**:139–145.

Littlefield, L. J. (1972) Development of haustoria of *Melampsora lini*. *Can. J. Bot.* **50**:1701–1703.

Littlefield, L. J., and M. C. Heath. (1979) *Ultrastructure of Rust Fungi*. Academic Press, New York. 277 pp.

Macko, V., R. C. Staples, Z. Yaniv, and R. R. Granados. (1976) Self-inhibitors of fungal spore germination. In *The Fungal Spore* (D. J. Weber and W. M. Hess, eds.). Wiley, New York. Pp. 73–100.

Macko, V., J. A. A. Renwick, and J. F. Rissler. (1978) Acrolein induces differentiation of infection structures in the wheat stem rust fungus. *Science* **199**:442–443.

Maheshwari, R., and A. C. Hildebrandt. (1967) Directional growth of the urediospore germ tubes and stomatal penetration. *Nature* **214**:1145–1146.

Maheshwari, R., P. J. Allen, and A. C. Hildebrandt. (1967a) Physical and chemical factors controlling the development of infection structures from urediospore germ tubes of rust fungi. *Phytopathology* **57**:855–862.

Maheshwari, R., A. C. Hildebrandt, and P. J. Allen. (1967b) The cytology of infection structure development in urediospore germ tubes of *Uromyces phaseoli* var. *typica* (Pers.) Wint. *Can. J. Bot.* **45**:447–450.

Marks, G. C., J. G. Berbee, and A. J. Riker. (1965) Direct penetration of leaves of *Populus tremuloides* by *Colletotrichum gloeosporioides*. *Phytopathology* **55**:408–412.

Martin, J. T., and B. E. Juniper. (1970) *The Cuticles of Plants*. St. Martin's Press, New York. 347 pp.

Mendgen, K. (1973) Feinbau der Infektionsstrukturen von *Uromyces phaseoli*. *Phytopathol. Z.* **78**:109–120.

Mendgen, K. (1978) Attachment of bean rust cell wall material to host and non-host plant tissue. *Arch. Microbiol.* **119**:113–117.

Mercer, P. C., R. K. S. Wood, and A. D. Greenwood. (1971) Initial infection of *Phaseolus vulgaris* by *Colletotrichum lindemuthianum*. In *Ecology of Leaf Surface Micro-organisms* (T. F. Preece and C. H. Dickinson, eds.). Academic Press, New York. Pp. 381–389.

Naito, N., M. Lee, and T. Tani. (1971) Inhibition of germination and infection structure formation of *Puccinia coronata* uredospores by plant growth regulators and antimetabolites. *Tech. Bull. Kagawa Univ.* **23**:51–56.

Onoe, T., T. Tani, and N. Naito. (1972) Scanning electron microscopy of crown rust appressorium produced on oat leaf surface. *Tech. Bull. Kagawa Univ.* **24**:42–47.

Parbery, D. G. (1963) Studies on graminicolous species of *Phyllachora* Fckl. I. Ascospores—Their liberation and germination. *Aust. J. Bot.* **11**:117–130.

Paus, F., and J. Raa (1973) An electron microscope study of infection and disease development in cucumber hypocotyls inoculated with *Cladosporium cucumerinum*. *Physiol. Plant Pathol.* **3**:461–464.

Politis, D. J., and H. Wheeler. (1973) Ultrastructural study of penetration of maize leaves by *Colletotrichum graminicola*. *Physiol. Plant Pathol.* **3**:465–471.

Preece, T. F., G. Barnes, and J. M. Bayley. (1967) Junctions between epidermal cells as sites of appressorium formation by plant pathogenic fungi. *Plant Pathol.* **16**:117–118.

Purdy, L. H. (1958) Some factors affecting penetration and infection by *Sclerotinia sclerotiorum*. *Phytopathology* **48**:605–609.

Rapilly, F., et B. Foucault. (1976) Premières études sur la rétention de spores fongiques par des épidermes foliaires. *Ann. Phytopathol.* **8**:31–41.

Robinson, P. M. (1973) Chemotropism in fungi. *Trans. Br. Mycol. Soc.* **61**:303–313.

Robinson, P. M., D. Park, and T. A. Graham. (1968) Autotropism in fungal spores. *J. Exp. Bot.* **19**:125–134.

Romig, R. W., and R. M. Caldwell. (1964) Stomatal exclusion of *Puccinia recondita* by wheat peduncles and sheaths. *Phytopathology* **54**:214–218.

Rowell, J. B., and C. R. Olien. (1957) Controlled inoculation of wheat seedlings with urediospores of *Puccinia graminis* var. *tritici. Phytopathology* **47**:650–655.

Russell, G. E. (1978) *Plant Breeding for Pest and Disease Resistance*. Butterworths, London. 485 pp.

Sharp, E. L., C. G. Schmitt, J. M. Staley, and C. H. Kingsolver. (1958) Some critical factors involved in establishment of *Puccinia graminis* var. *tritici. Phytopathology* **48**:469–474.

Stadler, D. R. (1952) Chemotropism in *Rhizopus nigricans*: The staling reaction. *J. Cell. Comp. Physiol.* **39**:449–474.

Staples, R. C. (1974) Synthesis of DNA during differentiation of bean rust uredospores. *Physiol. Plant Pathol.* **4**:415–424.

Staples, R. C. (1979) Unpublished data. Boyce Thompson Institute for Plant Research, Ithaca, N.Y.

Staples, R. C., and V. Macko. (1980) Formation of infection structures as a recognition response in fungi. *Exp. Mycol.* **4**:2–16.

Staples, R. C., A. A. App, and P. Ricci. (1975) DNA synthesis and nuclear division during formation of infection structures by bean rust uredospore germlings. *Arch. Microbiol.* **104**:123–127.

Staples, R. C., L. Laccetti, and Z. Yaniv. (1976) Appressorium formation and nuclear division in *Colletotrichum truncatum. Arch. Microbiol.* **109**:75–84.

Staub, T., H. Dahmen, and F. J. Schwinn. (1974) Light- and scanning electron microscopy of cucumber and barley powdery mildew on host and nonhost plants. *Phytopathology* **64**:364–372.

Van Burgh, P. (1950) Some factors affecting appressorium formation and penetrability of *Collectotrichum phomoides*. (Abstract) *Phytopathology* **40**:29.

Wheeler, H. (1977) Ultrastructure of penetration by *Helminthosporium maydis*. *Physiol. Plant Pathol.* **11**:171–178.

Wood, R. K. S. (1967) *Physiological Plant Pathology*. Blackwell Scientific Publications, Oxford. 570 pp.

Wynn, W. K. (1976) Appressorium formation over stomates by the bean rust fungus: response to a surface contact stimulus. *Phytopathology* **66**:136–146.

Wynn, W. K. (1979) Unpublished data. Department of Plant Pathology, University of Georgia, Athens, Ga.

Yang, S. L., and A. H. Ellingboe. (1972) Cuticle layer as a determining factor for the formation of mature appressoria of *Erysiphe graminis* on wheat and barley. *Phytopathology* **62**:708–714.

Yirgou, D., and R. M. Caldwell. (1968) Stomatal penetration of wheat seedlings by stem and leaf rusts in relation to effects of carbon dioxide, light, and stomatal aperture. *Phytopathology* **58**:500–507.

RECOGNITION IN *RHIZOBIUM*–LEGUME SYSTEMS

Bjørn Sölheim and Jack Paxton

Rhizobium/legume-fixed atmospheric nitrogen is by far the largest agricultural source of biologically fixed nitrogen. The degree of specificity in the symbiotic relationship between legumes and rhizobia varies to a great extent, but the importance of specificity is reflected in the fact that the taxonomy of rhizobia is based on the plants they are able to infect. Recognition reactions probably take place at an early stage in the infection process.

In the following sections we will discuss possible recognition reactions in the rhizosphere, the binding of bacteria to the root surface, the events leading up to root penetration, infection thread formation, and how these reactions relate to present agriculture. Several recent reviews have described the specificity of the infection process in great detail (Dart, 1974, 1977; Fåhraeus and Sahlman, 1977; Broughton, 1978; Dazzo, 1979, 1980a, b; Schmidt, 1979).

RHIZOBIAL GROWTH IN THE RHIZOSPHERE

Nutrients secreted from roots of legumes stimulate growth of microorganisms in the rhizosphere. Rhizobia in general seem to show a chemotactic response toward these secreted nutrients. Specificity in stimulation of growth of some rhizobial strains in the rhizosphere has been shown for some legumes, but this effect cannot be seen for others (Broughton, 1978).

Culture filtrates from compatible rhizobia cause curling and also branching of the root hairs of their hosts (Yao and Vincent, 1969), but several incompatible rhizobia also can cause these reactions. The chemical nature of the curling factor is not established, but there is evidence that one active component is a polysaccharide from the capsule of the bacterium (Dazzo and Hubbell, 1975b; Dazzo, 1978). Yao and Vincent (1976) suggest that there are at least two factors involved in root hair deformation, and that one might be a heat-stable protein. The curling factor could react with compounds secreted by the host roots and become more potent in the curling reaction (Solheim and Raa, 1973).

Turgeon et al. (1979) established in a study of early events in the infection of soybeans by *Rhizobium* that infections are initiated about 3 hr after inoculation. The initiation is in that region of the root that did not have root hairs at the time of inoculation. This delay between inoculation and the initiation of infection in soybean roots by *R. japonicum* could be eliminated by a 3-hr pretreatment of the roots with *R. japonicum* culture filtrates (Bauer et al., 1979). The active principle seems to be a capsular or exopolysaccharide of the symbiont which could also bind to soybean lectin. The work of Bauer et al. (1979) suggests that polysaccharides from the symbiont are capable of inducing emerging root hairs into an infectible state. Autoclaved polysaccharide and *N*-acetylgalactosamine, which is a potent hapten for soybean lectin, have the same effects. Their conclusion is that recognition and host range specificity in the symbiotic association between soybean and *R. japonicum* may require the triggering of appropriate host responses in differentiating root hairs by *R. japonicum* polysaccharide. It is not known whether the curling factor and this infection-conditioning polysaccharide are the same, but it seems clear that rhizobial polysaccharides, either at the cell surface or secreted in the rhizosphere, have very important biological effects on the root hairs of their hosts. It has also been shown (Bhuvaneswari and Bauer, 1978; Fjellheim and Solheim, 1979) that the plant roots contain compounds that alter the polysaccharide surface of the symbiont. Bhuvaneswari and Bauer (1978) found that some compatible rhizobia acquired soybean lectin receptors only when grown in the presence of soybean root exudates or in association with soybean seedling roots. They concluded that *R. japonicum* responded to the rhizosphere environment by synthesizing new cell-surface components. The symbiont might also develop lectin binding sites by another mechanism when growing in the rhizosphere. Fjellheim and Solheim (1979) have found that legume roots contain enzymes that specifically alter the capsular polysaccharide of compatible rhizobia.

It is possible that in the rhizosphere both partners in the symbiosis affect each other by secreting specific compounds. It is not known if this happens in a specific sequence, or if some of these reactions, in nature, take place after contact of the organisms. The curling and branching reaction and the infection conditioning could be caused by polysaccharides from compatible rhizobia that stimulate or initiate specific gene transcription of the affected host cell.

ADHERENCE OF RHIZOBIA TO LEGUME ROOTS

Polysaccharides extending from the surface of bacterial cells mediate adhesion, which determines particular locations of bacteria in most natural environments, ranging from soil particles to human teeth (Costerton et al., 1978). This adhesion could be specific and a major determinant in the initiation and progression of bacterial invasion of plants and animals.

Specific interactions between rhizobial cells and the host root seem to have taken place even before adherence of the bacteria to the root surface. It has been shown that all rhizobia can adhere to a certain extent to legume roots, but there also may be a selective mechanism of adherence which is specific for infective rhizobia. This specific mechanism found in the *R. trifolii*/*T. repens* system binds up to 10 times more bacterial cells to the root hair surface in 12 hr than the nonspecific interaction (Dazzo et al., 1976; Dazzo, 1980c). Adherence occurs both to root hairs and epidermal cells of the root (Dazzo et al., 1976). Chen and Phillips (1976), on the other hand, could not demonstrate specific binding of infective rhizobia to pea roots. There are reports of polar binding of the bacteria to the root surface in several host/rhizobia associations. This was first shown by Sahlman and Fåhraeus (1963) for clover roots. This polar binding might be due to the marked polarity of *Rhizobium* cells shown by Bohlool and Schmidt (1976) with *R. japonicum*. Fluorescent labeled soybean lectin bound specifically to the ends of the bacteria. The authors suggested that rhizobia may bind to a lectin on the root surface via its polar lectin binding site.

THE ROLE OF LECTINS IN BINDING AND RECOGNITION

The term lectin refers to proteins with multivalent binding sites for carbohydrates. Lectins may have several different biological functions in plants (Kauss and Glaser, 1974; Kauss, 1977).

Hamblin and Kent (1973) suggested that bean hemagglutinin is capable of binding rhizobia to roots of *Phaseolus vulgaris* at sites suitable for infection. They did not, however, discuss their results in relation to host/symbiont specificity. From studies of soybean lectin binding to different rhizobia Bohlool and Schmidt (1974) formulated the hypothesis that binding of infective rhizobia to host lectins was the specific determinant of host/symbiont infection. Dazzo and Hubbell (1975a), however, could not find a correlation between concanavalin A binding to rhizobia and their ability to nodulate jack bean. They (Dazzo and Hubbell, 1975b) presented a model for the clover/*R. trifolii* interaction, where the role of lectin was to cross-bridge between a common antigen on the bacterium and the root-cell walls. Later work has expanded this model (Bishop et al., 1977; Dazzo and Brill, 1977, 1978, 1979; Dazzo et al., 1976, 1978, 1979). Solheim (1975) has proposed a model for binding of *R. trifolii* to clover roots based on studies of the "curling factors" in rhizobial culture filtrates (Solheim and Raa, 1973) and on the effect of *Vicia* lectin on rhizobial adherence to *Vicia* roots (Solheim, 1975; Raa et al., 1977). In this model a high molecular weight "factor" from the host root, present in the rhizosphere, binds to the rhizobial surface. The bacteria are then able to bind to a receptor on the root hair surface. If the rhizosphere factor is a lectin, the major difference between the model of Dazzo and Hubbell (1975b) and that of Solheim (1975) is the release of the lectin. It has been shown (Fountain et al., 1977; Hwang et al., 1978) that soybean seed releases lectin when soaked in water. Lectin from roots of clover (Dazzo and Brill, 1977) and soybean (Jack et al., 1979a) is released into isotonic buffer washings. To investigate the involvement of lectins in the infection process it is necessary to get answers to certain key questions such as: What is the receptor for lectin on the bacterial surface? Is this receptor necessary for infection to take place? Is lectin in the plant necessary for the plant to become infected?

The origin and nature of the lectin receptors on the bacteria has been the subject of controversy. Lectin binding to rhizobial polysaccharides has been demonstrated in Ouchterlony plates by Brethauer (1977) and Kamberger (1979). Lectin has been shown to bind specifically to bacterial capsular polysaccharides (Bal et al., 1978; Calvert et al., 1978; Dazzo and Hubbell, 1975a), but also to the lipopolysaccharide (Kamberger, 1979; Wolpert and Albersheim, 1976). Planque and Kijne (1977) found that pea lectins bound to a glucan type polysaccharide in the cell wall of the symbiont, while Bauer et al. (1979) isolated an extracellular polysaccharide from *R. japonicum*, which bound specifically to soybean lectin and also conditioned the root hairs to become infectible. Mort et al. (1979) characterized a pentasaccharide as the repeating unit in the active *R. japonicum* extracellular polysaccharide. They concluded that substituents that were present in variable amounts on the polysaccharide might modify the biological activity of the polymer. It is possible that the host enzymatically removes specific groups (Fjellheim and Solheim, 1979). As discussed earlier, Bhuvaneswari and Bauer (1978) have shown that infective strains of *R. japonicum* that did not bind soybean lectin when grown in ordinary media, did bind lectin when soybean root exudates were added to the medium. For *R. japonicum* biphasic lectin binding curves were found (Bhuvaneswari et al., 1977), which suggest multiple carbohydrate receptors on bacteria. It is possible that rhizobia must acquire lectin binding sites before infection takes place.

Whether lectin is present at the actual infection sites is not known. Most work has been based on experiments with lectins isolated from seeds. There is evidence that the seed lectin in white clover also is found on the root (Dazzo and Brill, 1977, 1978; Dazzo et al. 1978). In developing soybean seedling roots there is a rapid decline of lectin levels (Pueppke et al., 1978; Hwang et al., 1978). The same has been found to be the case for the common lentil (Howard et al., 1972). The roots can be infected after there are no detectable traces of lectin left in the root. It is not possible to detect lectins on the surface or in extracts, of 1- and 2-week old roots of *Vicia hirsuta* (Van Touw and Solheim, 1978). Pull et al. (1978) quantitatively screened different soybean lines for the presence of the 120,000-dalton soybean lectin. Five lines of soybean whose seed lacked the lectin were identified. All these lines were effectively nodulated by several strains of *R. japonicum*, and they concluded that the 120,000-dalton soybean lectin probably was not required for the initiation of soybean–*Rhizobium* symbiosis. Jack et al. (1979b) examined the same five lines and found small amounts of lectin present in all of them. Some of the lectins were different from the 120,000-dalton soybean lectin, but all were immunochemically cross reactive with it. Except for white clover (Dazzo and Brill, 1977, 1979; Dazzo et al., 1978), it has not been shown that the lectins studied really are present on the roots during infection. In most cases the search for lectins on roots has not been done on roots infected by *Rhizobium*.

An explanation of the contradictory results on the presence of lectins at the infection site might be found in the results from Bauer's group at C. F. Kettering Research Laboratory (Bhuvaneswari et al., 1979; Turgeon et al., 1979; Bauer et al., 1979; Mort et al., 1979) (see also discussion on the rhizobial growth in the rhizosphere). They have reported that the initiation of infection in soybean by *Rhizobium* occurs only in short (15–30 μm), developing root hairs. A very early event in the initiation of infection can be triggered by extracellular polysaccharide from the symbiont. This polysaccharide seems to bind specifically to a lectin from the host. It is possible that this early event in the initiation of infectibility is an induction by the polysaccharide of the root hair and its adjacent cells to produce lectin, which takes part in a later event of the infection process. This might be an explanation of the observation that lectins usually are not found on axenically grown roots. In support of this hypothesis is the puzzling way nitrate regulates the infection process. The internal supply of NO_3^- in the root does not affect infection; it is the concentration in the rhizosphere that is important (Raggio et al., 1957). After washing away the NO_3^- initiation of infection starts immediately (Munns, 1968; Raggio et al., 1957). Therefore NO_3^- probably interferes with an early step in the infection process taking place on the outside of the root system. When the infection process has passed this step, NO_3^- will no longer block the infection process, even though Munns (1968) has reported an increase in infection thread abortion when NO_3^- is applied after initiation of infection. Dazzo and Brill (1978) have shown diminishing levels of lectin on root hairs grown in NO_3^-. If the uptake system in the root hair membrane for NO_3^- in some way is coupled with the receptor for the lectin inducing polysaccharide, increasing amounts of NO_3^- should gradually inhibit infection until all uptake sites for NO_3^- are saturated. There is no evidence that NO_3^- is inhibitory to rhizobia.

The root hair and a few adjacent cortical cells have enlarged nuclei and cytoplasmic content. The close relationship between the nucleus and infection thread initiation has been described (Fåhraeus, 1957; Fåhraeus and Sahlman, 1977; Haack, 1964; Nutman, 1959).

The involvement of lectins in the infection process of legumes by rhizobia has not been proved as a general mechanism. Some of the contradictory results may be explained by the hypothesis of induction or activation of lectin by components from the bacteria. We think it is important to stress that seed lectins, often assayed by hemagglutination, might not have anything to do with the infection process. One must look for lectins that react with the symbiont and are found at the infection site.

PENETRATION

Penetration of root hairs and root tissue by rhizobia has been studied by several investigators using both light- and electron microscopy (Broughton, 1978). These studies have confirmed Nutman's (1956) original hypothesis of invagination of the root hair wall at the infection site, and formation of an infection thread. This infection thread almost always originates from deformed and convoluted root hairs. The inward growth of the root hair cell wall possibly requires cell wall degrading enzymes such as pectinases and cellulases (Hubbell et al., 1978). A specific role for polygalacturonase has been suggested, but others have not been able to verify these results (Fåhraeus and Sahlman, 1977).

NODULE FORMATION

Very little is known about the mechanism of nodule formation. Cortical cells are induced to divide and rhizobia are released into the plant cells and differentiate into bacteroids (Dazzo, 1980b).

Signals from the host apparently derepress the nitrogen-fixing genes (NIF genes) in the bacteria. Probably various signals have to pass between host and symbiont before nitrogen fixation can take place. The bacteroids have to produce the heme component and the plant the protein component of the leghemoglobin (Beringer et al., 1979). Ineffective associations usually lack leghemoglobin. By biological symbiosis some plants can get the nitrogen needed for growth and crop production. The important question in agriculture is: Does this association *effectively* fix nitrogen from the atmosphere?

RELATIONSHIPS TO AGRICULTURE

Although the focus of this article is on recognition in legume nodulation, any changes in field manipulation of nodulation and our understanding of this process must be addressed in the context of present-day agriculture as a whole. Certainly international aid to agriculture cannot address single problems without considering implications for the whole system. Understanding the mechanism of recognition in *Rhizobium*–legume systems can allow insights into what governs the ability of nitrogen-fixing microorganisms to grow and compete in soils; especially the root rhizosphere. This in turn leads to a better understanding of the infection process and ultimately to the establishing of more, efficient nodules, which is the goal of most nodulation research. Establishing "better" rhizobial nodules on legumes in soils already containing infecting rhizobia is currently a problem. One approach other workers are investigating is the use of "nonnodulating" lines to allow nodulation by unique specific strains of *Rhizobium* spp. A fundamental understanding of the nodulation process from beginning to end is a valuable aid to intelligent applied research in improving field performance of legumes and other crops that depend on biological nitrogen fixation.

Symbiotic relationships such as nodulation and infection by mycorrhizal fungi (which increase the phosphate supply to plants), in addition to improving plant nutrition, may also be valuable in protecting crops from various pests and diseases. This can biologically reduce problems in producing crops and provide essential nutrients, including biologically fixed nitrogen, to crops that presently do not have this ability. "Synthetic" symbionts can be created and these may provide a breakthrough in biologically increasing plant mineral nutrition (Giles and Whitehead, 1977; Maier et al., 1978). Another possibility is creation of a promiscuous *Rhizobium* that can nodulate many hosts efficiently in the field.

Understanding biological fixation of atmospheric nitrogen includes understanding the recognition mechanisms in *Rhizobium*–legume interactions. And in a broader sense it includes biological nitrogen fixation in nodulating nonlegumes and nonnodulating plants such as rice, wheat, and maize. It may be possible to use biological nitrogen fixation for many crops in which traditionally this has not been considered possible. *Azotobacter* and *Spirillum* are examples of organisms that show promise in biologically fixing nitrogen for grasses (Dobereiner and Day, 1976).

Considerable work can be done in improving biological fixation of nitrogen in tropical agriculture. New strains of rhizobia are needed for improved biological nitrogen fixation under soil and climate types that are quite different than those for which most of the commercial rhizobia have been selected. Cheap, stable rhizobial inoculants are needed for tropical legumes, especially when introducing these legumes into new areas. And new strains of host plants, adapted to native rhizobia, can be developed.

Research into the genetic engineering of nitrogen fixation is demonstrating that it is possible to transfer NIF genes to many other organisms (Hollaender, 1977). Possibly this will lead to unique ways of supplying biologically fixed nitrogen to a wide range of plants. Crop rotation can also be increased to share biologically fixed nitrogen with other crops and at the same time reduce diseases and other problems.

Potential problems associated with attempts to improve crop performance include a human population increasing as fast or faster than crop production; no net gain is made in food availability. Rather than reducing human suffering, this problem as well as the competition for the world's finite natural resources, such as oil and minerals, is aggravated.

Another problem which cannot be ignored is the inherent instability induced in monocultures of crops pushed for ever increasing yields. A similar monoculture of the "best" symbionts could add undesirable instability to biological nitrogen fixation.

SUMMARY

The specificity of the infection process is probably governed by a series of events leading up to the formation of the infection thread. Some of these events have been studied in different rhizobia/legume systems. Sometimes results from one system do not fit with results from another system. A general infection mechanism for temperate rhizobia/legume systems probably is a series of recognition reactions in a certain sequence. If one is lacking, the infection might be stopped by a defense reaction in the host or by simply not providing the next, necessary signal in the process. Each particular *Rhizobium*/legume association has evolved its solution to avoid triggering a defense reaction. In some cases one or more recognition reactions might be completely bypassed. We think therefore that one can expect differences in the infection mechanisms in different rhizobia/legume systems, but that a general trend will be found.

We have summarized some of the possible events in a general infection model in the following:

Rhizosphere

1. Nonselective stimulation of growth of rhizobia and a chemotactic response towards the root.
2. Specific alteration of cell surface polysaccharides on the rhizobia by enzymes or specific growth factors from the host root.
3. Specific induction of synthesis of binding receptors (lectins) and necessary enzymes for infection in the host by polysaccharides from the symbiont.
4. The root hairs respond to the polysaccharides by a "curling reaction."
5. Specific binding of the bacteria to the root hair via a lectin–carbohydrate interaction.

Penetration

1. Formation of "Shepherds crook" and trapping of bacteria in "pockets" in the root hair wall.
2. Bacterial production of pectinase.

3. Invagination of the root hair cell wall.
4. Stimulation of cell wall synthesis and growth of the infection thread.

Nodule Formation

1. Stimulation of cortical cells to divide by signal(s) from the bacteria.
2. Release of bacteria into cortical cells.
3. Bacterial production of the heme component and plant production of the protein component of leghemoglobin.
4. Plant inducing bacteria to differentiate into bacteroid and derepression of the NIF gene.

Nitrogen Fixation

1. Recognition responses can take place up to and during nitrogen fixation.
2. The plant can closely regulate nodulation even during nitrogen fixation.

Biological nitrogen fixation makes an important contribution to modern agriculture, and this contribution can be increased. Developing a better understanding of biological nitrogen fixation and applying this knowledge could lead to substantial savings of petroleum energy now used to chemically fix atmospheric nitrogen.

PROJECTIONS OF ADVANCES IN THE NEXT 5 TO 10 YEARS

Lectins and rhizobial polysaccharides may be found to play important roles in recognition between legumes and rhizobia, but the recognition reaction leading to nodulation will be found to be very complex.

Improved strains of *Rhizobium* will be found and created which nodulate and fix nitrogen more efficiently including less H_2 evolution, but the lack of understanding of the soil ecosystem and recognition mechanisms in nodulation will hinder the use of these strains in agriculture. Other organisms, including mycorrhizal fungi, free-living bacteria, and algae will play an increasingly important role in commercial plant nutrition.

Artificial symbionts, including plant cells with nitrogenase created by molecular genetic engineering, will be tested under laboratory conditions and shown to have great promise in increasing crop production by biological means. Topical agriculture will expand and replace some temperate agriculture. Petroleum energy supplies will become even more expensive and biological nitrogen fixation will supply a significantly larger percentage of crop nitrogen (Evans and Barber, 1977; Hardy and Havelka, 1975).

REFERENCES

Bal, A. K., S. Shantharam, and S. Ratnam. (1978) Ultrastructure of *Rhizobium japonicum* in relation to its attachment to root hairs. *J. Bacteriol.* **133**:1393–1400.

Bauer, W. D., T. V. Bhuvaneswari, A. J. Mort, and G. Turgeon. (1979) The initiation of infections in soybean by *Rhizobium*. 3. *R. japonicum* polysaccharide pretreatment induces root infectibility. *Plant Physiol.* **63** (suppl.): Abstr. 745.

Beringer, J. E., N. Brewin, A. W. B. Johnston, H. M. Schulman, and D. A. Hopwood. (1979) The *Rhizobium*–legume symbiosis. *Proc. R. Soc. London, Ser. B* **204**:219–233.

Bhuvaneswari, T. V., and W. D. Bauer. (1978) Role of lectins in plant–microorganism interactions. III. Influence of rhizosphere/rhizoplane culture conditions on the soybean lectin-binding properties of rhizobia. *Plant Physiol.* **62**:71–74.

Bhuvaneswari, T. V., S. G. Pueppke, and W. D. Bauer. (1977) Role of lectins in plant–microorganism interactions. I. Binding of soybean lectin to rhizobia. *Plant Physiol.* **60**:486–491.

Bhuvaneswari, T. V., G. Turgeon, and W. D. Bauer. (1979) The initiation of infections in soybean by *Rhizobium*. 1. Localization of infectible root hairs. *Plant Physiol.* **63**(suppl.): Abstr. 743.

Bishop, P. E., F. B. Dazzo, E. R. Appelbaum, R. J. Maier, and W. J. Brill. (1977) Intergeneric transfer of genes involved in the *Rhizobium*–legume symbiosis. *Science* **198**:938–940.

Bohlool, B. B., and E. L. Schmidt. (1974) Lectins: A possible basis for specificity in the *Rhizobium*–legume root nodule symbiosis. *Science* **185**:269–271.

Bohlool, B. B., and E. L. Schmidt. (1976) Immunofluorescent polar tips of *Rhizobium japonicum*: Possible site of attachment or lectin binding. *J. Bacteriol.* **125**:1188–1194.

Brethauer, T. S. (1977) Soybean lectin binds to *Rhizobia* unable to nodulate soybean: Lectins may not determine host specificity in legume nodulation. M.S. Thesis, University of Illinois, Urbana, Ill.

Broughton, W. J. (1978) Control of specificity in legume–*Rhizobium* associations. *J. Appl. Bacteriol.* **45**:165–194.

Calvert, H. E., M. Lalonde, T. V. Bhuvaneswari, and W. D. Bauer. (1978) Role of lectin in plant–microorganism interactions. IV. Ultrastructural localization of soybean lectin binding sites on *Rhizobium japonicum*. *Can. J. Microbiol.* **24**:785–793.

Chen, A. T., and A. Phillips. (1976) Attachment of *Rhizobium* to legume roots as the basis for specific interactions. *Physiol. Plant.* **38**:83–88.

Costerton, J. W., G. G. Geesey, and K. J. Cheng. (1978) How bacteria stick. *Sci. Am.* **238**:86–95.

Dart, P. J. (1974) Development of root-nodule symbioses. I. The infection process. In *The Biology of Nitrogen Fixation* (A. Quispel, ed.). Elsevier/North-Holland, Amsterdam, Pp. 381–429.

Dart, P. J. (1977) Infection and development of leguminous nodules. In *A Treatise on*

Dinitrogen Fixation, Sect. III. *Biology* (R. W. F. Hardy and W. S. Silver, eds.). Wiley, New York. Pp. 367–472.

Dazzo, F. B. (1978) Personal communication. Michigan State University, Department of Microbiology and Public Health, East Lansing, Mich.

Dazzo, F. B. (1979) Lectins and recognition in the *Rhizobium*–legume symbiosis. *Am. Soc. Microbiol. News (Washington, DC)* **45:**238–240.

Dazzo, F. B. (1980a) Determinants of host specificity in the *Rhizobium*–clover symbiosis. In *Nitrogen Fixation.* Vol II (W. E. Newton and W. H. Orme–Johnsen, eds.). Univ. Park Press, Baltimore, Ma. Pp 165–187.

Dazzo, F. B. (1980b) Infection processes in the *Rhizobium*–legume symbiosis. In *Advances in Legume Science* (R. J. Summerfield and A. H. Bunting, eds.). Crown Publishing, London. Pp 49–59.

Dazzo, F. B. (1980c) Adsorption of microorganisms to roots and other plant surfaces. In *Adsorption of Microorganisms to Surfaces.* (G. Bitton and K. Marshall, eds.). Wiley, New York. Pp. 253–316.

Dazzo, F. B., and W. J. Brill. (1977) Receptor site on clover and alfalfa roots for *Rhizobium. Appl. Environ. Microbiol.* **33**:132–136.

Dazzo, F. B., and W. J. Brill (1978) Regulation by fixed nitrogen of host–symbiont recognition in the *Rhizobium*–clover symbiosis. *Plant Physiol.* **62**:18–21.

Dazzo, F. B., and W. J. Brill. (1979) Bacterial polysaccharide which binds *Rhizobium trifolii* to clover root hairs. *J. Bacteriol.* **137**:1362–1373.

Dazzo, F. B., and D. H. Hubbell. (1975a) Concanavalin A: Lack of correlation between binding to *Rhizobium* and specificity in the *Rhizobium*–legume symbiosis. *Plant Soil* **43**:713–717.

Dazzo, F. B., and D. H. Hubbell. (1975b) Cross reactive antigens and lectin as determinants of symbiotic specificity in the *Rhizobium*–clover association. *Appl. Microbiol.* **30**:1017–1033.

Dazzo, F. B., C. A. Napoli, and D. H. Hubbell. (1976) Adsorption of bacteria to roots as related to host-specificity in the *Rhizobium*–clover symbiosis. *Appl. Environ. Microbiol.* **32**:166–171.

Dazzo, F. B., W. E. Yanke, and W. J. Brill (1978) Trifoliin: A *Rhizobium* recognition protein from white clover. *Biochim. Biophys. Acta* **539**:276-286.

Dazzo, F. B., M. R. Urbano, and W. J. Brill. (1979) Transient appearance of lectin receptors on *Rhizobium trifolii. Curr. Microbiol.* **2**:15–20.

Dobereiner, J., and J. M. Day. (1976) In *Proceedings of the First International Symposium on Nitrogen Fixation*, Vol. 2 (W. E. Newton and C. J. Nyman, eds.). Washington State University Press, Pullman, Wash. Pp. 518–538.

Evans, H. J., and L. E. Barber. (1977) Biological nitrogen fixation for food and fiber production. *Science* **197**:332–339.

Fjellheim, K. E., and B. Solheim. (1979) Personal communication. Institute of Biology and Geology, University of Tromsø, Tromsø, Norway.

Fountain, D. W., D. E. Foard, W. D. Replogle, and W. K. Yang. (1977) Lectin release by soybean seeds. *Science* **197**:1185–1187.

Fåhraeus, G. (1957) The infection of clover root hairs by nodule bacteria studied by a simple glass slide technique. *J. Gen. Microbiol.* **16**:374–381.

Fåhraeus, G., and K. Sahlman. (1977) The infection of root hairs of leguminous plants by nodule bacteria. *Kungl. Vetenskaps samhällets i Uppsala årsbok* **20**:103–131.

Giles, K.L., and H. Whitehead. (1977) The localization of introduced *Azotobacter* cells within the mycelium of modified *Mycorrhiza* (*Rhizopogon*) capable of nitrogen fixation. *Plant Sci. Lett.* **10**:367–372.

Haack, A. (1964) Über den Einfluss der Knöllchen-bakterien auf die Wurzelhaare von Leguminosen und Nichtleguminosen. *Zentralbl. Bakteriol. Parasitenkd., Abt. II,* **117**:343–366.

Hamblin, J., and S. P. Kent. (1973) Possible role of phytohaemagglutinin in *Phaseolus vulgaris* L. *Nature* [New Biol.] **245**:28–30.

Hardy, R. W. F., and V. D. Havelka. (1975) Nitrogen fixation research: A key to world food? *Science* **188**:633–643.

Hollaender, A. (1977) Committee on genetic engineering for nitrogen fixation (A. Hollaender, Chairman). *Genetic Engineering for Nitrogen Fixation.* Plenum Press, New York.

Howard, I. K., H. J. Sage, and C. B. Horton. (1972) Studies on the appearance and location of haemagglutinins from a common lentil during the life cycle of the plant. *Arch. Biochem. Biophys.* **149**:323–326.

Hubbell, D. H., V. M. Morales, and M. Umalia-Garcia. (1978) Pectolytic enzymes in *Rhizobium. Appl. Environ. Microbiol.* **35**:210–213.

Hwang, D. L., W. K. Yang, and D. E. Foard. (1978) Rapid release of protease inhibitors from soybeans: Immunochemical quantitation and parallels with lectins. *Plant Physiol.* **61**:30–34.

Jack, M. A., E. L. Schmidt, and F. Wold. (1979a) Studies on the *in vivo* function of soybean lectin. *Fed. Proc., Fed. Am. Soc. Exp. Biol.* **38**:411 (Abstr. 965).

Jack, M. A., E. L. Schmidt, and F. Wold. (1979b) Personal communication by Finn Wold, Department of Biochemistry, University of Minnesota, St. Paul, Minn.

Kamberger, W. (1979) An Ouchterlony double diffusion study on the interaction between legume lectins and rhizobial cell surface antigens. *Arch. Microbiol.* **121**:83–90.

Kauss, H. (1977) The possible physiological role of lectins. In *Cell Wall Biochemistry Related to Specificity in Host–Plant Pathogen Interactions* (B. Solheim and J. Raa, eds.). Universitetsforlaget, Oslo, Norway. Pp. 347–359.

Kauss, H., and C. Glaser. (1974) Carbohydrate-binding proteins from plant cell walls and their possible involvement in extension growth. *FEBS Lett.* **45**:304–307.

Maier, R. J., P. E. Bishop, and W. J. Brill. (1978) Transfer from *Rhizobium japonicum* to *Azotobacter vinelandii* of genes required for nodulation. *J. Bacteriol.* **134**:1199–1201.

Mort, A. J., M. E. Slodki, R. D. Plattner, and W. D. Bauer. (1979) The initiation of infections in soybean by *Rhizobium*. 4. Molecular structure of biologically active *R. Japonicum* polysaccharides. *Plant Physiol.* **63**(suppl): Abstr. 746.

Munns, D. N. (1968) Nodulation of *Medicago sativa* in solution culture. III. Effects of nitrate on root hairs and infection. *Plant Soil* **29**:33–47.

Nutman, P. S. (1956) The influence of the legume in root–nodule symbiosis. A comparative study of host determinants and functions. *Biol. Rev. Cambridge Philos. Soc.* **31**:109–151.

Nutman, P. S. (1959) Some observations on root-hair infection by nodule bacteria. *J. Exp. Bot.* **10**:250–263.

Planque, K., and J. W. Kijne. (1977) Binding of pea lectins to a glycan type polysaccharide in the cell walls of *Rhizobium leguminosarum. FEBS Lett.* **73**:64–66.

Pueppke, S. G., K. Keegstra, A. L. Ferguson, and W. D. Bauer. (1978) The role of lectins in plant–microorganism interactions. II. Distribution of soybean lectin in tissue of *Glycine max* (L.) Merr. *Plant Physiol.* **61**:779–784.

Pull, S. P., S. G. Pueppke, T. Hymowitz, and J. H. Orf. (1978) Soybean lines lacking the 120,000-dalton seed lectin. *Science* **200**:1277–1279.

Raa, J., B. Robertsen, B. Solheim, and A. Tronsmo. (1977) Cell surface biochemistry related to specificity of pathogenesis and virulence of microorganisms. In *Cell Wall Biochemistry Related to Specificity in Host–Plant Pathogen Interactions* (B. Solheim and J. Raa, eds). Universitetsforlaget, Oslo, Norway. Pp. 11–30.

Raggio, M., N. Raggio, and J. G. Torrey (1957). The nodulation of isolated leguminous roots. *Am. J. Bot.* **44**:325–334.

Sahlman, K., and G. Fåhraeus. (1963) An electron microscope study of root-hair infection by *Rhizobium. J. Gen. Microbiol.* **33**:425–427.

Schmidt, E. L. (1979) Initiation of plant–root microbe interactions. *Annu. Rev. Microbiol.* **33**:355–376.

Solheim, B. (1975) A model of the recognition-reaction between *Rhizobium trifolii* and *Trifolium repens*, and possible role of lectins in the infection of legumes by rhizobia. *National Alliance Treaty Organization Conference on Specificity in Plant Diseases*, Advanced Study Institute, Sardinia.

Solheim, B., and J. Raa. (1973) Characterization of the substances causing deformation of root hairs of *Trifolium repens* when inoculated with *Rhizobium trifolii. J. Gen. Microbiol.* **77**:241-247.

Turgeon, G., T. V. Bhuvaneswari, and W. D. Bauer. (1979) The initiation of infections in soybean by *Rhizobium*. 2. Time course and cytology of the initial infection process. *Plant Physiol.* **63**(suppl.): Abstr. 744.

Van Touw, J. H., and B. Solheim. (1978) Personal communication. Institute of Biology and Geology, University of Tromsø, Tromsø, Norway.

Wolpert, J. S., and P. Albersheim. (1976) Host–symbiont interactions. I. The lectins of legumes interact with the O-antigen-containing lipopolysaccharides of their symbiont *Rhizobia. Biochem. Biophys. Res. Commun.* **70**:729–737.

Yao, P. Y., and J. M. Vincent. (1969) Host specificity in the root hair "curling factor" of *Rhizobium* spp. *Aust. J. Biol. Sci.* **22**:413–423.

Yao, P. Y., and J. M. Vincent. (1976) Factors responsible for the curling and branching of clover root hairs by *Rhizobium. Plant Soil* **45**:1–16.

NATURAL REGULATORS OF FUNGAL DEVELOPMENT

Edward J. Trione

Man is often criticized for using chemicals to control plant diseases, yet we are not original in the use of chemicals to control plant pathogens. Other forms of life have long employed a variety of compounds in attempts to control their own development and block the development of their competitors. Plants, fungi, bacteria, and insects are all known to affect one another chemically, and knowledge about these chemical interactions and their biological significance can lead to important new methods of controlling plant diseases.

There are many "natural regulators" of fungal development; for example, the physical environment (temperature, light, moisture, CO_2, O_2, etc.); allelopathic interactions between fungi and many other biological organisms; and roots of higher plants that develop symbiotic relationships with mycorrhizal fungi and greatly affect the development of these fungi. This chapter, however, will focus only on naturally occurring biochemical compounds that have been identified and shown to exert a profound influence on fungal development. The object of this review is to draw together information on the regulation of fungal development and attempt to relate it to new innovations in controlling plant pathogenic fungi. Several recent reviews on fungal hormones (Barksdale, 1969; Crandall et al., 1977; Crandall, 1978; van den Ende, 1976) and on endogenous germination inhibitors in fungi (Allen, 1976; Macko et al., 1976) may be consulted for additional details.

FUNGAL PHEROMONES

Allomyces

The water mold *Allomyces* is a phycomycete in the order Blastocladiales. This fungus has five free-swimming, unicellular stages and two mycelial stages. The three types of diploid swarm spores (zygotes, mitospores, and meiospores) are all attracted to nutrient substrates, particularly those containing the free amino acids L-leucine and L-lysine. The haploid male and female gametes, however, are not attracted to those amino acids.

Early studies (Machlis, 1958) clearly demonstrated that a chemotactic substance was produced by the female gametes, which attracted the male gametes. The male gametes were fast and active and could swim for several hours, whereas the female gametes were larger and slower. Thus, the release of a diffusible attractant from the slow female cells established a concentration gradient that attracted the male gametes.

To facilitate a study of the mating hormone of *Allomyces* it was necessary to have a clone that was essentially female in character to produce the hormone, and another clone essentially male in character to use in the bioassay. These female and male lines were developed from interspecific hybrids of *A. macrogynus* and *A. arbuscula* (Machlis. 1958).

The chemotactic agent was produced in very small amounts and was extracted and purified from large volumes of liquid culture media in which the female strain had grown. The bioassay was based on the attraction of motile male gametes to a dialysis membrane through which the test solution (female hormone) slowly diffused. Eventually about 2 g of pure hormone were obtained, and its structure was elucidated (Machlis et al., 1968).

The hormone sirenin was named after the temptingly beautiful female sirens of mythology, and was the first plant sex hormone to be characterized. The pure compound is a colorless, viscous, sesquiterpene with two double bonds, two hydroxyl groups, and a cyclopropane ring. It is active at 1×10^{-10} M.

Synthetic sirenin and a number of its analogs were tested, L-sirenin was found to be the natural hormone, and none of its synthetic analogs showed any chemotactic activity (Machlis, 1973). These results indicated that the allylic alcohol group is essential and that the stereochemistry of the side chain on the cyclopropane ring is of major importance for biological activity. Male gametes took up virtually all of the L-sirenin in a 5-nM solution in 20 min, and the sirenin could not be reextracted from the male gametes. This suggested that the hormone was irreversibly bound or metabolically altered. Sirenin was not inactivated by extracellular enzymes produced by the gametes.

Achlya

The genus *Achlya* belongs to a family of water molds, the Øomycetes. It is commonly found in shallow lakes where oxygen and food are plentiful. It

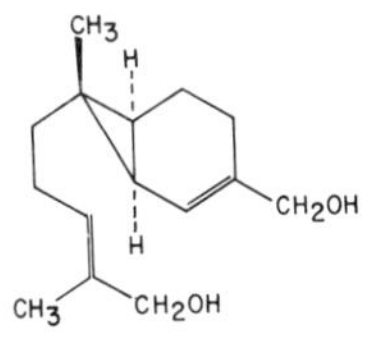

Sirenin

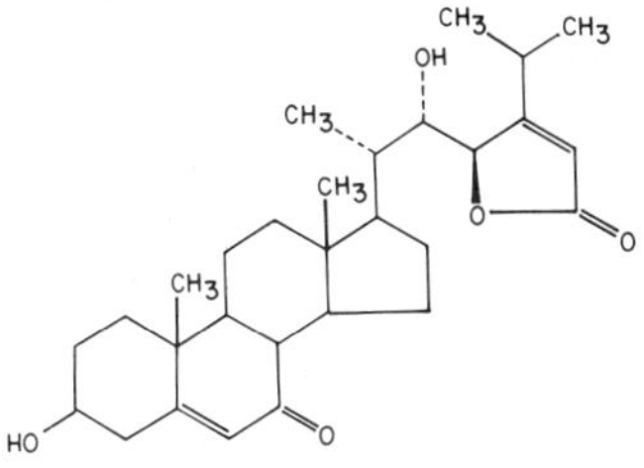

Antheridiol

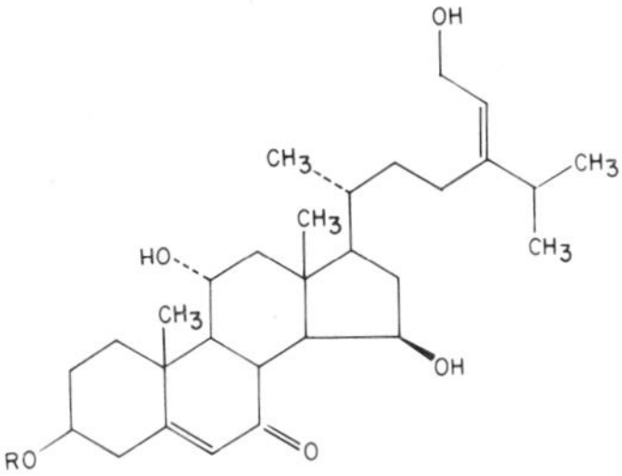

24(28) Dehydro-oogoniol-1

$R = (CH_3)_2\ CHCO$

Figure 1 The structures of the mating hormones of *Allomyces* and *Achlya*.

forms abundant asexual zoospores, but the zoospores do not mate as in the genus *Allomyces*. It also forms thick-walled sexual spores, the oospores, which are nonmotile and usually sink to the bottom of the lake and lie dormant for long periods before they germinate. Many species of *Achlya* are homothallic, but some are heterothallic (Barksdale, 1969).

When opposite mating types of hyphae are allowed to grow near one another, antheridial branches will develop on one culture and oogonial initials will be produced on the other culture. After many preliminary experiments and careful microscopic observations, the presence of mating hormones in *Achlya* was hypothesized (Raper, 1939). Evidence was obtained that a substance (hormone A) was produced continuously by the female mating type. Hormone A could be monitored because it induced profuse antheridial branching in the male culture. After the antheridial branching occurs, the antheridia produce hormone B, which induces the production of oogonial initials in the female cultures. As the oogonia increase in size they secrete larger amounts of hormone A, which direct the antheridial branches toward the oogonia. The antheridia and oogonia subsequently mate and oospores are formed.

Raper and Haagen-Smit (1942) attempted to isolate hormone A from 1400 liters of culture filtrate from the female strain. They obtained 2 mg of purified material, active at 10×10^{-12} g/ml, but were unable to identify it. In 1964 McMorris and Barksdale at the N.Y. Botanical Garden began large-scale experiments to isolate hormone A. They purified about 20 mg of hormone A from 1000 liters of culture filtrate and collaborated with chemists in determining the structure of the hormone, named antheridiol (Arsenault et al., 1968) (Figure 1). It was the first steroidal plant sex hormone to be identified and differs from the mammalian sex hormones in having a much longer side chain at C-17. The concentration of antheridiol determines, in the male strain, (1) the number of antheridial branches initiated, and (2) the time between application and the appearance of antheridial branches. Antheridiol is active at 6×10^{-12} g/ml (Barksdale, 1969).

In male strains the antheridial hyphae begin to produce hormone B about 2 hours after they have responded to hormone A. The oogonial initials begin to form about 12 hours after hormone B production begins. The number of oogonial initials produced increases as the concentration of hormone B increases. Purified hormone B preparations were found to contain 7 or more steroidal compounds, all closely related, that possess hormone B activity (McMorris, 1978). The structure shown in Figure 1 is by far the most active molecule in eliciting the formation of oogonia and was named dehydro-oogoniol. It is active at about 50 ng/ml, which is about 8000 times less active than hormone A. These two hormones, antheridiol and oogoniol, do not antagonize the action of the other and they have no known effect on other fungi.

The genus *Achlya* does not contain plant pathogenic species, but the sexual cycle of *Achlya* is very similar to that of *Pythium* and *Phytophthora*. It is interesting to consider the effects of sterols on these two important plant pathogenic genera. Pythiums and phytophthoras are unique in that they cannot synthesize sterol. Without sterol in the growth medium they remain vegetative, but with sterol added, growth is stimulated and sexual reproduction takes place. Fucosterol, sitosterol, and stigmasterol appear to be the most active in promoting antheridia and oogonia to develop with the subsequent formation of oospores (Elliot, 1977). Although these common sterols in the medium will bring about these effects, they are thought to function merely as precursors of the sexual hormones of the pythiums and phytophthoras. Preliminary evidence supports the hormonal regulation of sexual reproduction in phytophthoras (Ko, 1978). It has been hypothesized that the pheromones of *Pythium* and *Phytophthora* species are quite similar to the pheromones of *Achlya* and that these plant pathogenic fungi obtain the precursors of their sexual hormones from host plants (McMorris, 1978). Sterols are widespread in higher plants.

Mucorales

In the heterothallic species of the Mucorales there are only two mating types, that is, + or − lines. When the + or − lines are grown separately there are no distinguishing morphological differences between the two, but when the opposite mating types are grown close to each other zygophore formation is initiated. Early workers (Burgeff, 1924; Plempel, 1960, 1962, 1963; Plempel and Dawid, 1961) suggested that mating hormones were involved in the initiation of zygophores. Gooday (1968) and van den Ende (1968) successfully isolated and purified substances, trisporic acids (TSA) B and C, that induced zygophore production (Fig. 2). Either compound, TSA-B or TSA-C, could induce zygophores in both mating types, but 9-*cis*-trisporic acid B is the most active and is probably the "true" hormone (Bu'Lock et al., 1976).

If either the + or − line were grown alone, TSA production did not occur. If the two compatible mating lines were grown together or physically separated by a dialysis membrane then both + and − lines produced TSA-B and TSA-C; however, TSA production stopped immediately in both strains if the cultures were separated. Thus it was postulated that one or more small molecular weight substance(s) must signal each mating line of the presence of the opposite mating line. Sutter and co-workers (1970, 1973, 1974) demonstrated that +, as well as −, strains grown alone each produced a hormonal substance that stimulated the production of TSA in the opposite mating type. The structures of these prohormones are shown in Figure 2. The prohormones are produced continuously, but in small quantities (2–5 mg/liter), by both mating types. Once the prohormones are detected by the opposite mating type, TSA synthesis begins, and after TSA reaches a threshhold value zygophores are formed.

In the Mucorales, among the genera *Blakeslea, Mucor, Phycomyces,* and *Absidia*, interspecific matings are common (van den Ende, 1976). This suggests that the hormonal reactions described above are common to many species of these genera.

Yeasts

Rhodosporidium toruloides is the sexual stage of the yeast *Rhodotorula glutinis* (Banno, 1967). The haploid strains of *Rhodosporidium* (type *A* and *a*) are ovoid in shape, while the diploid and dicaryotic strains are filamentous (Abe et al., 1975). The opposite mating strains produce extracellular diffusible substances that induce the formation of mating tubes in the opposite mating-type cells (Abe et al., 1975). Type *A* cells produce constitutively an extracellular inducing substance, factor *A*, which is received by type *a* cells and stimulates *a* cells to form mating tubes and to secrete another inducing substance, factor *a*. Type *A* cells receive factor *a* and form mating tubes in response to it. The mating tubes of *A* and *a* cells elongate toward and recog-

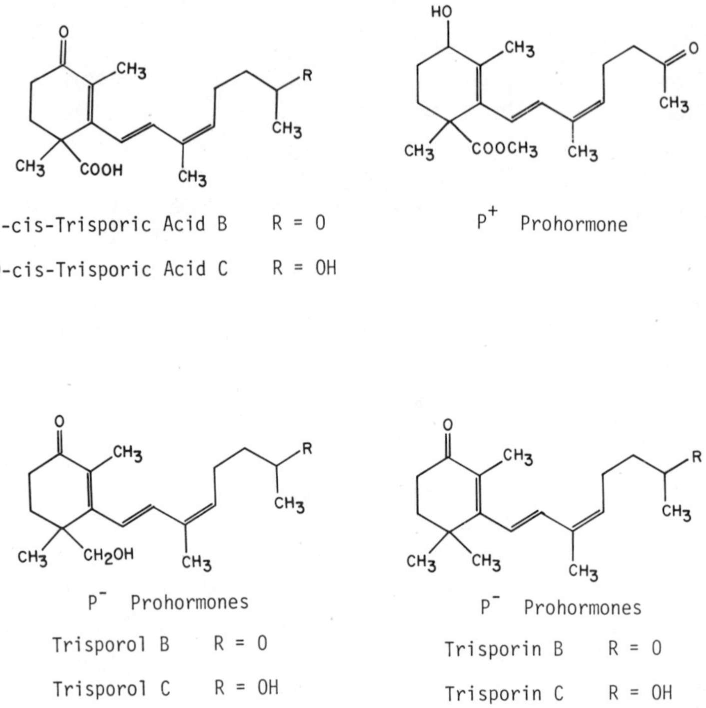
O
CH3
R
CH3
CH3
COOH
CH3
9-cis-Trisporic Acid B R = O
9-cis-Trisporic Acid C R = OH
HO
CH3
O
CH3
CH3
COOCH3
CH3
P+ Prohormone
O
CH3
R
CH3
CH3
CH2OH
CH3
P- Prohormones
Trisporol B R = O
Trisporol C R = OH
O
CH3
R
CH3
CH3
CH3
CH3
P- Prohormones
Trisporin B R = O
Trisporin C R = OH

Figure 2 The structures of the mating hormones of the Mucorales.

nize the opposite mating type and finally fuse with the mating tube or the cells. The structure of factor *A* has been determined (Kamiya et al., 1979) and found to be a peptide composed of 11 amino acid residues with a lipophilic group at the C-terminus (Figure 3). Factor *A* is active at 8 ng per ml (Kamiya et al., 1977). The farnesyl moiety is indispensable to hormone activity, because a synthetic peptide analogue without the farnesyl side chain had no biological activity (Tsuchiya et al., 1978).

The cells of haploid cultures of *Saccharomyces* species develop conjugation tubes when opposite mating types (*a* and α cells) are paired on an agar surface. When the tubes meet fusion occurs, a zygote is formed, and diploid cells may be budded from the zygote (Levi, 1956). The mating reaction does not require initial cell contact, for hormones transmit signals indicating the presence of the opposite mating type (Duntze et al., 1970). The hormone secreted by mating-type α cells, termed α factor, has been characterized as a linear oligopeptide which exists in two forms, one with 12 L-amino acids and the other with 13 (Stotzler and Duntze, 1976) (Figure 3).

The mating hormone secreted by *a* cells, termed *a* factor, has been more difficult to study due to the lack of a simple and reproducible bioassay system. However, partially purified preparations exhibiting *a*-factor activity

A factor from *Rhodosporidium*
H_2N-Tyr-Pro-Glu-Ilu-Ser-Trp-Thr-Arg-AsN-Gly-Cys—COOH
Farnesyl-OH

α factors from *Saccharomyces*
H_2N-Trp-His-Trp-Leu-GlN-Leu-Lys-Pro-Gly-GlN-Pro-Met-Tyr—COOH
H_2N-His-Trp-Leu-GlN-Leu-Lys-Pro-Gly-GlN-Pro-Met-Tyr—COOH
H_2N-Trp-His-Trp-Leu-GlN-Leu-Lys-Pro-Gly-GlN-Pro-Met(SO)-Tyr—COOH
H_2N-His-Trp-Leu-GlN-Leu-Lys-Pro-Gly-GlN-Pro-Met(SO)-Tyr—COOH

a factor from *Tremella*
H_2N-Glu-His-Asp-Pro-Ser-Ala-Pro-Gly-AsN-Gly-Tyr-Cys—COOH
Lipid

Figure 3 The structures of the mating hormones of yeast fungi.

have been obtained (Betz et al., 1977). Gel filtration studies suggested that the *a* factor had a molecular weight greater than 600,000, but this apparent molecular weight could have resulted from the aggregation of smaller subunits. Proteases destroy *a*-factor activity.

Haploid cells of the basidiomycete *Tremella mesenterica* normally grow as yeasts, but can produce filamentous conjugation tubes when sexually compatible strains are mixed. The conjugation tubes are induced by hormones that diffuse between the cells (Bandoni, 1965). The hormone produced by type *A* cells has been studied extensively. It has been purified and partially identified as a lipopeptide (Reid, 1974; Sakagami et al., 1978), and named Tremerogen A-10 (Figure 3). The lipid moiety attached to cysteine has not been structurally characterized, but it is not a farnesyl moiety, as found on the mating hormone for *Rhodosporidium* (Tsuchiya et al., 1978). Tremerogen A-10 is active at 2 ng/ml (Sakagami et al., 1978).

Other plant pathogenic fungi that have yeastlike stages and appear to have hormonally mediated mating reactions include many of the smut fungi, for example, *Ustilago* species, *Tilletia* species, and *Sorosporium* species.

INHIBITORS AND STIMULATORS OF FUNGAL SPORE GERMINATION

The germination of a spore is often the first crucial event in the development of a plant disease. To perform its role effectively a spore should germinate only in environments that provide a good opportunity for continued growth and development. The well-adapted spores have some built-in mechanisms for achieving that goal. One of the mechanisms is related to the concept of natural endogenous inhibitors of germination. In many mature fungal spores germination is regulated by reversible, static-type compounds so that germination occurs primarily in instances that favor outgrowth and survival of the organism. It is more difficult to envision an ecological role for endogenous germination stimulators, except in mating reactions of haploid spores of fungi, where precise timing and polarized growth is required.

Most of the research on endogenous inhibitors of fungal spore germination has focused on the uredospores of rust fungi. Physiological evidence indicated the presence of these inhibitors for more than 20 years (Allen, 1976) before they were identified from uredospores of the bean rust fungus, *Uromyces phaseoli* (Macko et al., 1970). Early evidence for the presence of endogenous inhibitors was obtained from:

1. The population effect, i.e., as more and more spores were crowded into a unit area their percentage germination was reduced.
2. Washing the spores with water reduced the population effect, and germination was enhanced.
3. Inhibitory substances were present in the water extract.
4. The inhibition was easily and quickly reversed (Allen, 1976).

Rust Fungi

A concerted effort was made to identify the germination inhibitors from rust uredospores. These substances were extracted from the spores with water, partitioned into ether and separated on TLC plates. The zones on the TLC plate were bioassayed against uredospores that had been presoaked in water to remove the natural inhibitor. The inhibitor was further purified by gas chromatography and identified as methyl-3,4-dimethoxycinnamate by mass spectrometry (Macko et al., 1970) (Fig. 4).

The endogenous germination inhibitor from wheat-stem rust uredospores was subsequently identified as methyl-3-methoxy-4-hydroxycinnamate (Macko et al., 1971) (Figure 4), also called methyl ferulate. Both the *cis* and *trans* isomers of methyl-3,4-dimethoxycinnamate were detected (Macko et al., 1970). It was later shown that only the *cis* isomer occurred naturally (Macko et al., 1972), and only the *cis* isomer was active as an inhibitor of spore germination (Allen, 1972).

Thus it appeared that a group of similar cinnamic acids function as germination inhibitors in the rust fungi. A number of other rust uredospores were studied, and it was found that methyl-3,4-dimethoxy-*cis*-cinnamate was the primary inhibitor present in most of the uredospores (Table 1). These inhibitors block only germination, they do not block germ-tube outgrowth or mycelial development; therefore, they must be present during the early stages of germination to be effective. The inhibitors are fungistatic, not fungicidal. They are active at 10^{-9} to 10^{-11} M, that is, in the same concentration range as the mating hormones of *Achlya* and *Allomyces*. These inhibitors are not specific for genus or species in the rust fungi, that is, they are active against a number of rust species.

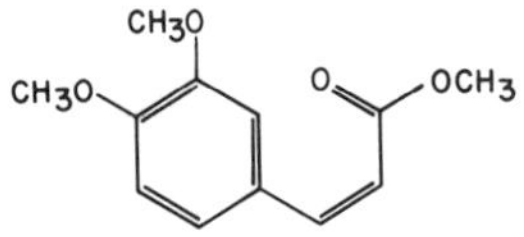

Methyl-cis-3,4-dimethoxycinnamate

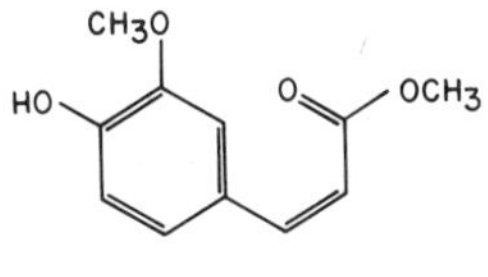

Methyl-cis-ferulate

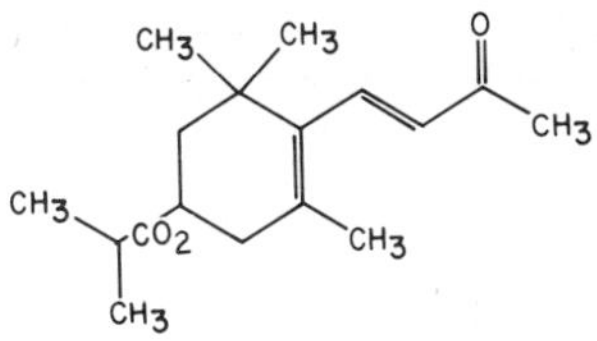

Quiesone, 5-isobutyroxy-β-ionone

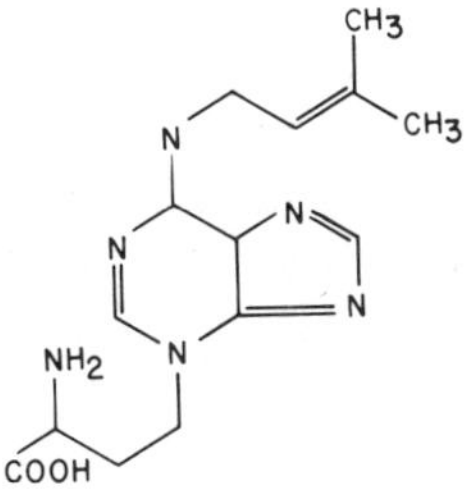

Discadenine

Figure 4 The structures of the inhibitors of fungal spore germination.

Rust uredospores contain several germination pores which upon germination are hydrolyzed prior to the emergence of the germ tube. The mode of action of the germination inhibitors appears to be the inactivation of enzymes responsible for hydrolyzing the germination pore plug (Allen, 1976). Removal of the inhibitor, by washing with water, lead to immediate digestion of the plug and germ tube emergence.

Table 1 Effects of Methyl-*cis*-3,4-dimethoxycinnamate on the Germination of Uredospores of Different Rust Fungi

Rust Species	Host	ED_{50} (ng/ml)	Natural Inhibitor	Reference
Uromyces phaseoli	Bean	2.72	MDMC[a]	Macko et al., 1972
Puccinia antirrhini	Snapdragon	0.23	MDMC[a]	Macko et al., 1972
P. arachidis	Peanut	0.08	MDMC[a]	Foudin and Macko, 1974
P. graminis tritici	Wheat	3.7	MF[b]	Macko et al., 1972
P. coronata	Oat	0.01	MDMC[a]	Macko et al., 1972
P. helianthi	Sunflower	1.65	MDMC[a]	Macko et al., 1972
P. striiformis	Wheat	4.0	MDMC[a]	Macko et al., 1977
Hemileia vastatrix	Coffee	Insensitive	Unknown	Stahmann et al., 1976
Melampsora lini	Flax	Insensitive	Unknown	Macko et al., 1972

[a]Methyl-*cis*-3,4-dimethoxycinnamate.
[b]Methyl-*cis*-ferulate.

Peronospora

A self-inhibitor of conidial germination of *Peronospora tabacina*, an obligate parasite causing the downy mildew of tobacco, was first reported by Shepherd and Mandryk (1962). Leppik et al. (1972) described the isolation and identification of the major endogenous inhibitor, and named it quiesone (Figure 4).

The β-ionone moiety of quiesone also inhibits conidial germination of *Peronospora*, ED_{50} 150 ng/ml, but it is not as potent as quiesone, ED_{50} 0.1 ng/ml (Leppik et al., 1972). Although β-ionone inhibits the germination of *Peronospora* spores, it is a potent stimulator of germination of dormant rust uredospores (Macko et al., 1976; French et al., 1977), where it counteracts the inhibitory effect of the natural endogenous methyl-*cis*-cinnamates. The biochemical bases of these β-ionone effects are unknown.

Compounds containing the ionone skeleton represent an important group of biologically active compounds, many of which are known to control developmental changes in organisms. One of the best known is abscisic acid, the plant hormones that exerts inhibitory effects on growth, regulates abscission of leaves, dormancy of buds and seeds, and other aspects of plant metabolism.

The quiesone molecule was synthesized by Mori (1973), and its biological and chemical properties were slightly different than those reported for the natural product which was isolated and identified by Leppik et al. (1972). Actually there was surprising agreement in the chemical spectral data, considering the small amount of natural product available for study. The synthetic compound, however, was reported to have an ED_{50} value 100 × higher than the natural product. Part of that large difference may be due to the fact that the synthetic product was a racemic mixture and the effect of *cis–trans* isomerism in the side chain was not tested. Only the *cis* isomers of the methyl cinnamate rust spore inhibitors are active. It is possible that some of the unnatural isomers in the synthetic product may compete with the natural inhibitors (for binding sites) and thus appear to reduce the inhibitory activity of the synthetic product (Leppik, 1979).

Dictyostelium

Spore germination is regulated by an endogenous inhibitor in the cellular slime mold *Dictyostelium discoideum*. Bacon et al. (1973) isolated an inhibitor from *D. discoideum* spores, identified it as *N,N*-dimethylguanosine (DMG), and reported that DMG at 50 μg/ml inhibited germination completely. However, Tanaka et al. (1974) found that DMG showed no inhibition of *D. discoideum* spores even at 300 μg/ml. The most potent inhibitor isolated from spores of *D. discoideum* is 3-(3-amino-3-carboxypropyl)-6-(3-methyl-2-butenylamino) purine (Fig. 4), given the name discadenine (Abe et al., 1976). It has an ED_{50} of 10 ng/ml.

Examples of self-inhibition involving decreased germination as the spore population is increased must be related to water-soluble or polar substances. This should not imply that all (or even most) of the germination inhibitors can be removed by washing spores with water. There may be many types of dormant spores that contain endogenous germination inhibitors that are not readily soluble in water or that possess permeability barriers to prevent leaching of the inhibitor. Such spores would not exhibit an inverse relationship between crowding and germination, because they probably have sufficient inhibitor within each spore to cause dormancy. Teliospores of the dwarf bunt fungus, *Tilletia controversa*, may be an example of the latter type (Trione, 1977). The self-inhibition phenomenon in fungal spores has been reported for 53 fungal species (Allen, 1976).

In addition to the germination inhibitors, fungal spores may also possess germination-promoting substances. Again, most of the research on this topic has focused on the uredospores of the rust fungi (Allen, 1976). Nonanal was identified as an endogenous germination stimulator of uredospores of *Puccinia graminia* v. *tritici* (French and Weintraub, 1957), and found subsequently in uredospores of *P. coronata, P. sorghi, P. recondita, P. striiformis, P. helianthi,* and *Uromyces phaseoli* (Rines et al., 1974). Methylheptenone, another endogenous germination stimulator, was found in uredospores of *P. graminis* v. *tritici* and *P. striiformis*. Nonanal appears to be the most potent of the germination stimulators and has enhanced the germination of 21 species of rust and smut fungi, including seven genera (French et al., 1978). In the rust fungi it has been found that the stimulators can reverse the action of the endogenous inhibitors and promote germination. The ratio and levels of endogenous stimulators and inhibitors may be responsible for the variation in germination characteristics of rust uredospores (Macko et al., 1977).

In addition to endogenous germination stimulators of fungal spores, natural exogenous stimulators from host plants probably play an important role in initiating germination of many types of fungal spores. Some of these compounds function by counteracting endogenous inhibitors, although there is no apparent structural relationship between the stimulator and the inhibitor. A large number of naturally occurring compounds from host plants have been reported to stimulate germination of fungal spores: for example, aldehydes, alcohols, ketones, esters, terpenes, and steroid saponins (Allen, 1976). It is noteworthy that many of the compounds that stimulate fungal spore germination also act as insect pheromones (French et al., 1978).

FORMATION OF INFECTION STRUCTURES

Many of fungal plant pathogens must develop specialized postgermination structures before penetration of the host can occur. In germinating uredospores of the rust fungi, these infection structures include appressorium, penetration peg, vesicle, and infection hyphae. Several mechanisms have been hypothesized to regulate or initiate this coordinated differentiation in the rust fungi, for example: (1) a volatile substance obtained from uredospores, (2) substances exuding from host cells, and (3) responses to the waxy crystals on the surface of the leaves (Allen, 1976). A water-soluble substance obtained from uredospores was recently purified, shown to induce infection structures in the wheat-stem rust fungus and identified as acrolein (2-propenal) (Macko et al., 1978). Many structurally related compounds were tested and similar bioactivity was found in short-chain aliphatic carbonyl compounds with conjugated double bonds. Acrolein, CH_2 CH CHO, was nearly 10-fold more potent, however, than any other compound tested.

SPOROGENIC SUBSTANCES

Light is a requirement for the formation of reproductive organs in many species of fungi. All the major groups of true fungi, as well as some myxomycetes, include light-sensitive species (Trione et al., 1966). With certain photosensitive fungi, the onset of sporulation is accompanied by the formation within the mycelium of substances which have a maximum absorption at 310 nm. These compounds, conventionally called "P-310s" (Trione and Leach, 1969), have been demonstrated to exist in many fungi, are nearly always correlated with the onset of sporulation, and have been found to induce sporulation in vegetative colonies of some species.

Trione et al. (1966) found at least three forms of P-310, which differed in their sporogenic activity when applied to actively growing colonies of *Ascochyta pisi* and *Pleospora herbarium* in amounts comparable to a normal hormonal activity level. An increase in sporulation was generally observed with increasing P-310 concentration up to a point, beyond which sporulation declined. The sporogenic effect was localized in young mycelium near the droplet sites. There was no evidence that the sporogens were translocated through the fungal colony to the older mycelium.

Favre-Bonvin et al. (1976) and Arpin and Favre-Bonvin (1977) published structural data of three compounds (Figure 5) named mycosporins (two from *Stereum hirsutum*, and one from *Botrytis cinerea*) that are chemically closely related to the P-310s from *Ascochyta pisi*. The structures of the three P-310s from *A. pisi* have not been fully determined. A P-310 compound and several related compounds have been identified from the sea anemone, *Palythoa tuberculosa* (Ito and Hirata, 1977; Takano et al., 1978). A chemically related compound (P-320) was found in the red alga, *Chondrus yendoi* (Tsujino et al., 1978). The function of the *Palythoa* and *Chondrus* compounds are unknown. It is interesting to note that the amino nitrogen of each of these seven P-310-like compounds is part of an amino-acid side chain attached to the chromophore, except that in mycosporin-1 the side chain is serinol, the reduced form of serine. None of the P-310-like compounds have been synthesized.

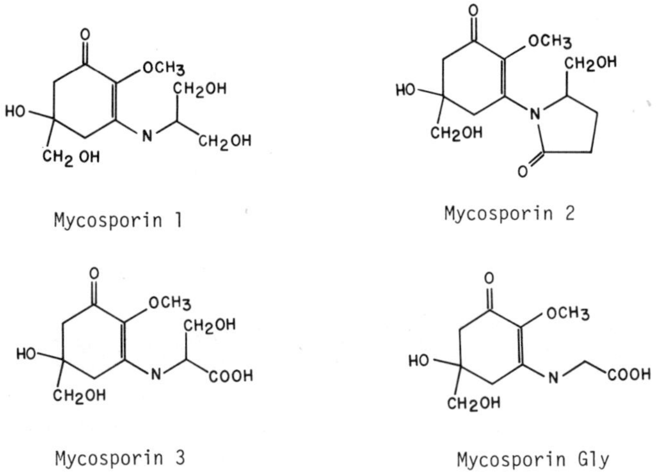

Mycosporin 1

Mycosporin 2

Mycosporin 3

Mycosporin Gly

Figure 5 The structures of P-310 compounds.

In addition to the work of Trione et al. (1966), the sporogenic property of P-310 was also demonstrated by the experiments of Tan and Epton (1974) and Dehorter (1976). According to these authors, P-310s, when added to the growing cultures, replaced the stimulus of the light by inducing, in the dark, the sporulation of *Botrytis cinerea* and *Nectria galligena*. However, experiments with *Helminthosporium dictyoides, Sclerotinia fructicola,* and *Botrytis cinerea* conducted by Vargas and Wilcoxson (1969), van den Ende and Cornelis (1970); and Hite (1973); respectively, did not demonstrate sporogenic activity of partially purified P-310 preparations.

In considering the conflicting results on the bioactivity of P-310 compounds, it should be noted that the literature references to P-310 are often ambiguous, since this general designation may refer to several, or to a series of compounds absorbing at 310 nm. In our purification procedure (Trione et al., 1966) each of the P-310 fractions, eluted from the third ion-exchange column, showed a single peak absorbing at 310 nm. Other researchers may not have purified their P-310 preparations to that degree. Thus, interpretation of experimental bioassay results must be made with caution for sporogenic activity can be influenced by P-310 type, concentration, and the presence of interfering substances. The natural role of P-310s in the mechanism of light-induced sporulation of fungi still remains to be clarified.

CONCLUSIONS

It is quite clear that various hormones or endogenous regulatory substances play important roles in developmental processes of fungi, but our knowledge of the occurrences of these hormones and their actions and interactions in fungal plant pathogens is rudimentary in the extreme. This lack of research on natural regulators of fungal development constitutes a major limitation to our understanding of potential new control methods. The information that is available from nonpathogenic fungi indicates that the chemical nature of fungal hormones varies nearly as much as the morphological traits between the different taxonomic groups. This wide variation in specific fungal hormones is in great contrast to the hormones of higher plants and higher animals, where the same hormone functions in a regulatory role in many dissimilar taxonomic groups. Although the diversity of chemical structures of fungal pheromones may be perplexing to the fungal physiologist, it could potentially be a great advantage to the plant pathologist. For it may be possible to exploit the specific regulatory control in particular genera of serious fungal pathogens by developing compounds that would be highly specific and potent against only the fungal genera for which they were designed. For example, the trisporic acid system of hormones appears to function in several species of the Mucorales but is ineffective against other fungi; similarly, the methyl-*cis*-cinnamates occur as endogenous germination inhibitors in many species of rust fungi but have not been shown to function in other fungi. Thus, although there is great variation in fungal hormones in general, there appears to be some common regulatory substances that control important physiological processes in very closely related fungi. It is quite probable that many of the *Phytophthora* species would be affected by the same or similar bioregulators; likewise, many of the smut fungi, for example, *Ustilago* and *Tilletia* species, probably respond to bioregulators common to these genera. If this hypothesis is correct, then it would not be necessary to study each fungal species individually, but rather to investigate the details of hormonal regulation in one model species that represents the larger group.

As the world becomes more concerned about the use of broad-spectrum, highly toxic pesticides to control insects and diseases, more research should be focused on the natural physiological and biochemical regulatory mechanisms of the most serious pest organisms. It does not seem to be widely recognized that basic biochemical and physiological studies on the regulation of growth and development of fungal pathogens can lead to significant new innovations in disease control. The great progress made in the last few years, however, in elucidating insect hormones and then using these highly specific and potent compounds for insect control demonstrates the relevance of this research approach. It should likewise be possible to develop highly specific and potent, multiple-inhibitor combinations to aid in better controlling fungal pathogens, without causing detrimental side effects on plants or animals. The naturally occurring sporogens and inhibitors of fungi can be altered by organic chemists to achieve compounds that are effective in controlling the growth and development of fungi. The impact of such highly specific and potent substances should assume greater significance in plant pathology.

REFERENCES

Abe, K., I. Kusaka, and S. Fukui. (1975) Morphological change in the early stages of the mating process of *Rhodosporidium toruloides*. *J. Bacteriol.* **122**:710–718.

Abe, H., M. Uchiyama, Y. Tanaka, and H. Saito. (1976) Structure of discadenine, a spore germination inhibitor from the cellular slime mold, *Dictyostelium discoideum*. *Tetrahedron Lett.* **42**:3807–3810.

Allen, P. J. (1972) Specificity of the cis-isomers of inhibitors of uredospore germination in the rust fungi. *Proc. Natl. Acad. Sci. USA* **69**:3497–3500.

Allen, P. J. (1976) Spore germination and its regulation. In *Encyclopedia of Plant Physiology*, Vol. 4 (R. Heitefuss and P. H. Williams, eds.). Springer-Verlag, New York. Pp. 51–85.

Arpin, N., and J. Favre-Bonvin. (1977) The mycosporins. *Int. Mycol. Congr. Abstr.* **1**:21.

Arsenault, G. P., K. Beimann, A. W. Barksdale, and T. C. McMorris. (1968) The structure of antheridiol, a sex hormone in *Achlya bisexualis*. *J. Am. Chem. Soc.* **90**:5635–5636.

Bacon, C. W., A. S. Sussman, and A. G. Paul. (1973) Identification of a self-inhibitor from spores of *Dictyostelium discoideum*. *J. Bacteriol.* **113**:1061–1063.

Bandoni, R. J. (1965) Secondary control of conjugation in *Tremella mesenterica*. *Can. J. Bot.* **43**:627–630.

Banno, I. (1967) Studies on sexuality of *Rhodotorula*. *J. Gen. Microbiol.* **13**:167–196.

Barksdale, A. W. (1969) Sexual hormones of *Achlya* and other fungi. *Science* **166**:831–837.

Betz, R., V. L. MacKay, and W. Duntze. (1977) *a*-Factor from *Saccharomyces cerevisiae:* Partial characterization of mating hormone produced by cells of mating type *a*. *J. Bacteriol.* **132**:462–472.

Bu'Lock, J. D., B. E. Jones, and N. Winskill. (1976) The apocarotenoid system of sex hormones and prohormones in Mucorales. *Pure Appl. Chem.* **47**:191–202.

Burgeff, H. (1924) Untersuchungen über Sexualität and Parasitismus bei Mucorinen. *Botanische Abhandlungen* **4**:1–135.

Crandall, M. (1978) Mating-type interactions in microorganisms. In *Receptors and Recognition*, Vol. 3 (P. Cuatrecasas and M. F. Greaves, eds.). Chapman and Hall, London, Pp. 47–100.

Crandall, M., R. Egel, and V. L. MacKay. (1977) Physiology of mating in three yeasts. *Adv. Microb. Physiol.* **15**:307–398.

Dehorter, B. (1976) Induction des peritheces de *Nectria galligena* Bres. par un photocompose mycelien absorbant a 310 nm. *Can. J. Bot.* **54**:600–604.

Duntze, W., V. L. MacKay, and T. R. Manney. (1970) *Saccharomyces cerevisiae:* A diffusible sex factor. *Science* **168**:1472–1473.

Elliot, Charles G. (1977) Sterols in fungi: Their functions in growth and reproduction. *Adv. Microb. Physiol.* **15**:121–173.

Favre-Bonvin, J., N. Arpin, and C. Brevard. (1976) Structure de la mycosporine (P. 310). *Can. J. Chem.* **54**:1105–1113.

Foudin, A. S., and V. Macko. (1974) Identification of the self-inhibitor and some germination characteristics of peanut rust uredospores. *Phytopathology* **64**:990–993.

French, R. C., and R. L. Weintraub. (1957) Pelargonaldehyde as an endogenous germination stimulator of wheat rust spores. *Arch. Biochem. Biophys.* **72**:235–237.

French, R. C., C. L. Graham, A. W. Gale, and R. K. Long. (1977) Structural and exposure time requirements for chemical stimulation of germination of uredospores of *Uromyces phaseoli. J. Agric. Food Chem.* **25**:84–88.

French, R. C., R. K. Long, F. M. Latterell, C. L. Graham, J. J. Smoot, and P. E. Shaw. (1978) Effect of nonanal, citral, and citrus oils on germination of conidia of *Penicillium digitatum* and *Penicillium italicum. Phytopathology* **68**:877–882.

Gooday, G. W. (1968) Hormonal control of sexual reproduction in *Mucor mucedo. New Phytol.* **67**:815–821.

Hite, R. E. (1973) Substances from *Botrytis cinerea* associated with sporulation and exposure to near-ultraviolet radiation. *Plant Dis. Rep.* **57**:760–764.

Ito, S., and Y. Hirata. (1977) Isolation and structure of a mycosporine from the zoanthid *Palythoa tuberculosa. Tetrahedron Lett.* **28**:2429–2430.

Kamiya, Y., A. Sakurai, S. Tamura, K. Abe, E. Tsuchiya, and S. Fukui. (1977) Isolation and chemical characterization of the peptidyl factor controlling mating tube formation in *Rhodosporidium toruloides. Agric. Biol. Chem.* **41**:1099–1100.

Kamiya, Y., A. Sakurai, S. Tamura, N. Takahashi, E. Tsuchiya, K. Abe, and S. Fukui. (1979) Structure of Rhodotorucine A, a peptidyl factor, inducing mating tube formation in *Rhodosporidium toruloides. Agric. Biol. Chem.* **43**:363–369.

Ko, W. H. (1978) Heterothallic phytophthora: Evidence for hormonal regulation of sexual reproduction. *J. Gen. Microbiol.* **107**:15–18.

Leppik, R. A. (1979) Personal communication. Division of Food Research. P.O. Box 12, Cannon Hill, QLD. 4170. Australia.

Leppik, R. A., D. W. Holloman, and W. Bottomley. (1972) Quiesone: An inhibitor of the germination of *Peronospora tabacina* conidia. *Phytochemistry* **11**:2055–2063.

Levi, J. D. (1956) Mating reaction in yeast. *Nature* **177**:753–754.

Machlis, L. (1958) Evidence for a sexual hormone in *Allomyces. Physiol. Plant.* **11**:181–192.

Machlis, L. (1973) The chemotactic activity of various sirenins and analogues and the uptake of sirenin by the sperm of *Allomyces. Plant Physiol.* **52**:527–531.

Machlis, L., W. H. Nutting, and H. Rapoport. (1968) The structure of sirenin. *J. Am. Chem. Soc.* **90**:1674–1676.

Macko, V., R. C. Staples, H. Gershon, and J. A. A. Renwick. (1970) Self-inhibitor of bean rust uredospores: Methyl 3,4-dimethoxycinnamate. *Science* **170**:539–540.

Macko, V., R. C. Staples, P. J. Allen, and J. A. A. Renwick. (1971) Identification of the germination self-inhibitor from wheat stem rust uredospores. *Science* **173**:835–836.

Macko, V., R. C. Staples, J. A. A. Renwick, and J. Pierone. (1972) Germination self-inhibitors of rust uredospores. *Physiol. Plant Pathol.* **2**:347–355.

Macko, V., R. C. Staples, Z. Yaniv, and R. Granados. (1976) Self-inhibitors of fungal spore germination. In *Form and Function in the Fungal Space* (D. J. Weber and W. M. Hess, eds.). Wiley, New York. Pp. 73–100.

Macko, V., E. J. Trione, and S. A. Young. (1977) Identification of the germination self-inhibitor from uredospores of *Puccinia striiformis. Phytopathology* **67**:1473–1474.

Macko, V., J. A. A. Renwick, and J. F. Rissler. (1978) Acrolein induces differentiation of infection structures in the wheat stem rust fungus. *Science* **199**:442–443.

McMorris, T. C. (1978) Sex hormones of the aquatic fungus *Achlya. Lipids* **13**:716–722.

Mori, K. (1973) Synthesis of *dl*-3-isobutyroxy-β-ionone and *dl*-dehydrovomifoliol. *Agric. Biol. Chem.* **37**:2899–2905.

Plempel, M. (1960) Die zygotropische Reaktion bei Mucorineen. I. *Planta* **55**:254–258.

Plempel, M. (1962) Die zygotropische Reaktion bei Mucorineen. III. *Planta* **58**:509–520.

Plempel, M. (1963) Die chemischen Grundlagen der sexual Reaktion bei Zygomyceten. *Planta* **59**:492–508.

Plempel, M., and W. Dawid. (1961) Die zygotropische Reaktion bei Mucorineen. II. *Planta* **56**:438–446.

Raper, J. R. (1939) Sexual hormones in *Achlya*. I. Indicative evidence for a hormonal coordinating mechanism. *Am. J. Bot.* **26**:639–650.

Raper, J. R., and A. J. Haagen-Smit. (1942) Sexual hormones in *Achlya*. IV. Properties of hormone A of *Achlya bisexualis. J. Biol. Chem.* **143**:311–320.

Reid, I. D. (1974) Properties of conjugation hormones (Erogens) from the basidiomycete *Tremella mesenterica. Can. J. Bot.* **52**:521–524.

Rines, H. W., R. C. French, and L. W. Daasch. (1974) Nonanal and 6-methyl-5-hepten-2-one: Endogenous germination stimulators of uredospores of *Puccinia graminis* var. *tritici* and other rusts. *J. Agric. Food Chem.* **22**:96–100.

Sakagami, Y., A. Isogai, A. Suzuki, S. Tamura, E. Tsuchiya, and S. Fukui. (1978) Amino acid sequence of tremerogen A-10, a peptidal hormone, inducing conjugation tube formation in *Tremella mesenterica* Fr. *Agric. Biol. Chem.* **42**:1301–1302.

Shepherd, C. J., and M. Mandryk. (1962) Auto-inhibitors of germination and sporulation in *Personospora tabacina. Trans. Br. Mycol. Soc.* **45**:233–244.

Stahmann, M. A., M. R. Musumeci, and W. B. C. Moraes. (1976) Germination of coffee rust uredospores and their inhibition by cinnamic acid derivatives. *Phytopathology* **66**:765–769.

Stotzler, D., and W. Duntze. (1976) Isolation and characterization of four related peptides exhibiting *a*-factor activity from *Saccharomyces cerevisiae. Eur. J. Biochem.* **65**:257–262.

Sutter, R. P. (1970) Trisporic acid synthesis in *Blakeslea trispora. Science* **168**:1590–1592.

Sutter, R. P., D. A. Capage, T. L. Harrison, and W. A. Kneen. (1973) Trisporic acid biosynthesis in separate plus and minus cultures of *Blakeslea trispora:* Identification by Mucor assay of two mating-type-specific components. *J. Bacteriol.* **114**:1074–1082.

Sutter, R. P., T. L. Harrison, and G. Galasko. (1974) Trisporic acid biosynthesis in *Blakeslea trispora* via mating-type-specific precursors. *J. Biol. Chem.* **249**:2282–2284.

Takano, S., D. Uemura, and Y. Hirata. (1978) Isolation and structure of two new amino acids, Palythinol and Palythene, from the zoanthid *Palythoa tuberculosa*. *Tetrahedron Lett.* **49**:4909–4912.

Tan, K. K., and H. A. S. Epton. (1974) Ultraviolet-absorbing compounds associated with sporulation in *Botrytis cinerea*. *Trans. Br. Mycol. Soc.* **63**:157–167.

Tanaka, Y., K. Yanagisawa, Y. Hashimoto, and M. Yamaguchi. (1974) True spore germination inhibitor of a cellular slime mold *Dictyostelium discoideum*. *Agric. Biol. Chem.* **38**:689–690.

Trione, E. J. (1977) Endogenous germination inhibitors in teliospores of the wheat bunt fungi. *Phytopathology* **67**:1245–1249.

Trione, E. J., and C. M. Leach. (1969) Light-induced sporulation and sporogenic substances in fungi. *Phytopathology* **59**:1077–1083.

Trione, E. J., C. M. Leach, and J. T. Mutch. (1966) Sporogenic substances isolated from fungi. *Nature* **212**:163–164.

Tsuchiya, E., S. Fukui, Y. Kamiya, Y. Sakagami, and M. Fujino. (1978) Requirements of chemical structure for hormonal activity of lipopeptidyl factors inducing sexual differentiation in vegetative cells of heterobasidiomycetous yeasts. *Biochem. Biophys. Res. Commun.* **85**:459–463.

Tsujino, I., K. Yabe, I. Sekikawa, and N. Hamanaka. (1978) Isolation and structure of a mycosporine from the red alga *Chondrus yendoi*. *Tetrahedron Lett.* **16**:1401–1402.

van den Ende, H. (1968) Relationship between sexuality and carotene synthesis in *Blakeslea trispora*. *J. Bacteriol.* **96**:1298–1303.

van den Ende, H. (1976) *Sexual Interactions in Plants*. Academic Press, New York, 186 pp.

van den Ende, G., and J. J. Cornelis. (1970) The induction of sporulation in *Sclerotinia fructicola* and some other fungi and the production of P-310. *Neth. J. Plant Pathol.* **76**:183–191.

Vargas, J. M., Jr., and R. D. Wilcoxson. (1969) Some effects of temperature and radiation on sporulation by *Helminthosporium dictyoides* on agar media. *Phytopathology* **59**:1706–1712.

II

HOST–PARASITE INCOMPATIBILITY

PATHOGEN-INDUCED CHANGES IN HOST ULTRASTRUCTURE

Richard M. Cooper

It is a truism that a knowledge of the workings of a system greatly eases the correction of any fault that arises within it. Accordingly, knowledge of ultrastructural aspects of pathogenicity is most likely to provide alternative approaches to problems of crop protection by providing insight and understanding of the nature of infection. It is my intention to select those aspects of recent ultrastructural studies that most advance our comprehension of mechanisms of pathogenicity and resistance, and where appropriate to indicate their relevance, present or future to disease control.

On the path to disease expression, a pathogen may pass through or into a succession of different plant parts including the external surfaces, cell walls, vascular system, and ultimately protoplasts. The resulting interactions with each of these will be considered in turn.

It is, of course, important to emphasize that many changes in infected tissues, including those at the ultrastructural level, will reflect spurious reactions or may be the result, rather than cause, of critical events during pathogenesis.

PLANT SURFACES

Surface features are especially amenable to examination by electron microscopy (EM) and are of particular use to plant breeders, as they are easy to identify phenotypically and less likely to select out new races of pathogens than "physiological" resistance.

Wax

Cuticular wax is often the first component encountered by aerial pathogens [Figure 1(*1*)]. Wax crystals appear to be altered or removed by some pathogens as evidenced by clear tracks left beneath germ tubes, hyphae, and appressoria on their removal from leaves (Kunoh et al., 1978; Sargent and Gay, 1977; Staub et al., 1974) and the disappearance of wax around colonies of an epiphytic yeast (McBride, 1972).

Kunoh et al. (1978) suggested that disintegration of wax might be a prerequisite for penetration of cereals by *Erysiphe graminis* as wax was only altered beneath germ tubes that breached the host cuticle. Others have also stressed the need for close contact between germ tubes and the cuticle for directed growth to find suitable penetration sites. Thus the regular lattice of wax crystals on wheat orientates germ tubes from *Puccinia graminis* f. sp. *tritici* uredospores parallel to the short axis of the leaf, thereby maximizing the probability of contacting stomata [Figure 1(*2*)] (Lewis and Day, 1972). The inability to remove surface wax and contact the cuticle undoubtedly contributes to the failure of certain pathogens to penetrate "nonhost" plants, viz., *Erysiphe cichoracearum* on barley (Staub et al., 1974) and *Uromyces phaseoli* on wheat (Wynn, 1976) and pea (Heath, this volume).

The exciting possibility of developing wax mutants for disease resistance was suggested by the claim that *E. graminis* only penetrates from "mature" appressoria, which require the unaltered wax layer for their formation—malformed appressoria (from which penetrations rarely arise) formed on dewaxed cuticles, reconstituted wax, or on plants with wax (eceriferum) mutations (Yang and Ellingboe, 1972).

Unfortunately, the potential of manipulating surface features that have such profound effects on orientation and penetration by pathogenic fungi

---

Figure 1 Interactions between pathogens and plant surfaces. *1*. Appressorium (a) formation from one cell of a germinating conidium (c) of *Colletotrichum graminicola* on a maize leaf, 6 hr after inoculation. Note intimate contact between electron-dense appressorial wall (OW) and wax rods (↑) on leaf surface. ×19,700. (From Politis and Wheeler, 1973.) *2*. Growth of germ tubes of *Puccinia graminis* f.sp. *tritici* parallel to the short axis of a wheat leaf. ×340. (From Lewis and Day, 1972.) *3*. Infection of a susceptible maize leaf (TmS cytoplasm) by *Helminthosporium maydis* race T, 6 hr after inoculation. Penetration usually occurred without appressorium formation and between epidermal cells (E), especially those adjacent to stomatal cells. Note disruption of host cytoplasm. ×3100. (From Wheeler, 1977.) *4*. Surface of cucumber cotyledon after removal of *Erysiphe cichoracearum* infection structure with gelatin 40 hr after inoculation. It is apparent that not only appressoria but also germ tubes and hyphae interact with the cuticular surface, as revealed by the deposits remaining on the surface and the location of epidermal penetration sites (↑). ×5000. (From Staub et al., 1974.) *5*. Infection of a broad bean-leaf epidermal cell by *Botrytis cinerea*. The germ tube (G) has produced an infection peg (P) that has penetrated the cuticle (c) and begun to alter the epidermal wall (E). [See Figure 3(*11*) for a later stage.] Note the extensive mucilage (M) around the germ tube and the absence of indentation or tearing of the cuticle during penetration. ×14,700. (From McKeen, 1974.)

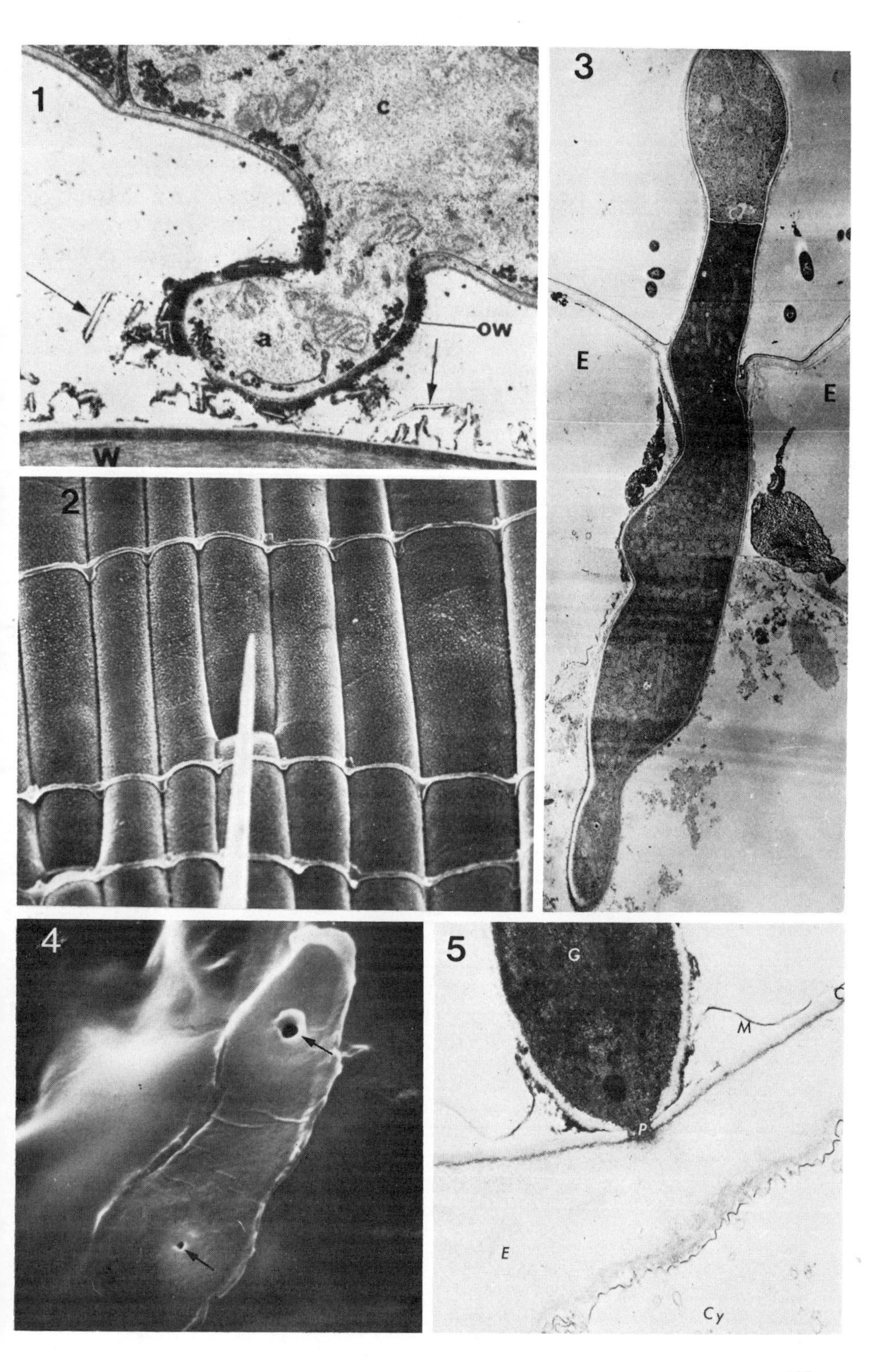
1
c
ow
a
W
2
3
E
E
4
5
G
M
C
P
E
Cy

(Heath, 1974a; Royle, 1975; Wynn, 1976) is seriously questioned by the following:

1. In spite of much reduced penetration the eventual susceptibilities to *E. graminis* of the wax mutants used by Yang and Ellingboe were unchanged from wild types.
2. Susceptibility of wheat cultivars was not determined by the ease of entry of stomata by *P. graminis* f. sp. *tritici* (Brown and Shipton, 1964).
3. Some pathogens do not require appressoria for penetration [Figure 1(*3,4*)] (Paus and Raa, 1973; Staub et al., 1974; Wheeler, 1977).
4. Although often delayed and/or reduced, penetration of nonhosts is often extensive (Akai et al., 1971; Heath, 1977; Maclean and Tommerup, 1979; Politis, 1976).

Nevertheless, such surface characteristics in combination with suitable cytoplasmic resistance could well contribute to field resistance; for example, as suggested for a cultivar of potato (Pimpernel), which exhibits field resistance against *P. infestans* and has reduced frequency of ES cells (epidermal cells adjacent to guard cells) through which the pathogen normally penetrates (Wilson and Coffey, 1978).

Cuticle

Whether penetration of the cuticle occurs by mechanical and/or enzymic means (Aist, 1976a) the final result is one of minimal damage to the host, even by necrotrophic pathogens such as *Rhizoctonia solani* which forms an extensive infection cushion on the host surface.

Penetration of cuticle and epidermal wall by many pathogens often results in a sharp clean pore without curled edges, sometimes performed by a blunt infection peg devoid of a cell wall, which suggests enzymic degradation, [Figures 1(*4,5*), 2(*6,7*), and 4(*19*)] (Ingram et al., 1975; McKeen, 1974;

Figure 2 Penetration of host cell walls. ***6.*** Cytochemical localization of lipase of *Bremia lactucae* during penetration of lettuce cotyledon epidermis. Lipase activity (↑) (revealed as electron-dense material) appears in the appressorium (A) during cuticle (c) digestion and is confined to the region of the appressorial wall (FW) in contact with the host surface. Host wall (HW). ×54,400. (From Duddridge and Sargent, 1978.) ***7.*** Penetration of a susceptible lettuce cultivar by *B. lactucae*. The nonmedian section through a penetration peg (P) reveals the fluid nature of the fungal wall extending between the apparently digested ends (↑) of host-wall microfibrils. Appressorium (A), epidermal cell (E). ×28,000. (From Ingram et al., 1975.) ***8.*** Growth between pine cells (cultured) by *Cronartium ribicola* (F) involves a splitting apart of walls (CW) of adjacent host cells. Vacuole (V). ×11,800. (From Robb et al., 1975b.) ***9.*** Region of penetration of a flax mesophyll cell (H) by *Melampsora lini*. Damage to the host cell wall (W) is minimal as evidenced by the indistinct boundary with the fungal wall. Haustorial mother cell (HMC). ×9700. (From Coffey et al., 1972b). ***10.*** Young haustorium of *Peronospora spinaciae* in spinach leaf cell. The host cell wall is invaginated and degraded and contributes to the sheath (fs) surrounding the haustorium (H). Fungal wall (W), host plasmalemma (p). ×10,900. (From Kajiwara, 1973.)

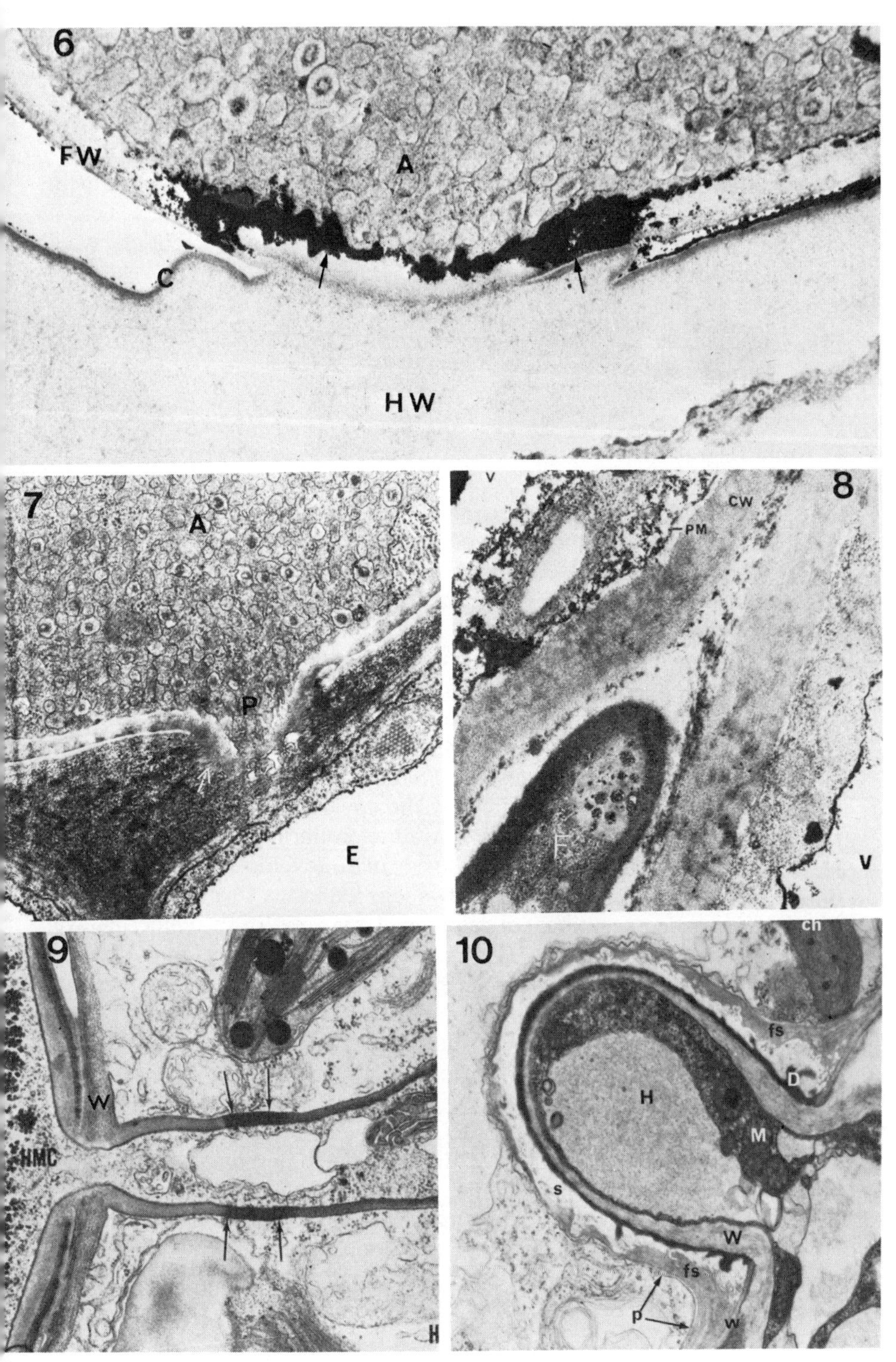
6
FW
A
C
HW
7
A
P
E
8
V
CW
PM
F
V
9
W
HMC
H
10
ch
fs
D
H
M
s
W
fs
p
w

Politis, 1976). Esterase activity has been detected by light microscopy at the tips of germ tubes of *Botrytis cinerea* and appressoria of *Venturia inaequalis* during cuticle penetration but not later (McKeen, 1974; Nicholson et al., 1972). The carefully controlled processes involved were revealed by recent cytochemical EM studies on infection of lettuce by *Bremia lactucae* (Duddridge and Sargent, 1978). Lipolytic enzyme activity appeared during penetration, localized precisely beneath the center of a group of apical vesicles in a small region of the appressorial wall immediately adjacent to cuticle, which swelled, distorted, and was later digested [Figure 2(*6*)].

CELL WALLS

Enzymic Degradation

Subsequent invasion necessitates repeated contact and breaching of host cell walls for which plant pathogens (and saprophytes) possess a battery of cell wall-degrading enzymes (CWDE) corresponding to most of the glycosidic linkages present. Various evidence suggests endopectic enzymes are critical factors in wall degradation, as they are the first CWDE produced *in vitro* and *in vivo* (they have been detected in ungerminated conidia) and are able to effect maceration (Bateman and Basham, 1976; Cooper, 1976, 1978).

Electron microscopy has revealed that, even during penetration of the epidermal wall, the distinction between biotrophic and necrotrophic fungi can be apparent. Biotrophs such as *B. lactucae* cause minimal degradation of wall polysaccharides, as evidenced by the apparent fluid state of the infection peg projecting between digested wall microfibrils [Figure 2(*7*)] (Ingram et al., 1975). The characteristic halos, which develop in epidermal walls during penetration by powdery mildews, has led to the suggestion that these biotrophs may be exceptional in degrading large areas of the epidermal wall. However, Sargent and Gay (1977), using electron-probe microanalysis, claimed that halos represent regions encrusted with opaline silica in association with cuticular lipid. Although silica has been implicated in disease resistance of cereals, it is unlikely in this case as it is deposited similarly during both resistant and susceptible reactions.

Subsequent growth by biotrophs is largely intercellular, which may involve a physical splitting apart of cell walls along the middle lamella [Figure 2(*8*)] (Robb et al., 1975b), but often the association is so intimate that fungal and host walls merge almost imperceptibly [Figure 2(*9*)] (Coffey et al., 1972b). An exception might be the marked wall changes induced in spinach by *Peronospora spinaciae* during haustorium formation, when the softened invaginated wall remained as part of the encapsulation surrounding the haustorium [Figure 2(*10*)] (Kajiwara, 1973; also see Figure 5(*27*), Wehtje et al., 1979).

In contrast, necrotrophs can cause profound wall changes even prior to penetration (Weinhold and Motta, 1973), and rapid swelling and degradation of the epidermal wall may surround penetration pegs, for example, *B. cinerea* [Figure 3(*11*)] (McKeen, 1974) and *Helminthosporium maydis* [Figure 7(*38*)] (Wheeler, 1977). Further colonization can involve disruption of the middle lamella, leading to tissue maceration, which is invariably accompanied by extensive protoplast damage (see the section on Endopectic Enzymes). For example, rotting of potato tubers by *Erwinia carotovora* results entirely from intercellular growth of bacteria, which secrete a low molecular weight pectate lyase able to cause disintegration of the middle lamella up to 500 μm from bacterial colonies [Figure 3(*12*)] (Fox et al., 1972). Similarly *Sclerotium rolfsii* causes extensive wall damage by secretion of copious, highly active CWDE and oxalic acid. The latter probably resides in hyphal microbodies and acts as a synergist to pectic enzymes or as a toxin in its own right [Figure 3(*13*)] (Hänssler et al., 1978).

Passage through or within host cell walls by other pathogens results in various degrees of wall alteration dependent on the host–parasite interaction and stage of infection [Figure 3(*14–16*)] (Ammon et al., 1974; Calonge et al., 1969; Heath and Wood, 1969; Hess, 1969; Paus and Raa, 1973, Pring, personal communication; Shimony and Friend, 1977). However, it is clear that many pathogens capable of producing highly active CWDE *in vitro,* for example, *Monilinia fructigena* and *Phytophthora infestans* often cause negligible wall degradation during invasion [Figure 3(*16*)]. A puzzling change of parasitic mode is exhibited during infection of bean by *Colletotrichum lindemuthianum*. This pathogen initially penetrates cell walls mechanically, inflicting no apparent damage to protoplasts during the first four days, then suddenly causes extensive wall breakdown and necrosis (Mercer et al., 1975).

Our ignorance of enzymic wall degradation *in vivo* may be lessened by ultrastructural localization of these important determinants of pathogenicity. Obviously this is dependent on the availability of suitable substrates with sufficiently electron-dense reaction products or methods involving deposition of heavy metal atoms at specific sites (Hislop and Pitt, 1976). The only attempt with a CWDE of a pathogen was localization of an arabinan-degrading enzyme (AF) in hyphae of *M. fructigena in vitro;* activity was detected in vacuoles and spherosome-like bodies, which appeared to migrate to the plasmalemma and secrete the enzyme by reverse pinocytosis [Figure 4(*17*)] (Hislop et al., 1974a). However, the technique failed to demonstrate AF in hyphae invading apple tissue. Similar structures have been observed in various fungi penetrating plant cell walls [Figure 4(*18,19*)].

Unfortunately suitable histochemical substrates are not available for pectic enzymes, although localization by light microscopy *in vitro* and *in vivo* with immunological methods (Hislop et al., 1974b) holds promise for future EM autoradiographical studies, which have already proved successful with endogenous cellulases in pea epicotyls (Bal et al., 1976).

Resistance

Any consideration of the manipulation of plant cell walls in a resistance program is complicated by the ultrastructural evidence that restriction of CWDE of pathogens is not necessarily a disadvantage, but may contribute to their success by avoiding undue host disruption (Cooper, 1978). The similar rates of penetration by fungi of resistant and susceptible cultivars (Shimony and Friend, 1975) support the chemical evidence that cell wall polymers are similar between varieties and are not responsible for "race-specific" resistance (Albersheim and Anderson-Prouty, 1975; Cooper, 1976). Walls of different species are also alike; thus these comments could be made for "nonhost" or "field" resistance. However, each host–parasite combination should be considered unique and warrants separate investigation; for example, we found that, when reacting CWDE of various pathogens with cell walls extracted from host and nonhost species in a few cases walls from an incompatible species were markedly inhibitory (unpublished data). Also a study by Wilson and Coffey (1978) suggested that the field resistance of a potato cultivar to *P. infestans* resulted partly from the epidermal wall providing a barrier to penetration.

Lignified and suberized tissues are usually inherently resistant to degradation and penetration, although a few specialized pathogens are capable of breaching such barriers directly, for example, *Gauemannomyces graminis* [Figure 4(*20*)] (Manners and Myers, 1975), or by exploiting natural weak points such as pits in xylem vessels [Figure 4(*21*)]. Variation in the distribution of such tissues can provide the basis for nonspecific resistance in some varieties (Royle, 1975).

These components can also accumulate in walls as a rapid response to

Figure 3 Penetration of host cell walls. ***11.*** Extensive degradation and swelling of epidermal cell walls during penetration of a broad bean leaf by *Botrytis cinerea*. Note constriction of the infection hypha (H) during passage through the cuticle (c), which has been pushed outward by the swollen host wall (W). Germ tube (G). ×6000. (From McKeen, 1974.) ***12.*** Maceration of potato tuber parenchyma by *Erwinia carotovora*. Bacteria (B) remain confined to the middle lamella, degradation of which results in separation of adjacent cells. Note swelling of remaining cell walls (W) and complete disruption of host cytoplasm. ×14,400. (From Fox et al., 1972.) ***13.*** Hyphal tip cell (P) of *Sclerotium rolfsii* within cell wall (HW) of *Phaseolus vulgaris*. The host cell wall has lost its fibrillar organization and host cytoplasm (HC) has become aggregated and electron opaque. Note the presence of hyphal microbodies (containing oxalate?) (↑). ×15,000. (From Hänssler et al., 1978.) ***14.*** Intercellular growth of *Macrophomina phaseolina* (F) in soybean roots. Note the extensive disintegration of the middle lamella (↑), electron-lucent areas (EL) in the host cell wall, and disrupted host cytoplasm (HC). ×18,900. (From Ammon et al., 1974.) ***15.*** Mycelium (M) of *Monilinia fructigena* growing intracellularly and intercellularly in adjacent cells of an infected pear fruit. Tissue was embedded in resin, sectioned and the face etched with sodium ethoxide before examination by scanning electron microscopy. ×500. (Courtesy of R. J. Pring, Long Ashton Research Station, England.) ***16.*** Hyphal tip (H) of *Monilinia fructigena* advancing within a cell wall (HW) of an infected pear fruit. The nonmedian section reveals host wall degradation is restricted to the immediate vicinity of the pathogen. ×8000. (From Byrde and Willetts, 1977.)

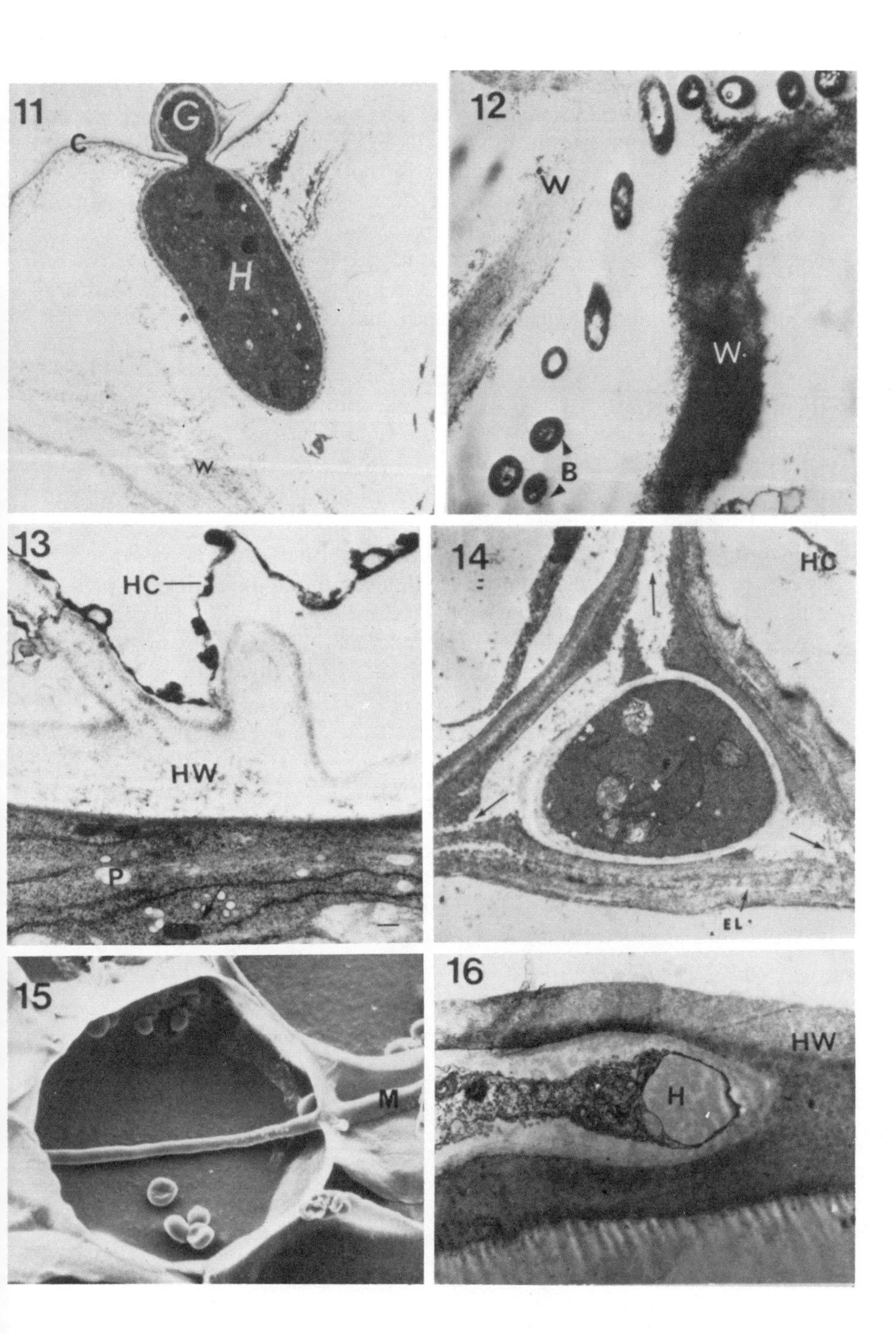
11
C
G
H
W
12
W
W
B
13
HC
HW
P
14
HC
EL
15
M
16
HW
H

invasion and form an effective barrier to penetration, for example, lignification of cereal leaves in response to incompatible pathogens (Ride and Pearce, 1979), suberization of cell walls of wounded potato tubers (Fox et al., 1972) or various cell wall changes in roots in response to vascular wilt pathogens, including suberization of the endodermis (Bishop, 1980). Although silicon is not a normal cell wall component of bean leaves, this element accumulates rapidly as electron-opaque deposits on mesophyll cell walls and appears to be responsible for the failure of potential haustorial mother cells to initiate haustoria (Heath, 1972, 1977, 1979).

Binding of Bacteria

Other recently discovered changes in plant cell walls in response to invasion have become of considerable interest with regard to resistance to incompatible and saprophytic bacteria.

Incompatible pathogens or saprophytic bacteria injected into intercellular spaces of leaves rapidly (1–4 hr) cause detachment of the cuticle (or pellicle) and erosion or swelling of the contacted mesophyll walls. This results in envelopment and immobilization of the bacteria, apparently by host cell wall components [Figure 5(*22,23*)] (Cason et al., 1978; Goodman et al., 1976b; Roebuck et al., 1978; Sequeira et al., 1977; Sing and Schroth, 1977). In contrast, virulent strains do not become attached and remain free to multiply in the intercellular fiuids.

It was suggested by Cason et al. (1978) that pectic enzymes may be involved in wall erosion; their action may expose the wall lectins probably responsible for binding bacterial lipopolysaccharide (LPS) (Goodman et al., 1976a; Sequeira, 1978). However, attachment of heat-killed cells and purified LPS in a similar manner implies that the pathogen only plays a passive role in this phenomenon (Graham et al., 1977; Sequeira et al., 1977).

This provides an interesting contrast to *Agrobacterium tumefasciens* and

---

Figure 4 Penetration of host cell walls. ***17.*** Localization of α-L-arabinofuranosidase (AF) in mycelium of *Monilinia fructigena*. A proportion of the activity is located between the plasmalemma (p) and hyphal wall (w) and in multivesicular bodies (VB). (*a*) ×27,500; (*b*) ×65,000. (From Hislop et al., 1974a.) ***18.*** Lomasome (L) in intercellular hypha (H) of *Monilinia fructigena* in infected pear fruit. Note increased staining of the host cell wall (HW) adjacent to the lomasome. Fungal wall (FW). ×45,000. (Courtesy of R. J. Pring, Long Ashton Research Station, England.) ***19.*** Penetration of cell wall (CW) in barley root by sporangium (s) of *Pythium graminicola*. Note the degradation of the host wall (HW) and the presence of three large lomasomes (L) where the penetrating hypha contacts the cell wall. ×31,600. (From McKeen, 1977.) ***20.*** Penetration of inner, thickened wall (EW) of wheat endodermal cell by *Gaeumannomyces graminis* (F). Note degradation of the wall layers impregnated with suberin (↑). ×7000. (From Manners and Myers, 1975.) ***21.*** Early stage of penetration (↑) by *Verticillium alboatrum* of a paired-pit between contiguous xylem vessels (Ve) in a tomato stem. Note the formation of an appressorium-like structure (A) on the pit membrane (P). Secondary wall (S). ×8900.

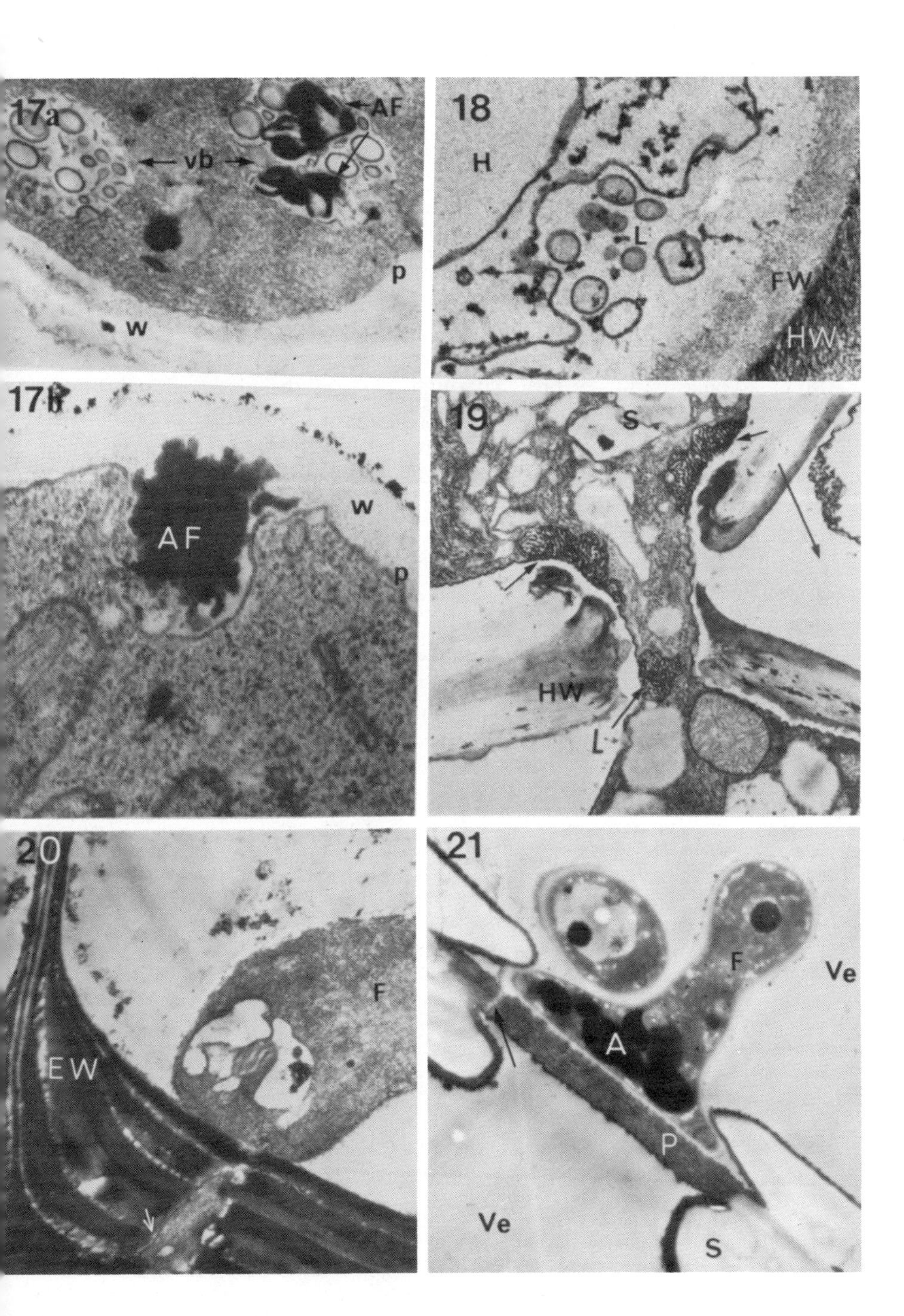
17a
AF
vb
p
w
18
H
L
FW
HW
17b
AF
w
p
19
S
HW
L
20
F
EW
21
F
Ve
A
P
Ve
S

related *Rhizobium* species for which binding to host walls is prerequisite for pathogenicity (Sequeira, 1978).

Wall Appositions

One of the earliest and most regular responses to penetration by pathogens (or wounding) is the rapid aggregation of host cytoplasm from which secretory vesicles (Zeyen and Bushnell, 1979) contribute to the formation of papillae or wall appositions in the paramural space directly beneath penetrating hyphae (Aist, 1976b). Although some of these appositions bear a striking resemblance to the host wall [Figure 5(*24*)] (Coffey et al., 1972b), most are quite distinct and may even contain entrapped membranous material [Figure 5(*25*)]. The appearance, size, and composition of these structures are diverse, even in the same host–parasite interactions (Bracker and Littlefield, 1973).

Papillae are resistant to attack by CWDE (Ride and Pearce, 1979) and are commonly considered to prevent or impede penetration by some pathogens as a result of correlation between the failure to establish a successful relationship and the presence of papillae [Figure 5(*25*)] (Bishop, 1980; Hohl and Stossel, 1976). However, their formation may be a reflection rather than a cause of resistance. The timing of papilla formation and host wall penetration can closely coincide. Thus interpretations made of EM images of fixed

---

Figure 5 Cell wall changes during pathogenesis. ***22.*** Envelopment of a cell of *Xanthomonas malvacearum* (Xm) by the swollen wall (CW) of a mesophyll cell in a resistant cotton cultivar, 10 hr after inoculation. Note the host cuticle (Cu) is continuous over the enveloped bacterium except for a rupture (R) through which fibrillar material is escaping into the intercellular space. Host cytoplasm (c) appears dense and coagulated. ×52,300. (From Cason et al., 1978.) ***23.*** Cells of an incompatible strain of *Pseudomonas solanacearum* (B) attached to the walls (W) of tobacco mesophyll cells. At 12 hr after infiltration the hypersensitive reaction is essentially complete with collapse of cells (thin arrows) and disorganization of organelles. Note the remaining pellicle (Pe) covering the bacterial cells and indications where bacteria were attached to the cell wall (thick arrows). ×8300. (From Sequeira et al., 1977.) ***24.*** Fibrillar collar (c) around the haustorial neck (HN) of *Melampsora lini* in a flax mesophyll cell. The nonmedian section suggests that the apposition is continuous with the host wall. Haustorial mother cell (MC). ×28,300. (From Coffey et al., 1972b.) ***25.*** Formation of a wall apposition (A) in response to attempted penetration by intercellular hypha of *Fusarium oxysporum* f. sp. *pisi* race 1 in root cortex of a resistant pea cultivar (Fek), 72 hr after inoculation. These structures were rarely penetrated and often contained senescing hyphae. Note degradation and swelling of host cell wall (HW). ×5800. (From Bishop, 1980.) ***26.*** Encased haustorium (H) of *Uromyces phaseoli* var. *vignae* in a resistant cultivar of cowpea (*Vigna sinensis*) 28 hr after inoculation. The callose-containing sheath (S) developed from the deposit which formed on the host cell wall at the time of penetration. Haustorial mother cell (MC). ×7700. (From Heath and Heath, 1971.) ***27.*** Infection of sunflower root epidermis by *Plasmopara halstedii*. Penetration by the zoospore (Z) results in localized degradation of the host wall (HW) and inward displacement of the wall, which contributes to the composition of the sheath (s) surrounding the infection vesicle (IV). ×11,100. (From Wehtje et al., 1979.)

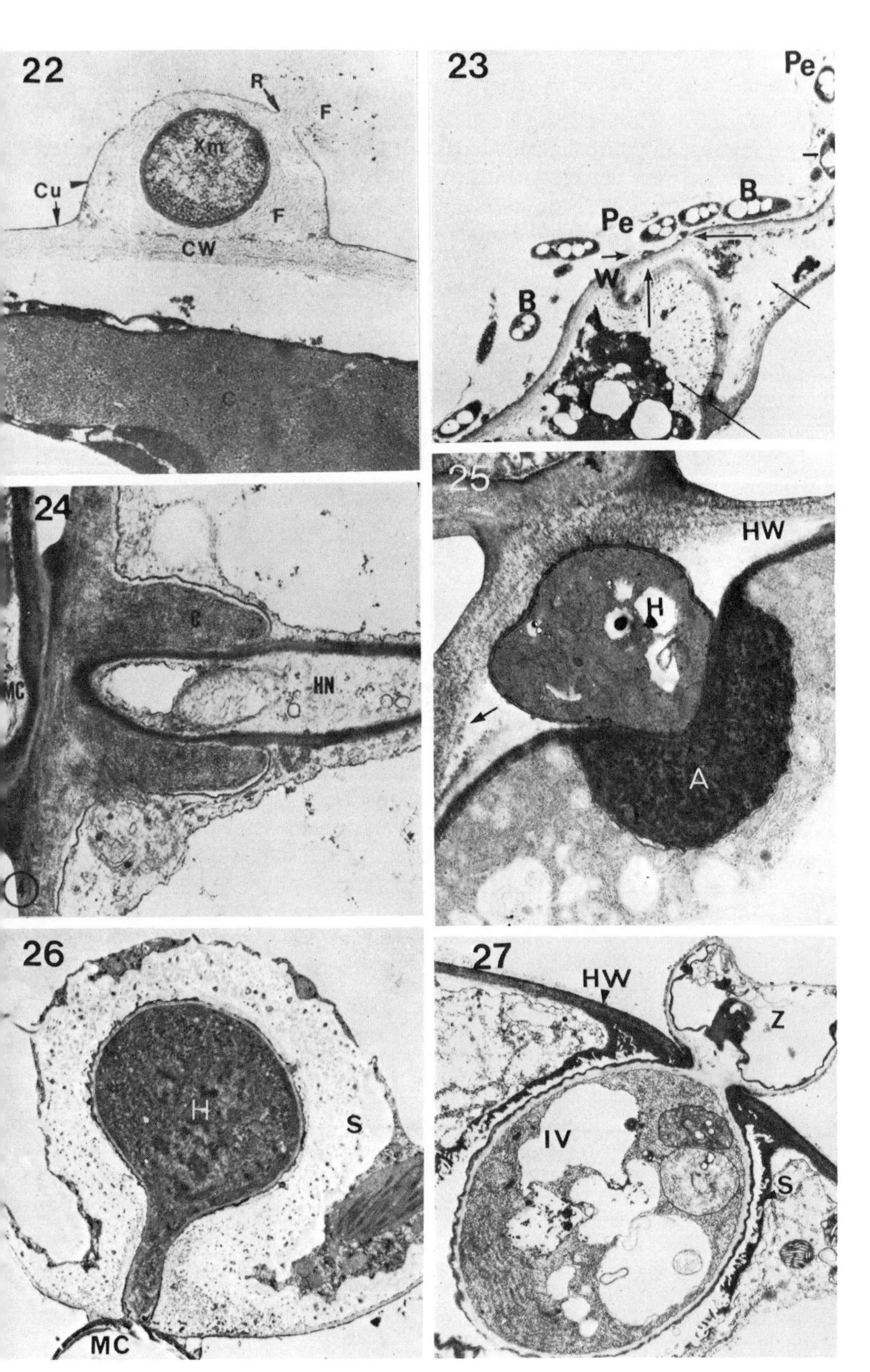
22
R
F
Xm
Cu
F
CW
23
Pe
B
Pe
W
B
24
C
HN
MC
25
HW
H
A
26
H
S
MC
27
HW
Z
IV
S

cells are speculative especially if serial sectioning is not performed (see Politis, 1976). More critical evidence has come from interference contrast light microscopy of living cells (Aist and Israel, 1977), which has suggested that papillae were usually formed too late to present a mechanical barrier to invasion of barley by *E. graminis* or kohlrabi by *Olpidium* zoospores.

Rust haustoria that form in nonhosts are often encased in an extension of the callose-containing wall apposition, and later become necrotic [Figure 5(*26*)] (Heath, 1972, 1977). However, such structures may only represent a response to intracellular hyphae damaged through some other host reaction, as suggested by their striking similarity to the encasements that form around necrotic rust haustoria in bean plants treated with the systemic fungicide oxycarboxin (Pring and Richmond, 1976). Apparently healthy haustoria of some oomycetes and intracellular hyphae of systemic smut fungi are often surrounded by extensive wall appositions [Figure 5(*27*)] (Fullerton, 1970; Hanchey and Wheeler, 1971; Wehtje et al., 1979), raising additional questions about the function of these structures.

XYLEM

Vascular wilt diseases are unique in that the symptoms are induced by microbial pathogens, which colonize and remain within xylem vessels [Figure 6(*28,30,31*)]. From this apparently hostile environment they multiply and induce symptoms of water stress, usually as a result of increased resistances of xylem to water flow.

Isolation and identification of those factors from pathogens responsible for the wilt syndrome could be of considerable value in screening cut stems for tolerance to wilt pathogens. This has already been suggested for *Verticillium*

Figure 6 Infection of xylem vessels by vascular wilt fungi. ***28.*** Scanning electron micrograph of hyphae of *Verticillium albo-atrum* in a xylem vessel (XV) of a susceptible tomato cultivar (Craigella). Colonization results from passive dispersal of propagules in the transpiration stream and lateral growth via paired pits between vessels [also see Figure 4(*21*)]. ×150. ***29.*** Scanning electron micrograph of xylem vessels occluded by gels (G) in uninoculated tap root of pea cultivar (Fek). Gels were induced by severing the root tip. ×650. (From Bishop, 1980.) ***30.*** Colonization of xylem vessels by *Fusarium oxysporum* f. sp. *pisi* in tap root of a resistant pea cultivar (Fek) 4 days after inoculation. Many of the vessels are occluded by gels (G) which appear to originate from pit membranes. Hyphae (H), xylem parenchyma (XP) pits (↑). ×2300. (From Bishop, 1980.) ***31.*** Hyphae (H) of *F. oxysporum* f. sp. *pisi* in stem xylem of a susceptible pea cultivar (Feltham First) 6 days after inoculation. Note the degradation of the primary wall (pw), which constitutes a high proportion of the vessel wall, especially in primary xylem. Secondary wall (SW). ×3700. (From Bishop, 1980.) ***32.*** Tylose (T) formation in xylem vessel (XV) (2nd stem internode) of a resistant tomato cultivar (Craigella 218) in response to invasion by *V. albo-atrum*. The longitudinal section reveals the complete occlusion of the vessel lumen by the tylose which results from the expansion of a contiguous parenchyma cell (XP) between the spiral secondary wall thickenings (↑). ×4000. (From Bishop, 1980.) ***33.*** Scanning electron micrograph of extensive tylosis in 2 xylem vessels of a tomato stem (cultivar Craigella). Tyloses formed as a rapid, localized (1–2 cm) response to wounding of xylem with a microneedle. ×500.

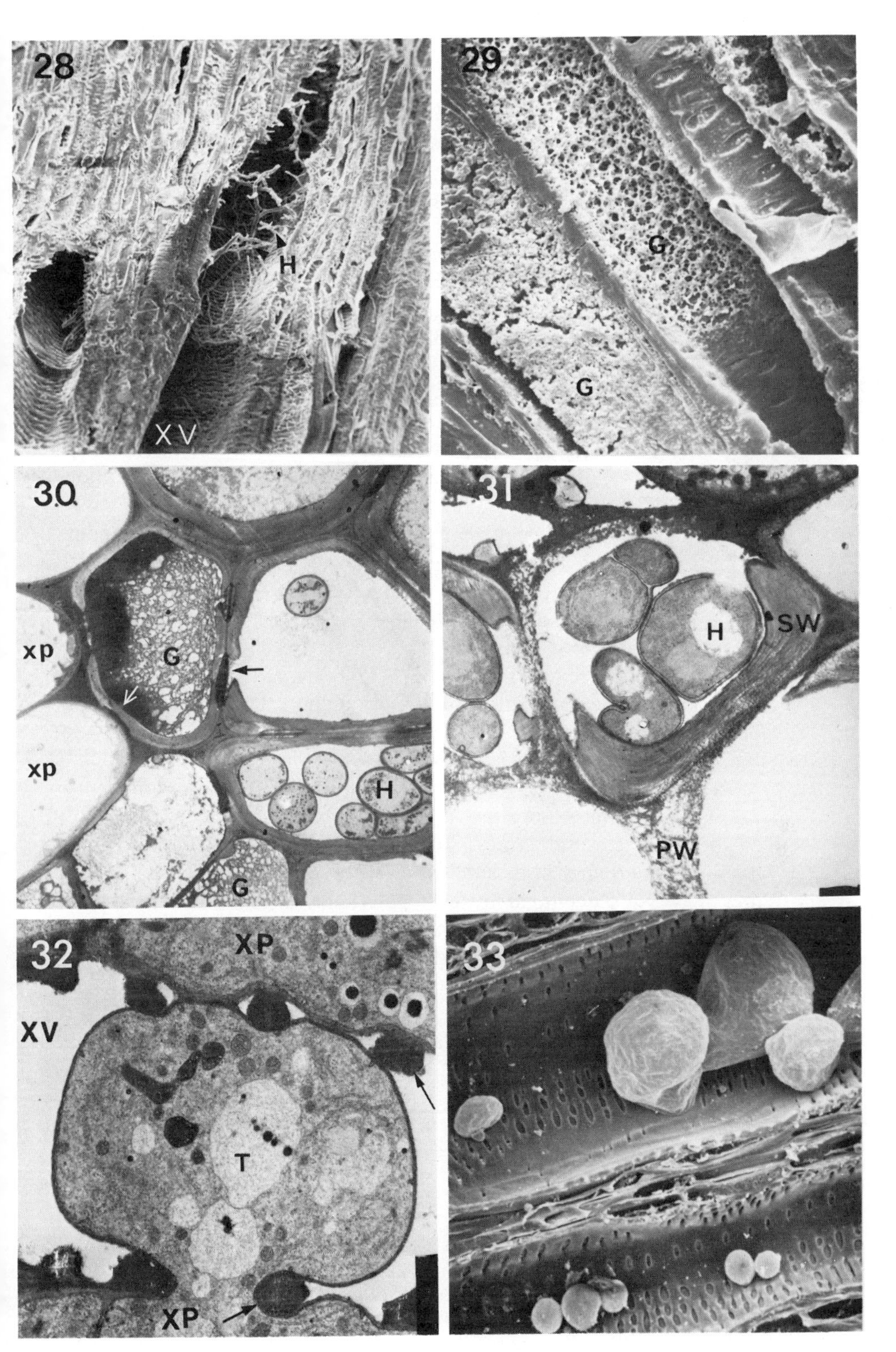
28
H
XV
29
G
G
30
xp
G
xp
H
G
31
H
SW
PW
32
XP
XV
T
XP
33

wilt of cotton and lucerne (Carr, 1970; Keen et al., 1972; Mussell and Strand, 1976); however, ultrastructural verification of their overall effects on plants is required because of the limited visual responses, especially wilting, of plants to microbial metabolites.

Recent EM studies have revealed some important features of susceptibility and resistance of vascular tissue, which nevertheless remain poorly understood. Infected vessels are often occluded by gels [Figure 6(*29,30*)], which can precede wilting or reduced vascular flow (Cooper and Wood, 1980; Robb et al., 1975a). They originate from the exposed primary wall-middle lamella complex at pit membranes (which constitute a high proportion of xylem walls, for example, 43% of banana vessels) apparently by a remarkable dilation of the membrane (Bishop, 1980; Van der Molen et al., 1977). This may involve the activity of fungal pectic enzymes (Cooper and Wood, 1980), which can be extensive in some wilt diseases, for example, *Fusarium* wilt of pea [Figure 6(*31*)] (Bishop, 1980; Wallis, 1977), although this is refuted by others (Moreau et al., 1978; Oullette, 1978).

Anomalously it has also been claimed that rapid gel formation represents a resistance response to microbial infection of vascular tissue by preventing the systemic dispersal of propagules (Van der Molen et al., 1977). Gels may indeed be a general response to stress, as occlusion of vessels occurs in several viral diseases (e.g., Favali et al., 1978) and as a wound reaction when tap roots of peas are severed [Figure 6(*29*)] (Bishop, 1980).

Confusion over the mechanism of induction and role in pathogenesis also concerns tyloses, which have a similar potential to gels in disease resistance or susceptibility [Figures 6(*32,33*) and 7(*34*)]. Although usually associated with resistance (e.g., Mollenhauer and Hopkins, 1976), there appear to be exceptions where these outgrowths of the xylem parenchyma may be involved in inducing water stress, for example, Dutch elm disease [Figure 7(*34*)] and *Verticillium* wilt of sunflower (Robb et al., 1979).

Water in vessels occluded by gels or tyloses is normally redistributed laterally via pits, but the almost universal occurrence in infected vessels of deposits coating pit membranes and vessel walls probably prevents or restricts water flow [Figure 7(*35*)] (Cooper and Wood, 1974; Corden and Chambers, 1966). The electron-dense material is probably phenolic (ferric positive) (Robb et al., 1979) phenylalanine-^{3}H-labeled (Favali et al., 1978) and may originate from injured or necrotic xylem parenchyma cells, which often contain "tannin-like" deposits in their vacuoles, which resembles the material lining vessels [Figure 7(*35*)] (Baur and Walkinshaw, 1974; Mueller and Beckman, 1976). Also similar structures result from adsorption to cell walls of phenolic material from necrotic cells in various other diseases (Brotzman et al., 1975; Lazarovits and Higgins, 1976; Robb et al., 1975b).

Xylem parenchyma cells react to invasion by forming a diffuse wall apposition against pit membranes known as the protective layer [Figure 7(*35*)]. Impregnation of this material with phenolics suggests a further barrier to water movement (Robb et al., 1979).

Another reaction to infection revealed by EM involves the agglutination of avirulent bacteria; this is responsible for the localization of *Erwinia amylovora* in xylem of apple petioles (Huang et al., 1975) and *Xanthomonas oryzae* in rice (Horino, 1976), which emphasizes that the xylem is far from "inert" as commonly supposed.

Xylem anatomy (vessel length, diameter, branching, contiguity [see Figure 7(*34*)], and end wall structure) can affect the success of these parasites, which relies on rapid axial and lateral dispersal throughout the vascular system before host localization responses become effective. Varietal resistance does not usually depend on differences in innate vascular structure (e.g., Beckman et al., 1976). However, resistance of clones of sugar cane to a bacterial disease (Teakle et al., 1978), varieties of alfalfa to *Corynebacterium insidiosum* (Cho et al., 1973), and elm hybrids to *Ceratocystis ulmi* (McNabb et al., 1970) are claimed to depend on vessel structures less amenable to colonization; this factor may be screened for by scanning EM in breeding programs. Also xylem formation and host susceptibility can be modified by treatment with chemicals such as trichlorophenyl acetic acid (Brener and Beckman, 1968; Venn et al., 1968).

PROTOPLASTS

Not all effects of infection on host cytoplasm are degenerative. Initially many stimulatory events may be associated with biotrophic parasitism; for example, increases in the volume of host cytoplasm (Mercer et al., 1975), smooth endoplasmic reticulum (ER) and ribosomes (Peyton and Bowen, 1963), mitochondria (Shimony and Friend, 1977), accumulation of starch in plastids [Figure 7(*36*)] (Coffey et al., 1972a). Even stimulatory changes to nuclei (Hadwiger and Adams, 1978) have been observed.

In contrast necrotrophs often cause severe disruption of host cytoplasm even during [Figures 1(*3*), 3(*11*), and 7(*38*)] (McKeen, 1974; Wheeler, 1977) or in advance (Weinhold and Motta, 1973) of penetration or invasion (Hess, 1969), changes which are normally associated with only the terminal stages of pathogenesis by biotrophs (Aist, 1976a).

Many pathogen-induced ultrastructural changes seem to be compatible with an acceleration of normal senescence (Butler and Simon, 1971; Goodman, 1972; Hislop et al., 1979; Ragetli et al., 1970). This often includes rapid deformation and disintegration of the chloroplast stroma and lamellae, degeneration of starch grains, and increase in the number of plastoglobuli. The ribosome population decreases and occasionally mitochondria also show early signs of alteration but are more resistant than other organelles. The ER often swells and vesiculates as do the dictyosomes. Breakdown of the tonoplast appears to accelerate cytoplasmic degeneration, which is in agreement with the evidence for its lysosomal character (Matile, 1975). The nucleus and plasmalemma are often the last recognizable components; the frequent persistence of the plasma membrane, even in severely damaged cells, is surprising considering the characteristic increase in cell permeability as an early response to invasion (Wheeler, 1976).

In view of the similarity of cellular responses to different forms of injury [pathogenesis, senescence and herbicides (Anderson and Thomson, 1973)] it is difficult to ascertain whether such effects are a direct result of pathogen activity or secondary, for example, various physiologically active compounds, such as ethylene, are generated during damage or infection, which can induce ultrastructural changes in plant cells (Freytag et al., 1977; Heath, 1974b).

Nevertheless, there are perhaps three fairly well-defined direct mechanisms of protoplast damage by pathogens for which there is ultrastructural evidence, for example, toxins, endopolygalacturonide hydrolases, and lyases, and hypersensitivity. These will be considered in turn.

Host-Selective Toxins

These remarkable low molecular weight compounds provide some of the rare examples where pathogenesis and specificity can be explained in relatively simple terms. Use of pure toxins has enabled unique studies of early critical physiological and ultrastructural changes in plants without the complications introduced by metabolic activities of living pathogens.

Figure 7 Effects of pathogens or their metabolites on host protoplasts. ***34.*** Scanning electron micrograph of T.S. of part of an elm (var Huntingdon) branch 6 days after inoculation with the aggressive strain of *Ceratocystis ulmi*. The majority of springwood vessels (SW) in the current growth ring became rapidly (after 2 days) occluded with tyloses (t). Note the ring-porous arrangement of vessels which is amenable to colonization by vascular wilt fungi by virtue of vessel contiguity, width, length, and superficial location. ×25. ***35.*** Area of xylem from a tomato stem 28 days after inoculation with *Verticillium albo-atrum*. The vessel secondary wall (SW) and pit membranes (↑) are coated with a thick, laminated, electron-dense material (m). Note the "tannin-like" deposits (d) in the vacuole (v) of the xylem parenchyma cell and the thick layer of host material (protection layer) (p) secreted beneath pit membranes bordering the affected vessel (Ve). ×37,500. ***36.*** A 5-day-old infection of sunflower rust (*Puccinia helianthi*). The plastids contain several large starch grains, which is a feature of this stage of infection. Haustorium (H), intercellular mycelium (IM). ×6000. (From Coffey et al., 1972a.) ***37.*** Effect of *Helminthosporium sacchari* toxin (helminthosporoside) on chloroplasts in susceptible sugar cane leaf cells. (*a*) Separation of lamellae (↑) 1 hr after toxin treatment. ×10,000. (*b*) After 12 hr most chloroplasts and other organelles are disrupted. Altered lamellae (L), cell wall (CW), remains of membrane (ME). ×10,800. (From Strobel et al., 1972.) ***38.*** Effects of *H. victoriae* toxin (victorin) or cortical cells in susceptible oat roots. 24 hr after exposure to toxin cell walls (CW) showed increased electron density and large numbers of lomasome-like structures (↑) formed next to the cell wall. At this stage the unit structure of the plasmalemma appears intact. ×6300. (From Hanchey and Wheeler, 1968.) ***39.*** Penetration by *Helminthosporium maydis* race T between guard (G) and subsidiary (s) cells of a maize leaf (normal (N) cytoplasm) 6 hr after inoculation. The cell wall of the subsidiary cell is swollen and distorted in areas (↑) well removed from the pathogen; cytoplasmic disruption probably results from wall degradation as the changes do not resemble the effects of T-toxin. ×2600. (From Wheeler, 1977.)

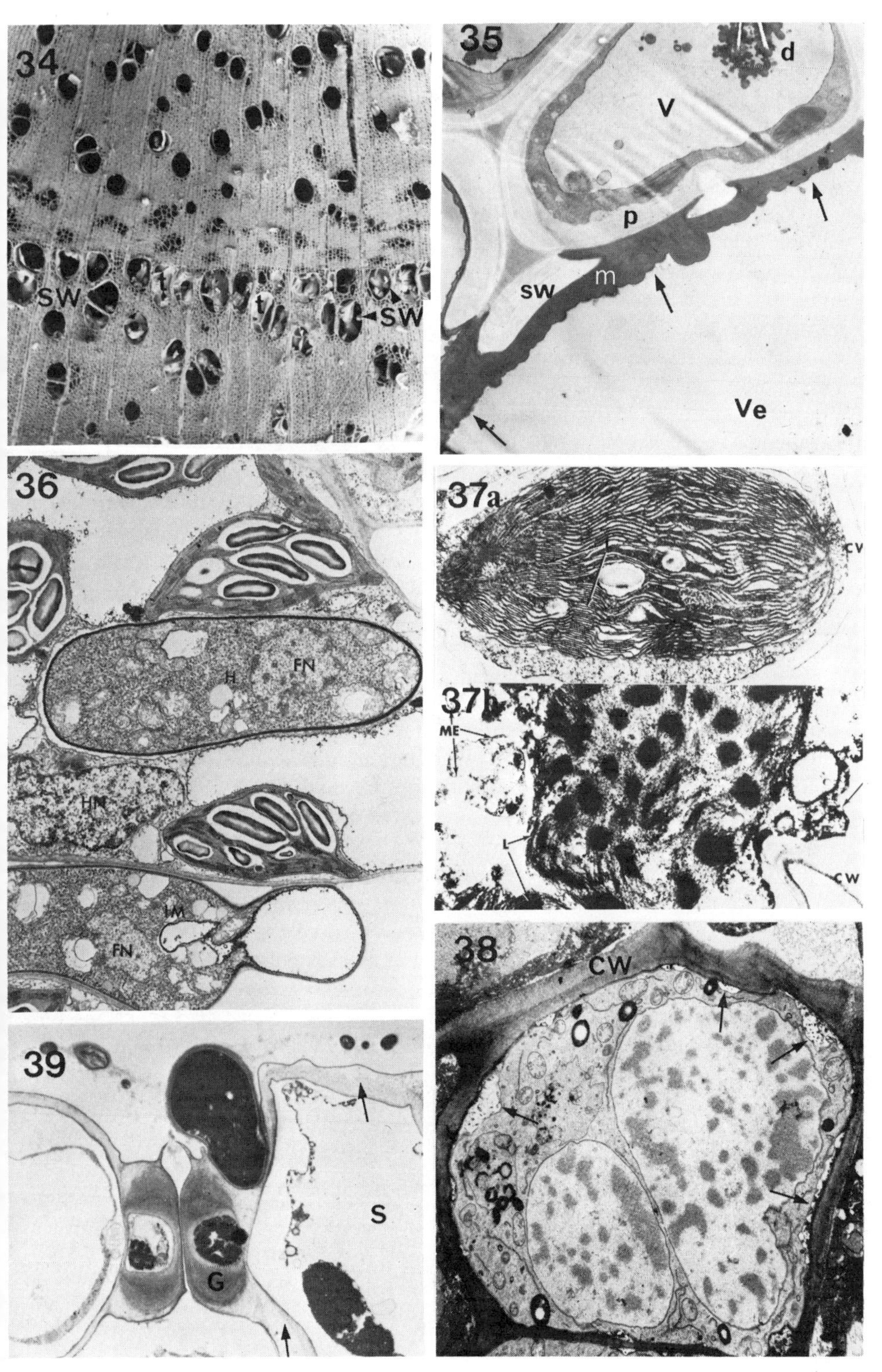
34
SW
t
t
SW
35
d
V
p
m
SW
Ve
36
H
FN
HN
IM
FN
37a
37b
ME
L
CW
38
CW
39
S
G

Treatment of plants with toxins have revealed in several cases effects closely resembling those during infection, although as many of these compounds cause a dramatic alteration in permeability of membranes the possibility of fixation artifacts should be borne in mind.

Helminthosporoside (ex. *Helminthosporium sacchari*) treatment of leaves of susceptible sugar-cane cultivars induces changes remarkably similar to those in infected leaves. After 1 hour, changes in chloroplast membranes are followed by severe cytoplasmic disruption, possibly as a result of damage to cellular membranes to which the toxin is known to bind [Figure 7(*37*)] (Strobel et al., 1972).

The lack of effects of victorin (ex. *H. victoriae*) on host cytoplasmic organelles and the rapid induction of electrolyte loss from susceptible oat tissue indicates that the plasmalemma is its site of action (Scheffer, 1976). Hanchey and Wheeler (1968, 1969) and Luke et al. (1966) showed that various ultrastructural changes were evident in cell walls, plasma membrane, Golgi and ER within 24 hours of exposure to the toxin [Figure 7(*39*)]. The expected lesions in the plasmalemma occurred only after long exposure to victorin. This may indicate that present EM techniques cannot detect subtle alterations to membranes at the molecular level. Indeed it is common for physiological changes (Park et al., 1976; Scheffer, 1976; Wheeler, 1977) and even external symptoms (Strobel et al., 1972) to precede detectable ultrastructural changes.

The toxin from *Alternaria kikuchiana* (AK-toxin) apparently resembles victorin in mode of action as it induced electrolyte loss and rapid (ca. 1 hr) alterations to the plasmalemma (invagination and vesiculation) of susceptible pear tissue as found in infected tissue, but had no effect on other membranes or organelles even after 6 hours. No changes to the unit membrane were detected (Park et al., 1976).

Although host-selective, the T-toxin of *H. maydis* race T may differ in its role from other selective toxins of *Helminthosporium* sp., which are essential to infection and colonization (Comstock and Scheffer, 1973). Wheeler (1977) demonstrated that necrosis of cells during penetration probably resulted from extensive wall degradation as changes were not typical of ultrastructural damage induced by the toxin [Figure 7(*38*)]; also reactions of susceptible and resistant cultivars could not be differentiated until subsequent colonization. These observations support previous genetic data of Yoder (1972, 1976), which suggested that the toxin may only be a secondary determinant of pathogenicity.

Particularly interesting is the highly selective ($1{:}10^4$) action of T-toxin (swelling and loss of respiratory control) on mitochondria from susceptible plants (Miller and Koeppe, 1971). This could account for the specificity and cytoplasmic inheritance of reaction to this disease. Also the extreme sensitivity of these organelles to the toxin (10^5 times more sensitive than induction of a leaf lesion (Wheeler, 1975) may be of potential benefit as an alternative to using intact plants in screening for resistant varieties; especially where a toxin is unstable and/or difficult to obtain in large quantities in a highly active form.

Although various studies have failed to provide evidence for early effects of T-toxin on mitochondria *in situ* (Wheeler, 1975), an ultrastructural study by Aldrich et al. (1977) appeared to confirm this organelle as the site of action. The toxin caused swelling and loss of matrix density in mitochondria of root cells after only 15 minutes. Nevertheless, in infected tissues the first detectable change was rupture of the tonoplast (6 hours) followed by general disorganization of membranes and organelles (24 hours) [Figure 8(*40*)] (White et al., 1973). Also see Earle, this volume.

Ultrastructural changes in infected tissues have also been mimicked by amylovorin, a toxin from *Erwinia amylovora* claimed to have host-selectivity to members of Rosaceae (Huang and Goodman, 1976). Plasmolysis of apple xylem parenchyma cells after 2 hours was followed by severe disruption of cytoplasm and organelles after 6 hours, as found in apple stems 48 hours after inoculation with bacteria [Figure 8(*41*)].

Nonselective Toxins

Although the role of these substances in pathogenesis is by definition difficult to establish, a few ultrastructural studies have contributed to an understanding of their possible activity.

Syringomycin, a polypeptide from *Pseudomonas syringae,* causes lysis of cellular membranes by a rapid detergent-like effect (Backman and De Vay, 1971); also a glycopeptide from *Corynebacterium sepedonicum* induced flaccidity in stems and leaves by destroying the integrity of cell membranes rather than plugging of xylem (Hess and Strobel, 1970; Strobel and Hess, 1968). The aggregation of ribosomes into helices by *P. phaseolicola* toxin (Lesemann and Rudolph, 1970) and marked changes to tobacco chloroplasts by *P. tabaci* toxin (Goodman, 1972) reflected by adverse effects on protein synthesis and photosynthesis, respectively, may also reveal their sites of action.

Similar studies on tentoxin from *Alternaria tenuis,* which induces chlorosis in seedlings of certain dicots, may have contributed to the first possible explanation of the molecular basis of toxin action and specificity. Electron microscopy revealed its apparent site of action as the chloroplast lamellae [Figure 8(*42*)] (Halloin et al., 1970), and subsequently the toxin was found to bind and inhibit a chloroplast ATPase involved in photophosphorylation of toxin susceptible species. (Steele et al., 1976).

In view of the potential in breeding programs of host-selective toxins for screening specific resistance (Day, 1974) or the possible use of nonselective toxins to reveal tolerance (Rudolph, 1976) it is desirable to understand their modes of action; in addition to revealing mechanisms of pathogenicity and resistance, identification of toxin-sensitive sites or receptors could alter our approach to detection of potential resistant varieties. Electron microscopy should be especially useful in conjunction with autoradiography to reveal the subcellular location of labeled toxins, previously only reported at the level of light microscopy (Backman and De Vay, 1971; Strobel and Hess, 1968).

Endopectic Enzymes

As previously mentioned, during infection extensive wall degradation is frequently if not invariably associated with cytoplasmic disruption. Use of homogeneous enzyme preparations from various microbial plant pathogens has clearly established the unique role of endopectic enzymes as the factors responsible for cell separation and associated protoplast damage (Bateman and Basham, 1976).

The mechanism of cell killing is generally considered to involve degradation of the pectic matrix of the wall, so that the weakened structure can no longer counteract the pressure exerted by the protoplast, resulting in a damaged plasmalemma. However, this may be an oversimplification as ion leakage, indicative of cell injury (Wheeler, 1976), occurs almost immediately when tissues are exposed to pectic enzymes (Hislop et al., 1979). Also the cell wall has been regarded as a component of the cell's lysosome system (Matile, 1975) enzymic degradation of which releases various enzymes with potential physiological activity, such as acid phosphatase, peroxidase, and IAA oxidase (Hislop et al., 1979; Mussell and Strand, 1976), although to date most attempts to establish the toxicity of these enzymes or their by-products have proved negative (Bateman and Basham, 1976; Hislop et al., 1979).

The only critical electron microscopy study using a purified pectic enzyme (Hislop et al., 1979) (*endo*-pectin lyase of *M. fructigena* on cultured apple cells) showed many features of tissues invaded by necrotrophic pathogens (e.g. Ammon et al., 1974; Calonge et al., 1969; Shimony and Friend, 1977): wall degradation, dilation and vesiculation of ER, retraction of the plas-

Figure 8 Effects of pathogens or their metabolites on host protoplasts. ***40.*** Effects of infection by *H. maydis* race T on mesophyll cells of a susceptible (TmS cytoplasm) maize cultivar. (*a*) The first detectable change after 6 hr was breakdown of the tonoplast (↑). Chloroplast (C), cell wall (W). ×17,100. (*b*) After 24 hr the plasma membrane (P) is broken, the plastid envelope (E) is disorganized, and the matrix within mitochondria (M) appears to be absent. ×12,400. (From White et al., 1973.) ***41.*** Changes in xylem parenchyma of apple stems induced by fire blight toxin (amylovorin) and *Erwinia amylovora* infection. (*a*) 4 hr after treatment with toxin, cells are plasmolyzed and organelles are disorganized. ×5500. (*b*) Similar changes are apparent in infected tissue after 48 hr. Note the aggregation of cytoplasm in cell B, and the remaining plasmodesmatal connections (↑). Cell walls (CW). ×5800. (From Huang and Goodman, 1976.) ***42.*** Effects of toxin from *Alternaria tenuis* (tentoxin) on plastid structure in cucumber cotyledons. (*a*) Juvenile plastids in 4-day-old plants. ×32,000. (*b*) As (*a*) but for toxin treatment during germination; note the absence of a lamellar system and the presence of starch grains (s). ×20,600. Chloroplast (c), granalamellae (g), mitochondrion (m). (From Halloin et al., 1970.)

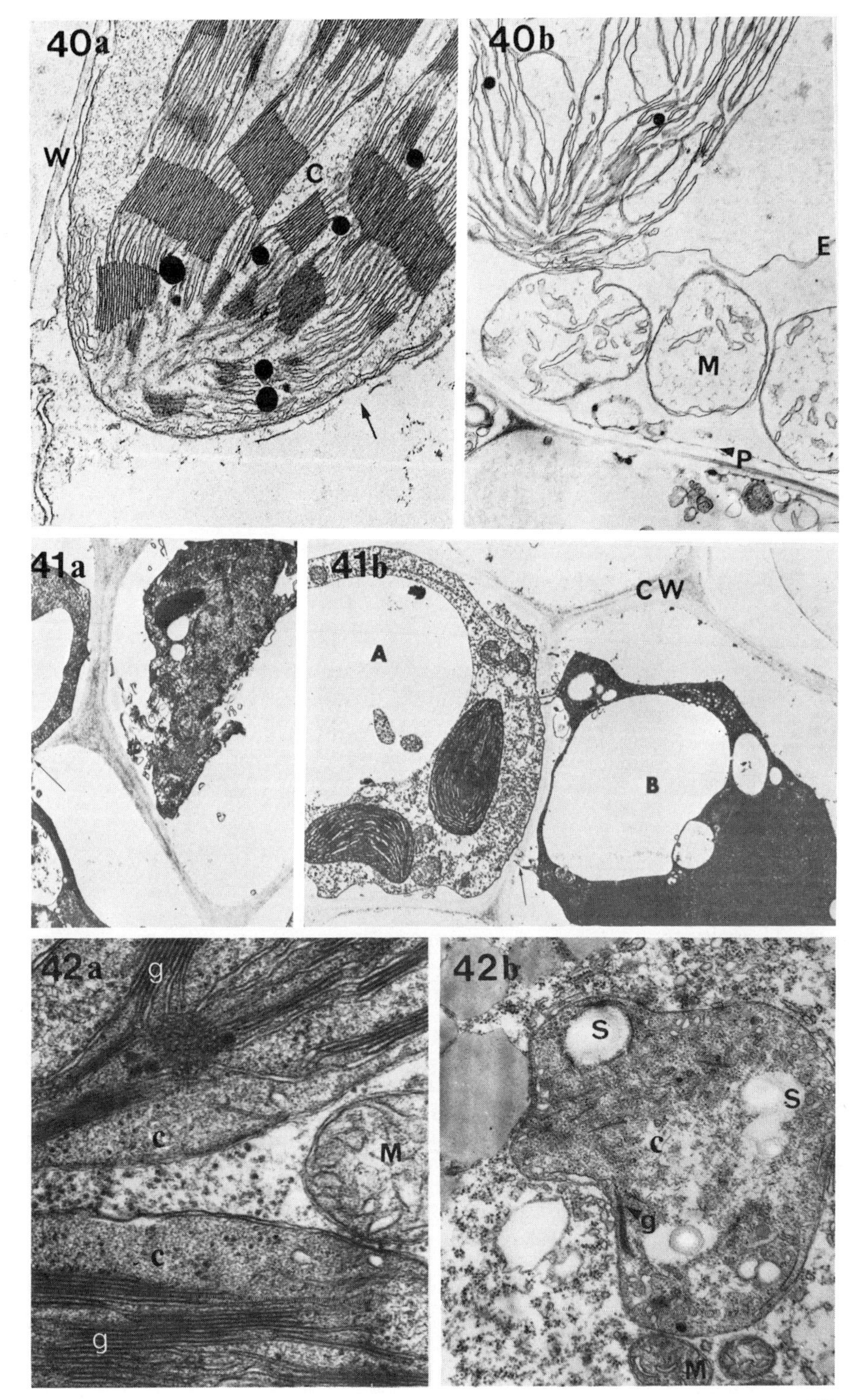
40a
W
C
40b
E
M
P
41a
41b
CW
A
B
42a
g
c
M
c
g
42b
S
S
c
g
M

malemma and pseudoplasmolysis [Figure 9(*43–46*)]. Cells with conspicuously altered walls were invariably "dead," but even many cells with apparently normal or only slightly degraded walls were often severely injured. This is highly significant in view of the common occurrence of visible wall degradation during infection by most plant pathogens. It is notable that the unit structure of the plasmalemma was apparently unaltered, although it stained less intensely and had swellings or strands extending to the altered wall [Figure 9(*45, 46*)]. Whether these represent points of rupture is not clear.

It is strange that what might be the most frequent mechanism of tissue damage during pathogenesis has been almost neglected at the ultrastructural level. Such studies are needed to clarify the existing wealth of biochemical and physiological information concerning protoplast damage by pectic enzymes.

During infection by some pathogens that degrade cell walls, membrane damage is often rapid and extensive [Figure 3(*11–14*)] (Fox et al., 1972; Hänssler et al., 1978; Hess, 1969; Mercer et al., 1975). This could result from protoplasts rupturing through weakened walls (although the effect of *M. fructigena* lyase suggests this as unlikely) or could obviously reflect the action of other toxic metabolites produced during infection. In one case membrane damage in pepper fruit coincided with the ability of culture filtrates of *Phytophthora palmivora* to degrade cellular membranes (Calonge et al., 1969). However, current evidence for the significance of phospholipases and proteinases of plant pathogens is scanty and in several cases the enzymes have shown no effect on plant tissue (Bateman and Basham, 1976).

HYPERSENSITIVE REACTIONS (HR)

Paradoxically this extreme, susceptible cellular reaction confers highly effective resistance to plants against incompatible races or species of mi-

---

Figure 9 Effects of pathogens or their metabolites on host protoplasts. ***43.*** Degradation by polygalacturonide from *Monilinia fructigena* has revealed the fibrillar texture of the cell wall (W) of cultured apple cells. Note pseudoplasmolysis of the protoplasts. ×42,800. ***44.*** Degradation of wall of cultured apple cell by pectin lyase from *Monilinia fructigena* is associated with a conspicuous dilation and vesiculation of the endoplasmic reticulum (er), swelling of mitochondria (m) and parts of the nuclear membrane (nm). ×17,400. ***45.*** The plasmalemma in an apple cell treated with pectin lyase has withdrawn from the modified wall (W) and appears normal except for the strands (↑) extending toward the cell wall. Note vesiculation of the endoplasmic reticulum (erv). ×20,800. ***46.*** High-power magnification of an er vesicle and plasmalemma extensions (↑) in cultured apple cells treated with pure pectin lyase from *Monilinia fructigena*. ×54,700. (From Hislop et al., 1979.) ***47.*** An early stage in the hypersensitive response of a resistant cultivar of cowpea to *Uromyces phaseoli* var *vignae*, 28 hr after inoculation. Both haustorium (H) and haustorial mother cell (MC) appear necrotic as does the collapsing host cell. A callose-like deposit (C) has formed in the adjacent healthy cell where it is in contact with the necrotic cell. ×5100. (From Heath and Heath, 1971.)

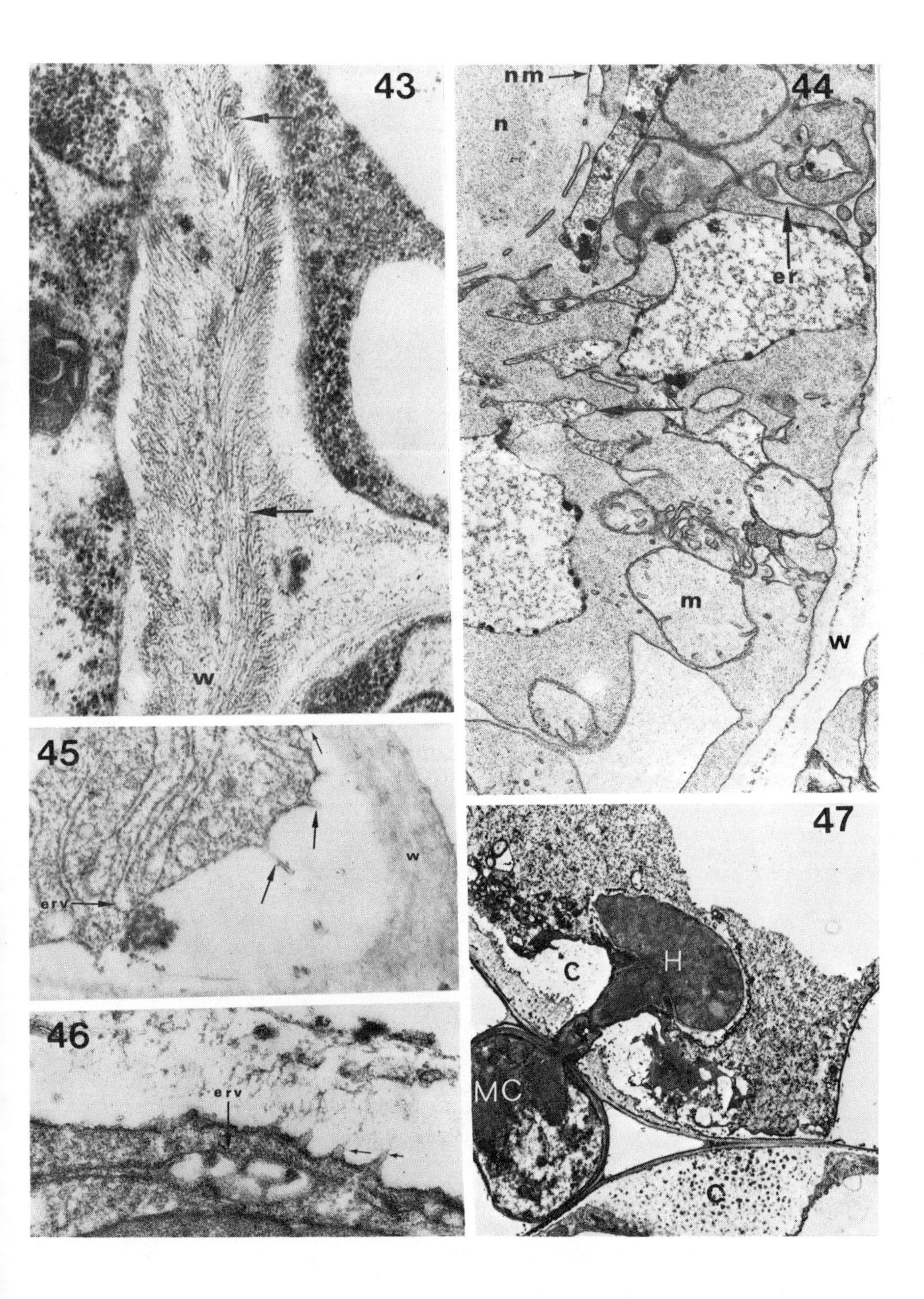
43
w
44
nm
n
er
m
w
45
erv
w
46
erv
47
C
H
MC
C

crobial pathogens. Although by definition such a response is not associated with extensive damage to crops, it warrants consideration here as ultrastructural comparisons with true compatible interactions may reveal the difference between these superficially similar reactions (e.g., respiration increase, electrolyte leakage, and browning) and indicate the subcellular sites of molecular exchange.

The main characteristic that separates the reactions is the rapidity of HR in response to invasion, which is apparent within several hours of inoculation rather than the several days that elapse before compatible cells become disrupted (e.g., Maclean et al., 1974; Sequeira, 1978). Somehow the hypersensitive reaction (HR) creates an unfavorable environment resulting in rapid decline in bacterial population or limitation of invading hyphae to one or a few cells. Various light microscope and electron microscopy studies have revealed that host cell necrosis precedes death of invading fungi by several hours, which strongly implicates HR as a cause not consequence of resistance [Figure 10(*48*)] (Jones et al., 1975; Maclean et al., 1974; Shimony and Friend, 1975).

Ultrastructural evidence suggests that HR is not simply a "rapid compatible reaction" but a unique response. Cellular necrosis associated with HR in various host–parasite combinations differs from that of damaged cells in compatible interactions by the rapid disruption of cellular membranes, including the plasmalemma, which, as already discussed, is usually one of the

---

Figure 10 Hypersensitive reactions. ***48.*** Primary vesicle (PV) of an incompatible race (W5) of *Bremia lactucae* in an epidermal cell (E_1) of lettuce cultivar Avondefiance, 4 hr after inoculation. The membranes and cytoplasm of the invaded cell are completely disrupted, whereas the cytoplasm of the pathogen and of the nonpenetrated cell (E_2) appear normal, except for the presence of crystal-containing microbodies (CM) in the latter. Appressorium (A), chloroplast (C), callose plug (K), host wall (HW). ×8700. (From Maclean et al., 1974.) ***49.*** Necrosis and collapse of root epidermal cells of a resistant sunflower cultivar in response to an encysted zoospore of *Plasmopara halstedii*. Serial sectioning revealed that the hypersensitive response occurred prior to penetration of host cells. Necrotic encysted zoospore (NZ), necrotic host cells (NC). ×25,000. (From Wehtje et al., 1979.) ***50.*** Reaction of *Phaseolus vulgaris* (cultivar KK, partially resistant) to *Colletotrichum lindemuthianum* (race δ), 6 d after inoculation. Part of a hypersensitive fleck involving several host cells showing electron-translucent cuticle (C) with attached moribund appressorium (A). The hyphae (H) and host cells appear to be dead. Note the reaction material (↑) in the cortical cell. ×2400. (From Mercer et al., 1974.) ***51.*** Reaction of *Phaseolus vulgaris* (cultivar Imuna, resistant) to *C. lindemuthianum* (race B), 65 hr after inoculation. Part of a hypersensitive cell (H) bounded by two apparently healthy epidermal cells (E) and a cortical cell (C). Note the characteristic electron-dense granules in the necrotic cell, which occlude the pit joining the cortical cell but do not appear to extend beyond the middle lamella of the surrounding cell wall, and the large volume of cytoplasm in the upper epidermal cell, which has formed reaction material (↑) adjacent to the necrotic cell. ×4300. (From Mercer et al., 1974.) ***52.*** A collapsed mesophyll cell (M) of *Phaseolus vulgaris* 24 hr after inoculation with an incompatible race of *Pseudomonas phaseolicola*. Bacterial cells (↑) are attached to the wall of the hypersensitive cell, which is partially degraded. The adjacent cell (C) shows extensive changes including accumulation of osmiophilic droplets, vesiculation, and retraction of plasmalemma (p). ×3300. (From Roebuck et al., 1978.)

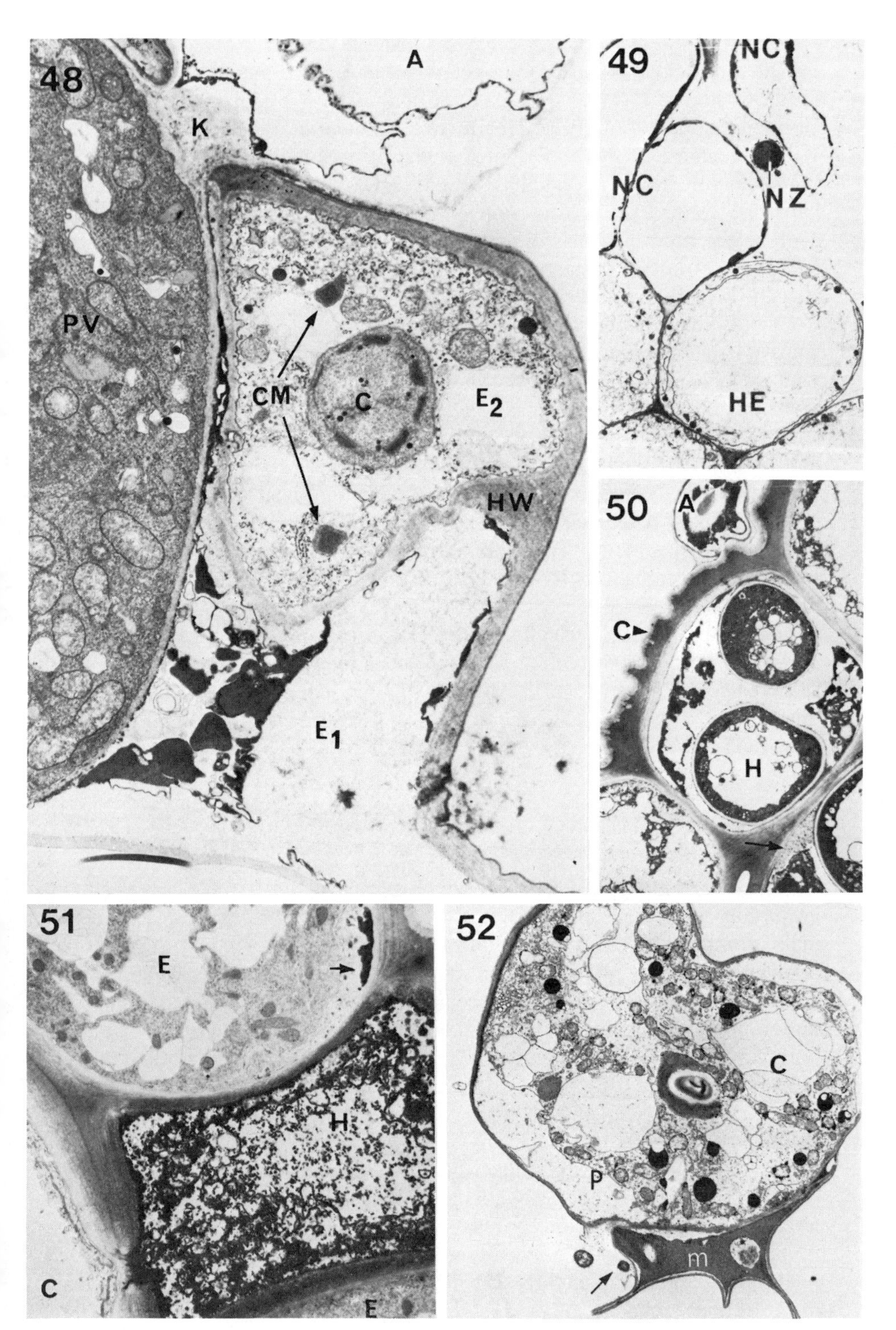
48
A
K
PV
CM
C
E2
HW
E1
49
NC
NC
NZ
HE
50
A
C
H
51
E
H
C
E
52
C
P
m

last factors to lose integrity in infected, toxin- or enzyme-treated cells [Figures 9(*47*) and 10(*48–52*)] (Cason et al., 1978; Goodman et al., 1976a; Goodman and Plurad, 1971; Heath, 1972; Huang et al., 1974; Jones et al., 1975; Maclean et al., 1974; Roebuck et al., 1978; Shimony and Friend, 1975).

In most infections by fungal pathogens, HR is initiated when germ tubes or haustoria contact the host plasmalemma, although there may be exceptions such as the rapid necrosis of root cells of a resistant cultivar of sunflower before penetration by encysting zoospores of *Plasmopara halstedii* [Figure 10(*49*)] (Wehtje et al., 1979).

Usually the timing and localized nature of HR [Figure 10(*48,51,52*)] suggests the reaction does not depend on toxic diffusible metabolites (as with many compatible reactions where damage can occur well in advance of invasion) but involves surface components of the pathogen. Observations such as these have led to many attempts to isolate from pathogens specific factors able to induce HR, which could have potential in disease control. Although various fractions (mainly glucans) from fungal cell walls have elicited "HR-like" responses (e.g., Anderson, 1978; Marcan et al., 1979), none have shown specificity. The major problem of distinguishing between true HR and nonspecific toxicity could be lessened by ultrastructural comparison of the cellular reactions.

In contrast, bacteria usually induce HR in incompatible hosts from their *extracellular* location in intercellular spaces once bound to the host cell wall [Figures 5(*22, 23*) and 10(*52*)]. Although the binding mechanism is becoming unraveled (Sequeira, 1978), the specificity factor, which presumably must travel via the cell wall, has yet to be isolated.

CONCLUSIONS

This chapter cites many cases in which ultrastructural studies have given important clues concerning critical events that occur during pathogenesis. A main stumbling block in most studies seems to be the lack of simultaneous application of EM, physiological, and biochemical techniques. Where such cooperation has been achieved, meaningful results have emerged. The interaction between bacterial LPS and host cell wall lectins is a prime example (Graham et al., 1977; Sequeira, 1978). The development and intelligent use of many alternative forms of control will rely on similar coordinated approaches. However, understanding the mechanisms behind such poorly defined concepts as tolerance and field resistance is only likely to follow further elucidation of well-studied host–parasite interactions in which susceptibility and resistance are genetically clearly defined. Nevertheless, for most diseases the basic mechanisms of pathogenicity and resistance have remained elusive. Possibly this is a reflection of our ignorance of recognition phenomena and specificity about which EM can only indicate possible sites of molecular exchange.

The most exciting prospects for EM concern the subcellular localization of putative determinants of pathogenicity or resistance. Perhaps understandably such technically difficult studies have been almost neglected, although they could yield essential information on the timing and location of key compounds during infection. One exception may be the visualization of host peroxidase and polyphenoloxidase activity (Mueller and Beckman, 1978). Thus in cases where the biosynthetic pathways are known, phytoalexins could be located by EM autoradiography after incorporating a radiolabeled precursor before inoculation. Evaluation of the role of phytoalexins in resistance depends on the timing and site of their accumulation relative to the invading pathogen as well as their relationship to host necrosis. Also the studies of Paxton et al. (1974) on the fine structure of living cells accumulating phytoalexin in response to an elicitor should be extended using the highly active glucan elicitors from fungal cell walls (Ayers et al., 1976). Radiolabeling of the elicitors may reveal their site of action and whether in fact they are unable to penetrate intact plant surfaces as suggested by accumulation of phytoalexins only in wounded tissues treated with elicitor.

Similarly toxins or CWDE could be easily obtained and labeled in culture and, after purification, applied to host tissue to reveal their location and effects on cells. However, the extremely variable microscopic responses even of adjacent cells to such compounds or to infection (Hislop et al., 1979; Robb et al., 1975a) complicates interpretation, as does the likelihood that microbial metabolites never act alone against plant tissues during infection.

Although ultrastructural studies have already made a substantial contribution to our understanding of pathogenesis, it remains to be seen whether such insight will enable a more rational approach to disease control.

ACKNOWLEDGMENTS

Grateful thanks are due to N. Weston and P. Clark for assistance with micrographs and to A. Rayner for critically reading the manuscript.

REFERENCES

Aist, J. R. (1976a) Cytology of penetration and infection. In *Encyclopedia of Plant Physiology*, Vol. 4 (R. Heitefuss and P. H. Williams, eds.). Springer-Verlag, New York. Pp. 198–221.

Aist, J. R. (1976b) Papillae and related wound plugs of plants. *Annu. Rev. Phytopathol.* **14**:145–163.

Aist, J. R., and H. W. Israel. (1977) Effects of heat shock inhibition of papilla formation on compatible host penetration by two obligate fungi. *Physiol. Plant. Pathol.* **10**:13–20.

Akai, S., O. Orino, M. Fukutomi, A. Nakata, H. Kunoh, and M. Shiraishi. (1971)

Cell wall reaction to injection and resulting change in cell organelles. In *Morphological and Biochemical Events in Plant–Parasite Interactions* (S. Akâi and S. Ouchi, eds.). Phytopathological Society of Japan, Tokyo. Pp. 329–347.

Albersheim, P., and A. J. Anderson-Prouty. (1975) Carbohydrates, proteins, cell surfaces and the biochemistry of pathogenesis. *Annu. Rev. Plant Physiol.* **26**:31–52.

Aldrich, H. C., V. E. Gracen, D. York, E. D. Earle, and O. C. Yoder. (1977) Ultrastructural effects of *Helminthosporium maydis* race T toxin on mitochondria of corn roots and protoplasts. *Tissue Cell* **9**:167–177.

Ammon, V., T. D. Wyllie, and M. F. Brown. (1974) An ultrastructural investigation of pathological alterations induced by *Macrophomina phaseolina* (Tassi) Goid in seedlings of soybean, *Glycine max* (L.) Merrill. *Physiol. Plant Pathol.* **4**:1–4.

Anderson, A. (1978) Initiation of resistant responses in bean by mycelial wall fractions from three races of the bean pathogen *Colletotrichum lindemuthianum*. *Can. J. Bot.* **56**:2247–2251.

Anderson, J. L., and W. W. Thomson. (1973) The effects of herbicides on the ultrastructure of plant cells. *Residue Rev.* **47**:167–189.

Ayers, A. R., B. Valent, J. Ebel, and P. Albersheim. (1976) Host–pathogen interactions. XI. Composition and structure of wall-released elicitor fractions. *Plant Physiol.* **57**:766–774.

Backman, P. A., and J. E. De Vay. (1971) Studies on the mode of action and biogenesis of the phytotoxin syringomycin. *Physiol. Plant Pathol.* **1**:215–233.

Bal, A. K., D. P. S. Verma, V. M. Byrne, and G. A. Maclachlin. (1976) Subcellular localization of cellulases in auxin-treated pea. *J. Cell Biol.* **69**:97–105.

Bateman, D. F., and H. G. Basham. (1976) Degradation of plant cell walls and membranes by microbial enzymes. In *Encyclopedia of Plant Physiology,* Vol. 4 (R. Heitefuss and P. H. Williams, eds.). Springer-Verlag, New York. Pp. 316–355.

Baur, P. S., and C. H. Walkinshaw. (1974) Fine structure of tannin accumulations in callus cultures of *Pinus elliotti* (slash pine). *Can. J. Bot.* **52**:615–619.

Beckman, C. H., G. E. Van der Molen, W. C. Mueller, and M. E. Mace. (1976) Vascular structure and distribution of vascular pathogens in cotton. *Physiol. Plant Pathol.* **9**:87–94.

Bishop, C. D. (1980) Ultrastructure of vascular wilt diseases. Ph.D. Thesis, University of Bath.

Bracker, C. E., and L. J. Littlefield. (1973) Structural concepts of host–pathogen interfaces. In *Fungal Pathogenicity and the Plant's Response* (R. J. W. Byrde and C. V. Cutting, eds.). Academic Press, New York. Pp. 159–317.

Brener, W. D., and C. H. Beckman. (1968) A mechanism of enhanced resistance to *Ceratocystis ulmi* in American elms treated with trichlorophenylacetate. *Phytopathology* **58**:555–561.

Brotzman, H. G., O. H. Calvert, J. A. White, and M. F. Brown. (1975) Southern corn leaf blight: Ultrastructure of host–pathogen association. *Physiol. Plant Pathol.* **7**:209–211.

Brown, J. F., and W. A. Shipton. (1964) Relationship of penetration to infection type when seedling wheat leaves are inoculated with *Puccinia graminis tritici*. *Phytopathology* **54**:89–91.

Butler, R. D., and E. W. Simon. (1971) Ultrastructural aspects of senescence in plants. *Adv. Gerontol. Res.* **3**:73–129.

Byrde, R. J. W., and A. J. Willetts. (1977) *The Brown Rot Fungi of Fruit: Their Biology and Control.* Pergamon Press, Oxford.

Calonge, F. D., A. H. Fielding, R. J. W. Byrde, and O. A. Akinrefon. (1969) Changes in ultrastructure following fungal invasion and the possible relevance of extracellular enzymes. *J. Exp. Bot.* **20**:350–357.

Carr, A. J. H. (1970) Plant pathology. In *Jubilee Report Welsh Plant Breeding Station 1919–1969.* University College of Wales, Aberystwyth. Pp. 168–182.

Cason, E. T., P. E. Richardson, M. K. Essenberg, L. A. Brinkerhoff, W. M. Johnson, and R. J. Venere. (1978) Ultrastructural cell wall alterations in immune cotton leaves inoculated with *Xanthomonas malvacearum*. *Phytopathology* **68**:1015–1021.

Cho, Y. S., R. D. Wilcoxson, and F. I. Frosheiser. (1973) Differences in anatomy, plant extract and movement of bacteria in plants of bacterial wilt resistant and susceptible varieties of alfalfa. *Phytopathology* **63**:760–765.

Coffey, M. D., B. A. Palevitz, and P. J. Allen. (1972a) Ultrastructural changes in rust infected tissues of flax and sunflower. *Can. J. Bot.* **50**:1485–1492.

Coffey, M. D., B. A. Palevitz, and P. J. Allen. (1972b) The fine structure of two rust fungi, *Puccinia helianthi* and *Melampsora lini*. *Can J. Bot.* **50**:231–240.

Comstock, J. C., and R. P. Scheffer. (1973) Role of host-selective toxin in colonization of corn leaves by *Helminthosporium carbonum*. *Phytopathology* **63**:24–29.

Cooper, R. M. (1976) Regulation of synthesis of cell wall-degrading enzymes of plant pathogens. In *Cell Wall Biochemistry Related to Specificity in Host–Plant Pathogen Interactions* (B. Solheim and J. Raa, eds.). Universitetsforlaget, Tromsø, Norway. Pp. 163–211.

Cooper, R. M. (1978) Factors affecting production and activity of cell wall-degrading enzymes in plant diseases. *International Congress of Plant Pathology, 3rd, Munich.* (Abstr.) P. 205.

Cooper, R. M., and R. K. S. Wood. (1974) Scanning electron microscopy of *Verticillium albo-atrum* in xylem vessels of tomato plants. *Physiol. Plant Pathol.* **4**:443–446.

Cooper, R. M., and R. K. S. Wood. (1980) The role of cell wall-degrading enzymes in vascular wilt diseases. III. Possible involvement of endopectin lyase. *Physiol. Plant Pathol.* **16**:285–300.

Corden, M. E., and H. L. Chambers. (1966) Vascular disfunction in *Fusarium* wilt of tomato. *Am. J. Bot.* **53**:284–287.

Day, R. P. (1974) *Genetics of Host–Parasite Interactions.* W. H. Freeman & Co., San Francisco, Calif.

Duddridge, J. A., and J. A. Sargent. (1978) A cytochemical study of lipolytic activity in *Bremia lactucae* Regel, during germination of the conidium and penetration of the host. *Physiol. Plant Pathol.* **12**:289–296.

Favali, M. A., G. G. Conti, and M. Bassi. (1978) Modifications of the vascular bundle ultrastructure in the "resistant zone" around necrotic lesions induced by tobacco mosaic virus. *Physiol. Plant Pathol.* **13**:247–251.

Fox R. T. V., J. G. Manners, and A. Myers. 1972) Ultrastructure of tissue disintegrations and host reactions in potato tubers infected by *Erwinia carotovora* var. *atroseptica*. *Potato Res.* **15**:130–145.

Freytag, A. H., J. D. Berlin, and J. C. Linden. (1977) Ethylene-induced fine structure alterations in cotton and sugarbeet radicle cells. *Plant Physiol.* **60**:140–143.

Fullerton, R. A. (1970) An electron microscope study of the intracellular hyphae of some smut fungi. *Aust. J. Bot.* **18**:285–292.

Goodman, R. N. (1972) Phytotoxin-induced ultrastructural modifications of plant cells. In *Phytotoxins in Plant Diseases* (R. K. S. Wood, A. Ballio, and A. Graniti, eds.). Academic Press, New York. Pp. 311–329.

Goodman, R. N., and S. B. Plurad. (1971) Ultrastructural changes in tobacco undergoing the hypersensitive reaction caused by plant pathogenic bacteria. *Physiol. Plant Pathol.* **1**:11–15.

Goodman, R. N., P. Y. Huang, J. S. Huang, and V. Thaipanich. (1976a). Induced resistance to bacterial infection. In *Biochemistry and Cytology of Plant–Parasite Interaction* (K. Tomiyama, J. M. Daly, I. Uritani, H. Oku, and S. Ouchi, eds.). Elsevier, New York. Pp. 25–42.

Goodman, R. N., P. Y. Huang, and J. A. White. (1976b) Ultrastructural evidence for immobilization of an incompatible bacterium, *Pseudomonas pisi*, in tobacco leaf tissue. *Phytopathology* **66**:754–764.

Graham, T. L., L. Sequeira, and T. R. Huang. (1977) Bacterial lipopolysaccharides as inducers of disease resistance in tobacco. *Appl. Environ. Microbiol.* **34**:424–432.

Hadwiger, L. A., and M. J. Adams. (1978) Nuclear changes associated with the host-parasite reaction between *Fusarium solani* and peas. *Physiol. Plant Pathol.* **12**:63–72.

Halloin, J. M., G. A. De Zoeten, G. Gaard, and J. C. Walker. (1970) The effects of tentoxin on chloroplast synthesis and plastid structure in cucumber and cabbage. *Plant Physiol.* **45**:310–314.

Hanchey, P., and H. Wheeler. (1968) Pathological changes in ultrastructure: Effects of victorin on oat roots. *Am. J. Bot.* **55**:53–61.

Hanchey, P., and H. Wheeler. (1969) Pathological changes in ultrastructure: False plasmolysis. *Can. J. Bot.* **47**:675–678.

Hanchey, P., and H. Wheeler. (1971) Pathological changes in ultrastructure: Tobacco roots infected with *Phytophthora parasitica* var. *nicotianae*. *Phytopathology* **61**:33–39.

Hänssler, G., D. P. Maxwell, and M. D. Maxwell. (1978) Ultrastructure and acid phosphatase cytochemistry of hyphae of *Sclerotium rolfsii* in hypocotyl of *Phaseolus vulgaris*. *Phytopathol. Z.* **92**:157–167.

Heath, M. C. (1972) Ultrastructure of host and non-host reactions to cowpea rust. *Phytopathology* **62**:27–38.

Heath, M. C. (1974a) Light and electron microscope studies of the interactions of host and non-host plants with cowpea rust—*Uromyces phaseoli* var. *vignae*. *Physiol. Plant Pathol.* **4**:403–414.

Heath, M. C. (1974b) Ethylene production and changes in chloroplast ultrastructure in rust-infected cowpea leaves. *Proc. Am. Phytopathol. Soc.* Abstr.: 136.

Heath, M. C. (1977) A comparative study of non-host reactions with rust fungi. *Physiol. Plant Pathol.* **10**:73–88.

Heath, M. C. (1979) Partial characterization of the electron-opaque deposits formed in the non-host plant, French bean, after cowpea rust infection. *Physiol. Plant Pathol.* **15**:141–148.

Heath, M. C., and I. B. Heath. (1971) Ultrastructure of an immune and a susceptible reaction of cowpea leaves to rust infection. *Physiol. Plant Pathol.* **1**:277–287.

Heath, M. C., and R. K. S. Wood. (1969) Leaf spots induced by *Ascochyta pisi* and *Mycosphaerella pinodes*. *Ann. Bot.* **33**:657–670.

Hess, W. M. (1969) Ultrastructure of onion roots infected with *Pyrenochaeta terrestris,* a fungus parasite. *Am. J. Bot.* **56**:832–845.

Hess, W. M., and G. A. Strobel. (1970) Ultrastructure of potato stems infected with *Corynebacterium sepedonicum*. *Phytopathology* **60**:1428–1431.

Hislop, E. D., and D. Pitt. (1967) Sub-cellular organization in host–parasite interactions. In *Encyclopedia of Plant Physiology,* Vol. 4 (R. Heitefuss and P. H. Williams, eds.). Springer-Verlag, New York. Pp. 389–412.

Hislop, E. C., V. M. Barnaby, C. Shellis, and F. Laborda. (1974a) Localization of α-L-arabinofuranosidase and acid phosphatase in mycelium of *Sclerotinia fructigena*. *J. Gen. Microbiol.* **81**:79–99.

Hislop, E. C., C. Shellis, A. H. Fielding, F. J. Bourne, and J. W. Chidlow. (1974b) Antisera produced to purified extracellular pectolytic enzymes from *Sclerotinia fructigena*. *J. Gen. Microbiol.* **83**:125–143.

Hislop, E. C., J. P. R. Keon, and A. H. Fielding. (1979) Effect of pectin lyase from *Monilinia fructigena* on viability, ultrastructure and localization of acid phosphatase of cultured apple cells. *Physiol. Plant Pathol.* **14**:371–381.

Hohl, H. R., and P. Stossel. (1976) Host–parasite interfaces in a resistant and a susceptible cultivar of *Solanum tuberosum* inoculated with *Phytophthora infestans:* Tuber tissue. *Can. J. Bot.* **54**:900–912.

Horino, O. (1976) Induction of bacterial leaf blight resistance by incompatible strains of *Xanthomonas oryzae* in rice. In *Morphological and Biochemical Events in Plant–Parasite Interactions* (S. Akai and S. Ouchi, eds.). Phytopathological Society of Japan, Tokyo. Pp. 43–55.

Huang, P. Y., and R. N. Goodman. (1976) Ultrastructural modifications in apple stems induced by *Erwinia amylovora* and the fire blight toxin. *Phytopathology* **66**:269–276.

Huang, J., P. Y. Huang, and R. N. Goodman. (1974) Ultrastructural changes in tobacco thylakoid membrane protein caused by bacterially induced hypersensitive reaction. *Physiol. Plant Pathol.* **4**:93–97.

Huang, P. Y., J. S. Huang, and R. N. Goodman. (1975) Resistance mechanisms of apple shoots to an avirulent strain of *Erwinia amylovora*. *Physiol. Plant Pathol.* **6**:283–287.

Ingram, D. S., J. A. Sargent, and I. C. Tommerup. (1975) Structural aspects of infection by biotrophic fungi. In *Biochemical Aspects of Plant–Parasite Relationships* (J. Friend and D. R. Threlfall, eds.). Academic Press, London.

Jones, D. R. W. G. Graham, and E. W. B. Ward. (1975) Ultrastructural changes in pepper cells in an incompatible interaction with *Phytophthora infestans*. *Phytopathology* **65**:1274–1285.

Kajiwara, T. (1973) Ultrastructure of host–parasite interface between downy mildew fungus and spinach. *Shokubutsu Byogai Kenkyu* **8**:167–178.

Keen, N. T., M. Long, and D. C. Erwin. (1972) Possible involvement of a pathogen-produced protein-lipopolysaccharide complex in *Verticillium* wilt of cotton. *Physiol. Plant Pathol.* **2**:317–331.

Kunoh, H., T. Tsuzuki, and H. Ishizaki. (1978) Cytological studies of early stages of powdery mildew in barley and wheat. IV. Direct ingress from superficial primary germ tubes and appressoria of *Erysiphe graminis hordei* on barley leaves. *Physiol. Plant Pathol.* **13**:327–333.

Lazarovits, G., and V. J. Higgins. (1976) Ultrastructure of susceptible, resistant and immune reactions of tomato to races of *Cladosporium fulvum*. *Can. J. Bot.* **54**:235–249.

Lesemann, D., and K. Rudolph. (1970) Die Bildung von Ribosomen—Helices unter dem Einfluss des Toxins von *Pseudomonas phaseolicola* in Blättern von Mangold (*Beta vulgaris L.*). *Z. Pflanzenphysiol.* **62**:108–115.

Lewis, B. G., and J. R. Day. (1972) Behavior of germ tubes of *Puccinia graminis tritici* in relation to fine structure of wheat leaf surfaces. *Trans. Br. Mycol. Soc.* **58**:139–145.

Luke, H. H., H. E. Warmke, and P. Hanchey. (1966) Effects of the pathotoxin victorin on ultrastructure of root tissue and leaf tissue of *Avena* species. *Phytopathology* **56**:1178–1183.

Maclean, D. J., and I. C. Tommerup. (1979) Histology and physiology of compatibility and incompatibility between lettuce and the downy mildew fungus, *Bremia lactucae* Regel. *Physiol. Plant Pathol.* **14**:291–312.

Maclean, D. J., J. A. Sargent, I. C. Tommerup, and D. S. Ingram. (1974) Hypersensitivity as the primary event in resistance to fungal parasites. *Nature* **249**:186–187.

Manners, J. G., and A. Myers. (1975) The effect of fungi (particularly obligate pathogens) on the physiology of higher plants. *Symp. Soc. Exp. Biol.* **29**:279–296.

Marcan, H., M. G. Jarvis, and J. Friend. (1979) Effect of methyl glycosides and oligosaccharides on cell death and browning of potato tuber discs induced by mycelial components of *Phytophthora infestans*. *Physiol. Plant Pathol.* **14**:1–9.

Matile, P. H. (1975) The lytic component of plant cells. In *Cell Biology Monographs*, Vol. 1 (M. Alfert et al., eds.). Springer-Verlag, New York.

McBride, R. P. (1972) Larch leaf waxes utilized by *Sporobolomyces roseus in situ*. *Trans. Br. Mycol. Soc.* **58**:329–331.

McKeen, W. E. (1974) Mode of penetration of epidermal cell walls of *Vicia faba* by *Botrytis cinerea*. *Phytopathology* **64**:461–467.

McKeen, W. E. (1977) Growth of *Pythium graminicola* in barley roots. *Can. J. Bot.* **55**:44–47.

McNabb, H. S., H. M. Heybroek, and W. L. McDonald. (1970) Anatomical factors in resistance to Dutch elm disease. *Neth. J. Plant Pathol.* **76**:196–204.

Mercer, P. C., R. K. S. Wood, and A. D. Greenwood. (1974) Resistance to anthracnose of French bean. *Physiol. Plant Pathol.* **4**:291–306.

Mercer, P. C., R. K. S. Wood, and A. D. Greenwood. (1975) Ultrastructure of the parasitism of *Phaseolus vulgaris* by *Colletotrichum lindemuthianum. Physiol. Plant Pathol.* **5**:203–214.

Miller, R. J., and D. E. Koeppe. (1971) Southern corn leaf blight: Susceptible and resistant mitochondria. *Science* **173**:67–69.

Mollenhauer, H. H., and D. L. Hopkins. (1976) Xylem morphology of Pierce's disease-infected grapevines with different levels of tolerance. *Physiol. Plant Pathol.* **9**:95–100.

Moreau, M., A. M. Catesson, M. Péresse, and Y. Czaninski. (1978) Dynamique comparée des réactions cytologiques du xylème de l'Oeillet en présence de parasites vasculaires. *Phytopathol. Z.* **91**:289–306.

Mueller, W. C., and C. H. Beckman. (1976) Ultrastructure and development of phenolic-storing cells in cotton roots. *Can. J. Bot.* **54**:2074–2082.

Mueller, W. C., and C. H. Beckman. (1978) Ultrastructural localization of polyphenoloxidases and peroxidase in roots and hypocotyls of cotton seedlings. *Can. J. Bot.* **56**:1579–1587.

Mussell, H., and L. Strand. (1976) Pectic enzymes: Involvement in pathogenesis and possible relevance to tolerance and specificity. In *Cell Wall Biochemistry Related to Specificity in Host–Plant Pathogen Interactions* (B. Solheim and J. Raa, eds.). Universitetsforlaget, Tromsø, Norway, PP. 31–70.

Nicholson, R. L., J. Kuć, and E. B. Williams. (1972) Histochemical demonstration of transitory esterase activity in *Venturia inaequalis. Phytopathology* **62**:1242–1247.

Oullette, G. B. (1978) Fine structural observations on substances attributable to *Ceratocystis ulmi* in American elm and aspects of host cell disturbances. *Can. J. Bot.* **56**:2550–2566.

Park, P., M. Fukutomi, S. Aki, and S. Nishimura. (1976) Effect of the host-specific toxin from *Alternaria kikuchiana* on the ultrastructure of plasma membranes of cells in leaves of Japanese pear. *Physiol. Plant Pathol.* **9**:167–174.

Paus, F., and J. Raa. (1973) An electron microscope study of infection and disease development in cucumber hypocotyls inoculated with *Cladosporium cucumerinum. Physiol. Plant Pathol.* **3**:461–464.

Paxton, J., D. J. Goodchild, and I. A. M. Cruickshank. (1974) Phaseollin production by live bean endocarp. *Physiol. Plant Pathol.* **4**:167–171.

Peyton, G. A., and C. C. Bowen. (1963) The host–parasite interface of *Peronospora manshurica* on *Glycine max. Am. J. Bot.* **50**:789–797.

Politis, D. (1976) Ultrastructure of penetration by *Colletotrichum graminicola* on highly resistant oat leaves. *Physiol. Plant Pathol.* **8**:117–122.

Politis, D. J., and H. Wheeler. (1973) Ultrastructural study of penetration of maize leaves by *Colletotrichum graminicola. Physiol. Plant Pathol.* **3**:465–471.

Pring, R. J., and D. V. Richmond. (1976) An ultrastructural study of the effect of oxycarboxin on *Uromyces phaseoli* infecting leaves of *Phaseolus vulgaris. Physiol. Plant Pathol.* **8**:155–162.

Ragetli, H. W., M. Weintraub, and E. Lo. (1970) Degeneration of leaf cells, resulting from starvation after excision. I. Electron microscope observations. *Can. J. Bot.* **48**:1913–1922.

Ride, J. P., and R. B. Pearce. (1979) Lignification and papilla formation at sites of attempted penetration of wheat leaves by non-pathogenic fungi. *Physiol. Plant Pathol.* **15**:79–92.

Robb, J., L. Busch, and B. C. Lu. (1975a) Ultrastructure of wilt syndrome caused by *Verticillium dahliae*. I. In chrysanthemum leaves. *Can. J. Bot.* **53**:901–903.

Robb, J., A. E. Harvey, and M. Shaw. (1975b) Ultrastructure of tissue cultures of *Pinus monticola* infected by *Cronartium ribicola*. I. Prepenetration host changes. *Physiol. Plant Pathol.* **5**:1–8.

Robb, J., J. D. Brisson, L, Busch, and B. C. Lu. (1979) Ultrastructure of wilt syndrome caused by *Verticillium dahliae*. VII. Correlated light and transmission electron microscope identification of vessel coatings and tyloses. *Can. J. Bot.* **57**:822–834.

Roebuck, P., R. Sexton, and J. W. Mansfield. (1978) Ultrastructural observations on the development of the hypersensitive reaction in leaves of *Phaseolus vulgaris* cv. Red Mexican inoculated with *Pseudomonas phaseolicola* (race 1). *Physiol. Plant Pathol.* **12**:151–157.

Royle, D. J. (1975) Structural features of resistance to plant diseases. In *Biochemical Aspects of Plant–Parasite Relationships* (J. Friend and D. R. Threlfall, eds.). Academic Press, London. Pp. 161–163.

Rudolph, K. (1976) Non-specific toxins. In *Encyclopedia of Plant Physiology*, Vol. 4 (R. Heitefuss and P. H. Williams, eds.). Springer-Verlag, New York. Pp. 270–315.

Sargent, C., and J. L. Gay. (1977) Barley epidermal apoplast structure and modification by powdery mildew contact. *Physiol. Plant Pathol.* **11**:195–205.

Scheffer, R. P. (1976) Host-specific toxins in relation to pathogenesis and disease resistance. In *Encyclopedia of Plant Physiology*, Vol. 4 (R. Heitefuss and P. H. Williams, eds.). Springer-Verlag, New York. Pp. 247–269.

Sequeira, L. (1978) Lectins and their role in host–pathogen specificity. *Annu. Rev. Phytopathol.* **16**:453–481.

Sequeira, L., G. Gaard, and G. A. De Zoeten. (1977) Interaction of bacteria and host cell walls: Its relation to mechanism of induced resistance. *Physiol. Plant Pathol.* **10**:43–50.

Shimony, C., and J. Friend. (1975) Ultrastructure of the interaction between *Phytophthora infestans* and leaves of two cvs. of potato (*Solanum tuberosum* L.) Orion and Majestic. *New Phytol.* **74**:59–65.

Shimony, C., and J. Friend. (1977) The ultrastructure of the interaction between *Phytophthora infestans* (Mont.) de Bary and tuber discs of potato (*Solanum tuberosum* L.) cv. King Edward. *Physiol. Plant Pathol.* **11**:243–249.

Sing. V. O., and M. N. Schroth. (1977) Bacteria–plant cell surface interaction: Active immobilization of saprophytic bacteria in plant leaves. *Science* **197**:759–761.

Staub, T., H. Dahmen, and F. J. Schwinn. (1974) Light- and scanning-electron microscopy of cucumber and barley powdery mildew on host and non-host plants. *Phytopathology* **64**:364–372.

Steele, J. A., T. F. Uchytil, R. D. Durbin, P. Bhatnagar, and D. H. Rich. (1976) Chloroplast coupling factor 1: A species-specific receptor for tentoxin. *Proc. Natl. Acad. Sci. USA* **73**:2245–2283.

Strobel, G. A., and W. M. Hess. (1968) Biological activity of a phytotoxic glycopeptide produced by *Corynebacterium sepedonicum. Plant Physiol.* **43**:1673–1688.

Strobel, G. A., W. M. Hess, and G. W. Steiner. (1972) Ultrastructure of cells in toxin-treated and *Helminthosporium sacchari*-infected sugarcane leaves. *Phytopathology* **62**:339–345.

Teakle, D. S., J. M. Appleton, and D. R. L. Steindl. (1978) An anatomical basis for resistance of sugar-cane to ratoon stunting diseases. *Physiol. Plant Pathol.* **12**:83–91.

Van der Molen, G. E., C. H. Beckman, and E. Rodehorst. (1977) Vascular gelation: A general response phenomenon following infection. *Physiol. Plant Pathol.* **11**:95–100.

Venn, K. O., V. M. G. Nair, and J. E. Kuntz. (1968) Effects of TCPA on oak sapwood formation and the incidence and development of oak wilt. (Abstr.) *Phytopathology* **58**:1071.

Wallis, F. M. (1977) Ultrastructural histopathology of tomato plants infected with *Corynebacterium michiganense. Physiol. Plant Pathol.* **11**:333–342.

Wehtje, G., L. J. Littlefield, and D. E. Zimmer. (1979) Ultrastructure of compatible and incompatible reactions of sunflower to *Plasmopara halstedii. Can. J. Bot.* **57**:315–323.

Weinhold, A. R., and J. Motta. (1973) Initial host responses in cotton infection by *Rhizoctonia solani. Phytopathology* **63**:157–162.

Wheeler, H. (1974) Cell wall and plasmalemma modifications in diseased and injured plant tissues. *Can. J. Bot.* **52**:1005–1009.

Wheeler, H. (1975) The role of phytotoxins in specificity. In *Specificity in Plant Diseases* (R. K. S. Wood and A. Graniti, eds.). Plenum Press, London. Pp. 217–235.

Wheeler, H. (1976) Permeability alterations in diseased plants. In *Encyclopedia of Plant Physiology,* Vol. 4 (R. Heitefuss and P. H. Williams, eds.). Springer-Verlag, New York. Pp. 413–429.

Wheeler, H. (1977) Ultrastructure of penetration by *Helminthosporium maydis. Physiol. Plant Pathol.* **11**:171–178.

White, J. A., O. H. Calvert, and M. F. Brown. (1973) Ultrastructural changes in corn leaves inoculated with *Helminthosporium maydis* race T. *Phytopathology* **63**:296–300.

Wilson, V. E., and M. D. Coffey. (1978) Cytological observations on field resistance to potato blight. *Ann. Appl. Biol.* **89**:298–302.

Wynn, W. K. (1976) Appressorium formation over stomates by the bean rust fungus: Response to a surface contact stimulus. *Phytopathology* **66**:136–146.

Yang, S. L., and A. H. Ellingboe. (1972) Cuticle layer as a determining factor for the formation of mature appressoria of *Erysiphe graminis* on wheat and barley. *Phytopathology* **62**:708–714.

Yoder, O. C. (1972) Host-specific toxins as determinants of successful colonization by fungi. In *Phytotoxins in Plant Diseases* (R. K. S. Wood, A. Ballio, and A. Graniti, eds.). Academic Press, London. Pp. 457–463.

Yoder, O. C. (1976) Evaluation of the role of *Helminthosporium maydis* race T in southern corn leaf blight. In *Biochemistry and Cytology of Plant–Parasite Interaction* (K. Tomiyama, J. M. Daly, I. Uritani, H. Oku, and S. Ouchi, eds.). Elsevier, New York. Pp. 16–24.

Zeyen, R. J., and W. R. Bushnell. (1979) Papilla response of barley epidermal cells caused by *Erysiphe graminis:* Rate and method of deposition determined by microcinematography and transmission electron microscopy. *Can. J. Bot.* **57**:898–913.

INDUCTION OF HOST PHYSICAL RESPONSES TO BACTERIAL INFECTION: A RECOGNITION PHENOMENON

Luis Sequeira

Plants, in the course of evolution, have developed mechanisms to prevent multiplication of the numerous microorganisms that gain access to the internal tissues. The presence of large numbers of natural openings, such as stomata, allows microorganisms ready access to internal plant tissues. Because of their motility and small size, bacteria frequently invade the substomatal chambers when there is sufficient free water to provide a continuous avenue through the stomatal opening. Although the intercellular fluid is relatively rich in nutrients (Klement, 1965), relatively few bacteria are able to multiply extensively in the intercellular spaces. This constitutes a general form of resistance, which is extremely important for survival of the plant species, but which is poorly understood. It depends on morphological and biochemical changes that are induced once the potential pathogen comes in contact with the host cell wall. The induction of these responses, therefore, may depend on specific recognition of the surface molecules that interact.

For several years we have been studying the interaction of strains of *Pseudomonas solanacearum,* an important wilt-inducing pathogen, and mesophyll cells in tobacco and potato leaves. The evidence we have accumulated has led to the hypothesis that recognition between host and parasite is effected by a three-membered system that includes (1) the lipopolysaccharide (LPS) component of the outer bacterial membrane, (2) the extracellular polysaccharide (EPS) that coats the bacterial cell, and (3) a host cell wall lectin that binds specifically to certain saccharides of the bacterial LPS or EPS. This chapter reviews the evidence that has led to this concept for recognition of *P. solanacearum* in potato and tobacco, and presents more recent information that suggests that the phenomenon may be of general occurrence in other bacteria–plant interactions. Also the potential usefulness of our understanding of recognition in the design of novel methods of disease management is discussed.

RESPONSES OF MESOPHYLL CELLS TO BACTERIAL INFECTION

It has been established that infiltration of tobacco leaves with high concentrations of heterologous bacteria, that is, bacteria pathogenic to other plant species, results in a hypersensitive response (HR) (Klement and Goodman, 1968). This reaction is characterized by rapid collapse of the host cell, reduction in bacterial populations, and containment of the pathogen in the infiltrated area. In contrast, homologous bacteria, that is, bacteria pathogenic to tobacco, multiply rapidly in the intercellular spaces, but there is no overt response of the leaf tissues until 48 hr later, when the tissues become necrotic and the bacterium invades the adjoining tissues.

Examination of the relationship between homologous or heterologous bacteria and the tobacco mesophyll cell wall reveals marked differences in the host response. Only the heterologous bacteria become closely appressed to the host cell wall; within 2–3 hr after infiltration, the bacteria are surrounded by a thin, fibrillar pellicle, presumably of cuticular origin, that separates from the host cell wall surface. By 7 hr after infiltration, at a time when the HR is irreversible, the bacteria become surrounded by granular and fibrillar materials that are electron dense. The bacteria become attached and then enveloped, singly or in groups. Bacteria become entrapped in the interstices between host cells, and it is in these regions that the host response is most marked. However, envelopment occurs throughout the surface of the cell and, in cross-section, the sites of bacterial envelopment appear as small blisters (Sequeira et al., 1977). By means of scanning electron microscopy, Huang and Van Dyke (1978) have shown that the plant cell wall often becomes depressed at the site of bacterial attachment, and that the pellicle develops from projections at the edge of the depression.

The material that surrounds the bacteria appears to be of host origin. At the site of bacterial attachment, and extending for a considerable distance on either side, the host plasmalemma becomes invaginated, and numerous vesicles accumulate on the inner side of the cell wall. It is thought that the material that surrounds the bacteria is released from these vesicles.

In contrast with the reaction of the host to heterologous bacteria, the homologous bacteria are not attached or enveloped. They multiply freely in the intercellular fluid and cause partial disintegration of the cell wall. There is no apparent change in host cell organelles during the first 24 hr after infiltration. The cell wall breakdown presumably is the result of the activity of pectolytic and cellulolytic enzymes that are released by many plant pathogenic bacteria.

The envelopment responses that we have described are typical of tobacco leaves infiltrated with *P. pisi* (Goodman et al., 1976), avirulent or heterologous strains of *P. solanacearum* (Sequeira et al., 1977), heat-killed bacteria (Sequeira et al., 1977), or saprophytic bacteria (Sequeira et al., 1977; Goodman et al., 1976). The envelopment process per se does not lead to the HR, but there is considerable evidence that very close contact of the bacterium and the cell wall is necessary for induction of the HR (Stall and Cook, 1979).

The attachment and envelopment of bacteria may be a general defense reaction that rids plants of unwanted invaders. What evidence is there that attachment and envelopment are of common occurrence in plants other than tobacco? We have observed the same phenomenon in potato leaves infiltrated with avirulent strains of *P. solanacearum,* in cucumber leaves inoculated with saprophytic bacteria, and in soybean leaves inoculated with avirulent strains of *Xanthomonas phaseoli* var. *sojensis* (Fett and Sequeira, unpublished). Victoria (1977) reported the same phenomenon in corn leaves infiltrated with strains of *Erwinia carotovora.* Sing and Schroth (1977) reported envelopment of saprophytic bacteria in bean leaves, and, more recently, envelopment of *P. syringae* in bean pod tissues has been described (Daub and Hagedorn, 1979). Therefore, there is little question that attachment and envelopment of bacteria occur in many different plant species. That the phenomenon may be involved in specific resistance was reported by Cason et al. (1978), who observed attachment of *X. malvacearum* only in leaves of cotton varieties that carry specific genes for resistance.

The significance of the attachment and envelopment process has been questioned by M. N. Schroth and collaborators, who had initially reported that saprophytic bacteria were entrapped on cell surfaces of bean leaves (Sing and Schroth, 1977). A reexamination of this phenomenon led to the conclusion that the entrapment is an artifact caused by loosening of the microfibrillar elements; when the tissues dry out, bacteria or inert materials are trapped behind the fibrillar material. Homologous pathogens reverse this process because they cause water to flow into the site of entrapment (Alosi et al., 1978). Although there is little question that bacteria become trapped in the interstices between mesophyll cells as the water film dries out, and some loosening of the cuticular film of these cells may occur upon infiltration, the entrapment phenomenon is general throughout the exposed surfaces of mesophyll cells and is an active process, accompanied by extensive morphological and physiological changes in the cellular organelles. Also, we could not detect entrapment of inert materials, such as polystyrene balls (0.1 μm diam). In addition, strains of race 2 of *P. solanacearum,* which produce copious amounts of extracellular polysaccharide and thus maintain the leaf tissues watersoaked for several hours, are rapidly attached and enveloped.

The system we have described involves the attachment of heterologous bacteria to plant cell walls. It is evident that with certain bacteria, it is the homologous, rather than the heterologous, strains that attach to cell walls. With *Agrobacterium tumefaciens* and *Rhizobium* spp., there is considerable evidence that attachment to the host cell wall is a definite prerequisite for infection by the compatible forms (Lippincott and Lippincott, 1969; Bogers,

1972; Dazzo, 1978). Because these two genera are closely related and cause growth alterations of the host tissues, it seems possible that in the course of evolution, certain strains have adapted to, and taken advantage of, the attachment process (Sequeira, 1978). Unlike other bacteria, both crown gall and root nodule bacteria are capable of rapid multiplication after attachment to the host cell wall.

THE MOLECULAR BASIS FOR HOST–PATHOGEN RECOGNITION

The response of tobacco mesophyll cells to homologous and heterologous bacteria must be based on a recognition phenomenon; only certain bacteria are attached and enveloped, and induce the HR. In many other biological systems, recognition is based on the complementary interaction of surface macromolecules, generally involving a carbohydrate containing molecule in one member, and a protein capable of recognizing specific carbohydrate moieties (a lectin) in the other. We have investigated the nature of the macromolecules that interact in the case of *P. solanacearum* infiltrated into tobacco or potato leaf tissues.

Pathogen Components Involved in Recognition

The outer membrane of gram-negative bacteria contains lipopolysaccharide (LPS), which is composed of a complex heteropolysaccharide and a covalently bound lipid A (Lüderitz et al., 1971). In addition, all plant pathogenic bacteria produce extracellular polysaccharide (EPS), which is composed of simple sugars and uronic acids linked by O-glycosidic bonds to form complex oligosaccharides. The LPS may play the major role in the recognition phenomenon that follows the interaction of the bacterium with the host cell wall. Although EPS appears to play an important role in modifying or altering this phenomenon under certain conditions, there are no structural differences in the EPS of smooth strains of *P. solanacearum* that differ in their ability to induce the HR in tobacco. Lack of EPS, however, appears to be important in allowing rapid recognition, attachment, and immobilization of many rough (B-1) forms of the bacterium.

The attachment of bacteria to plant cell walls is mediated by LPS components. Differences in HR-inducing abilities of different strains of *P. solanacearum* can be correlated with differences in LPS composition. Whatley et al. (1979) have shown by several methods that there are clear differences in LPS composition that correlate directly with the ability of the host to respond to the pathogen. The major variations among 11 avirulent strains of *P. solanacearum* were in the relative amounts of xylose, rhamnose, and glucose. The LPS of strains that caused the HR had low xylose/glucose and rhamnose/glucose ratios; the reverse was true for most strains that did not cause the HR. When the LPS of all strains was separated by SDS polyacrylamide gel electrophoresis, distinct differences in apparent molecular weight became evident. The HR-inducing strains had a fast-moving band close to the dye front; those of the non-HR-inducing strains (with one exception) had a major slow-moving band.

Further evidence for the structural differences in the LPS of HR-inducing and -noninducing strains of *P. solanacearum* was provided by Hendrick et al. (1979). A bacteriophage (CH 154), isolated from a lysogenic strain of P. *solanacearum,* was capable of lysing smooth isolates (such as K60), but not the rough variants (such as B-1). The phage was inactivated by LPS from the smooth strains, but not by that from the rough strains. The evidence suggests that part of the O-specific antigen of LPS is necessary for attachment of the phage; mutations that alter the length of this chain apparently result in loss of these phage-attachment sites, but also uncover sites for attachment to the tobacco cell wall and/or induction of the HR. At present, the phage is being used as a probe to examine the changes in LPS structure that are essential for induction of the HR.

These results indicate that rough LPS is an important component of the recognition system that results in induction of the HR. In most of the rough strains that we have studied, EPS is not produced, or is produced in very small amounts. Although EPS is essential for pathogenesis, because loss of EPS is invariably accompanied by loss of virulence, it does not appear to play a primary role in recognition, contrary to our original hypothesis (Sequeira and Graham, 1977). First, most of the strains of race 2 of *P. solanacearum,* which are pathogenic to banana but not to tobacco, produce copious amounts of EPS but induce the HR in tobacco leaves. Examination of the LPS of several of these strains indicates that it is of the "rough" type (Cantrell, unpublished), thus confirming that the correlation between this type of LPS and induction of the HR holds for fluidal strains as well as for the rough variants. Second, numerous attempts to inhibit induction of the HR by adding a wide range of concentrations of purified EPS before, during, or after infiltration with strain B-1 were unsuccessful.

Host Components Involved in Recognition

Considerable evidence has accumulated on the role of carbohydrate-binding proteins (lectins) in recognition systems in animals (Hughes, 1975). These lectins generally interact with glycoprotein constituents of the outer membrane of animal cells. In plants, the presence of a massive cell wall prevents direct contact of bacteria with the host plasmalemma, at least during the initial stages of the interaction. The host cell wall, however, contains proteins, and some of these have been shown to have lectin activity (Lamport, 1979). A chitin-binding lectin, for example, can be extracted from tobacco leaves by the simple expedient of infiltrating the leaves with saline and centrifuging the leaf tissues to recover the solution. Because cytoplasmic enzymes cannot be detected in this solution, and lectin constitutes the major portion of the protein, it is unlikely that there is significant contamination by cytoplasmic components (Graham and Sequeira, unpublished). This lectin is very similar to potato lectin and has affinity for internal *N*-acetylglucosamine groups, which are components of fungal as well as bacterial cell walls.

The hapten specificity of potato and tobacco lectins suggested that they night be involved in specific recognition of avirulent, HR-inducing strains of *P. solanacearum*. This hypothesis requires that such lectins be located at or near the host cell wall. Preliminary results with fluorescent antibody techniques suggest that these lectins are present on the cell walls of potato and tobacco stem tissues. Because antibodies specific for purified lectin were isolated by means of affinity chromatography before they were used to treat tissue sections, there is little question that the lectin, or its antigenic constituents, are present on cell walls (Leach et al., 1978). Potato lectin, however, contains approximately 50% carbohydrate, made up mostly of arabinogalactans that are common components of plant cell walls (Lamport, 1979). If these carbohydrates are effective antigens, fluorescent antibody techniques may not be appropriate for localization of potato lectin, because fluorescence would be obtained wherever arabinogalactans are present. This question cannot be resolved until the antigenic properties of both protein and carbohydrate components of the lectin can be examined separately.

That a cell wall lectin might be involved in recognition of avirulent forms of *P. solanacearum* was first proposed by Sequeira and Graham (1977) based on results of agglutination assays of a wide range of strains exposed to potato lectin. All avirulent isolates agglutinated strongly with the lectin; all virulent isolates either failed to agglutinate or agglutinated only weakly. The avirulent forms formed typical rough colonies on ordinary media and were slimeless. Because addition of EPS prevented their agglutination, it was thought that EPS played an important role in preventing attachment of virulent cells to host cell walls, thus allowing them to multiple freely in the intercellular spaces.

Both LPS and EPS from *P. solanacearum* bind to potato lectin *in vitro* and a simple assay was devised to study the binding characteristics of these substances. In this assay lectin was immobilized on cellulose nitrate filter membranes without apparent loss of activity; neither LPS nor EPS were retained by the filter. Thus the amount of radioactively labeled ligand retained by lectin on the filter could be used as a measure of the affinity of the lectin for the ligand (Duvick et al., 1979). Although EPS binds to the lectin rapidly and quantitatively, it does so only as relatively low ionic strength; in contrast, binding of LPS to the lectin is unaffected by the ionic strength of the buffer, except at very high molarity (Duvick, unpublished). These facts suggest that, at the normal ionic strength in the intercellular fluids of tobacco leaves, binding of EPS to the lectin would be very weak. These findings explain our inability to prevent the HR reaction by adding EPS to leaves infiltrated with avirulent forms of *P. solanacearum*. In contrast, LPS readily inhibits the HR under the same conditions (Graham et al., 1977).

The interactions of rough and smooth LPS with potato lectin are of interest. In general, rough LPS binds much more strongly than smooth LPS, particularly in the undissociated form. In the dissociated form (after treatment with sodium deoxycholate), the differences in binding of the two types of LPS to lectin were not as striking, but still were about fourfold. These differences may be more significant *in vivo* because most of the *in vitro* binding of the smooth LPS can be accounted for by the binding of lectin to the lipid A, a component of LPS which may not be readily accessible in its normal location within the outer membrane of the bacterium.

Although the data that we have discussed above support the concept that a cell wall lectin acts as the receptor for bacterial LPS, an important inconsistency should be mentioned. The binding of lectin and LPS is hapten-reversible; chitin oligomers effectively inhibit agglutination of bacteria or cells or binding of LPS by lectin *in vitro*. When chitin oligomers were infiltrated into tobacco leaves, at a wide range of concentrations, induction of the HR by B-1 cells was not prevented. We have no explanation for this phenomenon, but we are exploring the possibility that chitin oligomers are degraded by plant enzymes.

That the system we have described for the tobacco *P. solanacearum* may be of general significance in bacterial interactions with plants is illustrated by recent work with a lectin from corn seed (Woods et al., 1979). This lectin strongly agglutinated 16 of 25 strains of soft-rotting *Erwinia* spp. that were avirulent or of low virulence, but agglutinated only two of 17 highly virulent strains. Of 21 strains of *Erwinia stewartii*, all slimeless, avirulent strains were strongly agglutinated by corn lectin, and all virulent, slime-forming strains were not agglutinated (A. Woods, unpublished information). With the latter pathogen, the pattern appears to be similar to that of *P. solanacearum* in tobacco.

BACTERIAL LIPOPOLYSACCHARIDES AS INDUCERS OF DISEASE RESISTANCE

It has been known for some time that the introduction of living (Novacky et al., 1973), heat-killed (Lozano and Sequeira, 1970), or saprophytic (Klement and Goodman, 1968) bacteria into the intercellular spaces of tobacco leaves results in the induction of disease resistance. Unlike the HR, this resistance is nonspecific, light-dependent, and, under some conditions, becomes systemic throughout the plant (Sequeira, 1979). Efforts have been made to determine the nature of the inducer and of the mechanism of resistance.

Evidence from work with *P. solanacearum* and *E. rubrifaciens* indicates that lipopolysaccharides or protein–lipopolysaccharide complexes are the inducers of resistance in tobacco (Graham et al., 1977; Mazzucchi et al.,

1979). In the case of *P. solanacearum* chemical modifications of LPS from both smooth and rough forms suggest that both acylation of lipid A and intact core–lipid A linkage are required for inducer activity. Deacylation of LPS by mild alkaline saponification, which releases fatty acids from lipid A, caused a total loss of activity. Similarly, mild acid hydrolysis, which breaks the linkage between lipid A and the core oligosaccharide of the LPS, resulted in total loss of inducer activity (Graham et al., 1977). The loss of activity may have been associated with the highly insoluble nature of the intact lipid A portion.

It has been suggested that systemic resistance, like the HR, results from a recognition system in which cell wall lectins are involved (Sequeira, 1979). The lipid A (solubilized by conjugation with bovine serum albumin) binds to potato lectin (Sequeira and Graham, 1977) and intact LPS becomes tightly associated with the host cell wall (Graham et al., 1977). The evidence for lectin involvement, however, is not strong. The requirement that fatty acids remain intact in lipid A parallels the activity of LPS (endotoxin) in animals (Nowotny, 1969) and suggests that interactions of lipid A with cellular membranes are important determinants of the resistant response.

Although antibacterial compounds appear in the intercellular fluids of tobacco leaves at about the same time that disease resistance becomes evident (Rathmell and Sequeira, 1975), the mechanism by which bacterial LPS induces these changes is not known. Specific peroxidase changes are associated with the binding of bacteria, or LPS, to the host cell wall (Nadolny and Sequeira, 1980), but such changes are not directly associated with the resistant response.

GENERAL CONCLUSIONS

1. Knowledge of the molecular basis of the recognition phenomena that lead to resistant responses in plants should be very useful in the design of alternative means of disease control. The use of compounds that elicit the plant's own defense response, theoretically, should be the most effective means for biological control of plant diseases.
2. Induction of the HR, which appears to depend on the interaction of specific saccharides on the bacterial LPS with cell wall lectins, may not be the most desirable goal for the control of bacterial diseases. The localized nature of the response, and the collapse of host cells that results, militate against the use of this potential method of disease control.
3. The finding that the lipid A portion of LPS, in soluble form, is an effective inducer of systemic, nonspecific resistance suggests a possible approach to practical applications. Since this segment of LPS is formed of repeating units of acylated glucosamine moieties linked by phosphodiester bonds or O-glycosidic linkages (Lüderitz et al., 1971), it should be possible, by careful enzymatic digestion of this component, to isolate fragments that

are biologically active, yet of sufficiently small molecular weight to be taken up by plant leaves or roots.

4. The main constraints to progress in this approach to disease control are (a) the lack of adequate, quantitative assays for induction of resistance, (b) the limited knowledge of the cell wall structure of bacterial plant pathogens, (c) the lack of adequate proof of the nature and location of the receptors for the inducers, and (d) our complete ignorance as to the transducer system that carries the information from the receptor to the active site where the resistance mechanism is activated. Although progress has been made in these different areas, information is not being generated at a rate fast enough to insure that practical applications will be made within the foreseeable future.

REFERENCES

Alosi, M. C., D. C. Hildebrand, and M. N. Schroth. (1978) Entrapment of bacteria in bean leaves by films formed at air–water interfaces. (Abstr.) *Phytopathol. News* **12**:197.

Bogers, R. J. (1972) On the interaction of *Agrobacterium tumefaciens* with cells of *Kalanchoe daigremontiana.* In *Proceedings of the Third International Conference on Plant Pathogenic Bacteria* (H. P. Maas Geesteranus, ed.) Wageningen, The Netherlands. Pp. 239–250.

Cason, E. T., P. E. Richardson, M. K. Essenberg, L. A. Brinkerhoff, W. M. Johnson, and R. J. Venere. (1978) Ultrastructural cell wall alterations in immune cotton leaves inoculated with *Xanthomonas malvacearum. Phytopathology* **68**:1015–1021.

Daub, M., and D. Hagedorn. (1979) Resistance of Phaseolus line WBR 133 to *Pseudomonas syringae. Phytopathology* **69**:946–951.

Dazzo, F. B. (1978) Adsorption of microorganisms to roots and other plant surfaces. In *Adsorption of Microorganisms to Surfaces* (G. Bitton and K. C. Marshall, eds.) Wiley, New York.

Duvick, J. P., L. Sequeira, and T. L. Graham. (1979) Binding of *Pseudomonas solanacearum* surface polysaccharides to plant lectin *in vitro*. (Abstr. #742) *Plant Physiol.* **63**(suppl.):134.

Goodman, R. N., R. Y. Huang, and J. A. White. (1976) Ultrastructural evidence for immobilization of an incompatible bacterium, *Pseudomonas pisi,* in tobacco leaf tissue. *Phytopathology* **66**:754–764.

Graham, T. L., L. Sequeira, and T. R. Huang. (1977) Bacterial lipopolysaccharides as inducers of disease resistance in tobacco. *Appl. Environ. Microbiol.* **34**:424–432.

Hendrick, C. A., M. H. Whatley, N. Hunter, M. A. Cantrell, and L. Sequeira. (1979) The hypersensitive response in tobacco: A phage capable of differentiating HR- and non-HR-inducing *Pseudomonas solanacearum*. (Abstr. #741) *Plant Physiol.* **63**(suppl.):134.

Huang, J-S., and C. G. Van Dyke. (1978) Interaction of tobacco callus tissue with

Pseudomonas tabaci, P. pisi, and *P. fluorescens. Physiol. Plant Pathol.* **13**:65–72.

Hughes, C. (1975) Cell surface membranes of animal cells as the sites of recognition of infectious agents and other substances. In *Specificity in Plant Diseases* (R. K. S. Wood and A. Graniti, eds.). Plenum Press, New York. Pp. 77–99.

Klement, Z. (1965) Method of obtaining fluid from the intercellular spaces of foliage and the fluid's merit as a substrate for phytobacterial pathogens. *Phytopathology* **55**:1033–1034.

Klement, Z., and R. N. Goodman. (1968) The hypersensitive reaction to infection by bacterial plant pathogens. *Annu. Rev. Phytopathol.* **5**:17–44.

Lamport, D. T. A. (1979) Structure and function of plant glycoproteins. In *Biochemistry of Plants,* Vol. 3 (J. Preiss, ed.) Academic Press, New York.

Leach, J. E., M. A. Cantrell, and L. Sequeira. (1978) Localization of potato lectin by means of fluorescent antibody techniques. (Abstr.) *Phytopathol. News* **12**:197.

Lippincott, B. B., and J. A. Lippincott. (1969) Bacterial attachment to a specific wound site as an essential stage in tumor initiation by *Agrobacterium tumefaciens. J. Bacteriol.* **97**:620–628.

Lozano, J. C., and L. Sequeira. (1970) Prevention of the hypersensitive reaction in tobacco leaves by heat-killed cells of *Pseudomonas solanacearum. Phytopathology* **60**:875–879.

Lüderitz, O., O. Westphal, A. M. Staub, and H. Nikaido. (1971) Isolation and chemical and immunological characterization of bacterial lipopolysaccharides. In *Microbial Toxins,* Vol. 4 (G. Weinbaum, S. Kadis, and S. J. Ajl, eds.). Academic Press, New York. Pp. 145–233.

Mazzucchi, U., C. Bazzi, and P. Pupillo. (1979) The inhibition of susceptible and hypersensitive reactions by protein–lipopolysaccharide complexes from phytopathogenic pseudomonads: Relationship to polysaccharide antigenic determinants. *Physiol. Plant Pathol.* **14**:19–30.

Nadolny, L., and L. Sequeira. (1980) Increases in peroxidase activities are not directly involved in induced resistance in tobacco. *Physiol. Plant Pathol.* **16**:1–8.

Novacky, A., G. Acedo, and R. N. Goodman. (1973) Prevention of bacterially induced hypersensitive reaction by living bacteria. *Physiol. Plant Pathol.* **3**:133–136.

Nowotny, A. (1969) Molecular aspects of endotoxic reactions. *Bacteriol. Rev.* **33**:72–98.

Rathmell, W. G., and L. Sequeira. (1975) Induced resistance in tobacco leaves: The role of inhibitors of bacterial growth in the intercellular fluid. *Physiol. Plant Pathol.* **5**:65–73.

Sequeira, L. (1978) Lectins and their role in host–pathogen specificity. *Annu. Rev. Phytopathol.* **16**:453–481.

Sequeira, L. (1979) The acquisition of systemic resistance by prior inoculation. In *Recognition and Specificity in Plant Host–Parasite Interactions* (J. M. Daly and I. Uritani, eds.). Japan Science Society Press, Tokyo. Pp. 231–251.

Sequeira, L., and T. L. Graham. (1977) Agglutination of avirulent strains of *Pseudomonas solanacearum* by potato lectin. *Physiol. Plant Pathol.* **11**:43–54.

Sequeira, L., G. Gaard, and G. A. de Zoeten. (1977) Attachment of bacteria to host cell walls: Its relation to mechanisms of induced resistance. *Physiol. Plant Pathol.* **10**:43–50.

Sing, V. O., and M. N. Schroth. (1977) Bacteria–plant cell surface interactions: Active immobilization of saprophytic bacteria in plant leaves. *Science* **197**:759–761.

Stall, R. E., and A. A. Cook. (1979) Evidence that bacterial contact with the plant cell is necessary for the hypersensitive reaction but not the susceptible reaction. *Physiol. Plant Pathol.* **14**:77–84.

Victoria, J. I. (1977) Resistance in corn (*Zea mays* L.) to bacterial stalk rot in relation to virulence of strains of *Erwinia chrysanthemi.* Ph.D. Thesis, University of Wisconsin, Madison, Wis. 179 pp.

Whatley, M. H., N. Hunter, M. A. Cantrell, K. Keegstra, and L. Sequeira. (1979) Bacterial lipopolysaccharide structure and the induction of the hypersensitive response in tobacco. (Abstr. #740) *Plant Physiol.* **63**(suppl.):134.

Woods, A., N. Hunter, L. Sequeira, and A. Kelman. (1979) Lectin activity isolated from corn seed. (Abstr. #739) *Plant Physiol.* **63**(suppl.):134.

EVALUATION OF THE ROLE OF PHYTOALEXINS

Noel T. Keen

Phytoalexin production is one of the conferral mechanisms of natural disease resistance in plants, but there is ambiguity concerning how widespread it is. Phytoalexins have been demonstrated in at least 100 plant species representing 21 families, but some efforts, including Deverall's (1977) and my own unpublished work with *Cucumis sativus,* failed to detect phytoalexins after challenge of plant tissue with microorganisms. This may indicate that such plants do not use phytoalexins as a defense mechanism or that we are simply missing them. I searched for phytoalexins in the classic flax–flax rust system, using traditional extraction techniques without initial success. However, upon devising a modified extraction procedure, two inducibly formed chemicals, coniferyl aldehyde (Figure 1) and coniferyl alcohol, were readily recovered from leaves inoculated with incompatible races of *Melampsora lini* (Keen and Littlefield, 1979). Several lines of evidence suggested that the chemicals were phytoalexins involved in cessation of fungus growth in resistant genotypes, but without the proper extraction technique they would not have been isolated. The previous suspicion by some that phytoalexins are not produced by members of the Gramineae has been disproven with the isolation and identification of the momilactones from rice (Cartwright et al., 1977), ρ-1,4-hydroxybenzoquinone and ρ-1,4-benzoquinone from barley (Evans and Pluck, 1978), and of the avenalumins from oats by Tani's group (1979).

Phytoalexin production has been associated with the expression of both general and specific disease resistance (Table 1 and see Keen and Bruegger, 1977). The large number of experimental approaches taken and the mounting number of plant taxa known to produce phytoalexins minimize the possibility that they are coincidentally produced at the same time as some elusive, unknown defense mechanism. Why would so many plants make large quan-

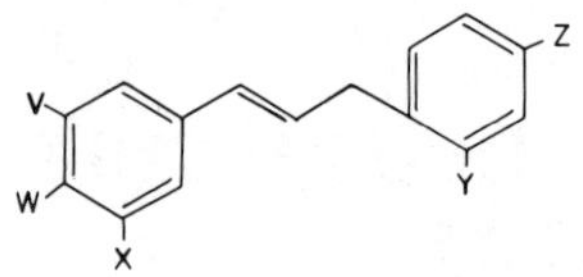
V
W
X
Y
Z

	V	W	X	Y	Z
Resveratrol	OH	H	OH	H	OH
4-prenylresveratrol	OH	Prenyl	OH	H	OH
Oxyresveratrol	OH	H	OH	OH	OH
4-prenyloxyresveratrol	OH	Prenyl	OH	OH	OH
Pterostilbene	OCH_3	H	OCH_3	H	OH

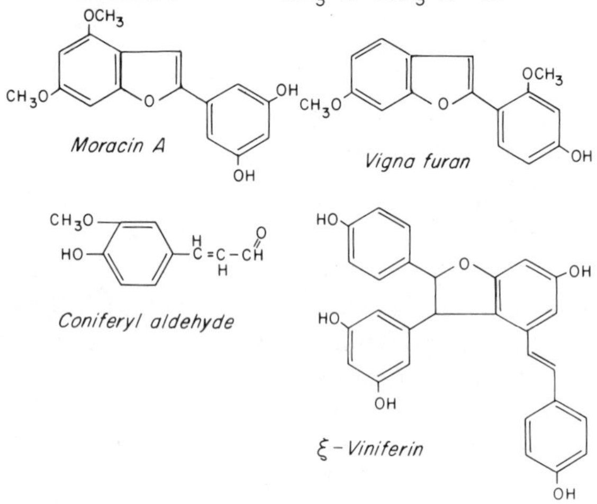
OCH3
CH3O
O
OH
OH
Moracin A
CH3O
O
OCH3
OH
Vigna furan
CH3O
HO
C=C-CH
H
H
O
Coniferyl aldehyde
HO
O
OH
HO
OH
OH
ξ - Viniferin

Figure 1 Recently described phytoalexins.

tities of relatively sophisticated antibiotic chemicals in response to parasite challenge for no reason? There are, however, a few examples of successful pathogens that appear to escape inhibition by plant phytoalexins. This is occasionally used as a detracting argument for phytoalexin-mediated defense, but, as will be discussed in a later section, there is nothing inexplicable, since analogous escape mechanisms are well known among certain pathogens of higher animals.

Additional tests of the relation between phytoalexin production and disease defense in a wider variety of plants would be desirable. One proposal is to select mutants of a normally sensitive pathogen that are resistant to a phytoalexin(s) *in vitro*. If resistant mutants could be isolated, they should produce a more susceptible host response than the wild-type pathogen. However, as will be discussed later, such mutants have not been obtained and there is good reason not to expect their ready isolation. A more plausible approach would be to select supersensitive pathogen strains and see if these gave a more resistant plant reaction. Another appealing genetic test of phytoalexin involvement with resistance would involve the selection of mu-

Table 1. A Summary of Some of the Data Linking Phytoalexins to the Expression of Disease Resistance in Plants[a]

1. Compatible pathogens are frequently tolerant of or chemically inactivate the phytoalexins of their host plant but closely related nonpathogens are not (Cruickshank, 1963)
2. Pretreatment of normally susceptible plants with agents such as UV light that elicit phytoalexin production render the plants resistant to subsequent inoculation (Bridge and Klarman, 1973)
3. Alterations in the environment that deleteriously affect phytoalexin production by the host with lesser or no effects on the parasite have been shown to decrease resistance (Bell and Presley, 1969; Murch and Paxton, 1977)
4. Metabolic inhibitors of various kinds and preinoculation heat treatments frequently block phytoalexin production and resistance (Chamberlain and Gerdemann, 1966; Yoshikawa et al., 1978); both resistance and phytoalexin production are regained by heat-treated plants after about 3 days
5. In the studied gene-for-gene systems, substantial phytoalexin production specifically occurs only with incompatible combinations of host and parasite (Keen and Littlefield, 1979; Yoshikawa et al., 1978)
6. Resistance alleles leading to rapid cessation of parasite growth give more rapid accumulation of phytoalexins in the infection court than alleles observed to restrict pathogen development more slowly (Keen and Littlefield, 1979)
7. Careful studies have established that highly toxic concentrations of phytoalexins are formed at precisely the time and cellular location in which development of an incompatible pathogen race ceases (Yoshikawa et al., 1978)
8. Coinoculation of compatible and incompatible races of a pathogen generally results in restricted development of both, concomitant with the accumulation of high levels of phytoalexins (Paxton and Chamberlain, 1967)
9. Application of purified phytoalexins to the infection site of a normally compatible plant pathogen results in incompatibility (Chamberlain and Paxton, 1968) and removal of phytoalexins results in increased compatibility (Klarman and Gerdemann, 1963)
10. Limited data suggest that specific elicitors obtained from various races of pathogens in gene-for-gene systems give the same specificity for phytoalexin production on incompatible and compatible plant genotypes as the living pathogens (Anderson, 1979; Bruegger and Keen, 1979; Keen, 1978)

[a] The references are not comprehensive, but exemplary only.

tant lines of a plant species that were unable to produce phytoalexins in response to a suitable challenge. The deficient mutants should be susceptible to all those organisms against which phytoalexins confer resistance in the wild-type plant. This has not yet been attempted, in part due to the technical problems of obtaining suitable haploid cell lines, screening for deficient mutants, and obtaining plant regeneration.

There are strong indications of phytoalexin involvement in the conferral of single gene-mediated resistance (Keen and Bruegger, 1977), but it is not yet clear whether phytoalexins are a causal mechanism for polygenic or "horizontal" resistance (Robinson, 1976; Van der Plank, 1975). However, as pointed out by Ellingboe (1975) and Nelson (1978), it is most plausible to consider this type of resistance the same as "vertical" resistance—that is, a collection of individual monogenes, each contributing small but additive effects toward restriction of pathogen development. If this reasoning is correct, the substantial precedent for phytoalexin association with single-gene resistance makes it likely that at least some of the genes in a "horizontal" resistance complex would also be expressed through phytoalexin production.

PHYTOALEXINS AND THE HYPERSENSITIVE REACTION

The hypersensitive reaction (HR) is a widely occurring mechanism of disease resistance. The only viable mechanistic explanations for conferral of hypersensitive resistance are inducibly formed chemical barriers (phytoalexins) and inducibly formed structural barriers. Former suggestions, such as permeability changes or nutritional deprivation due to host cell death, are at odds with considerable experimental data as general mechanisms. The evidence for involvement of induced structural barriers is more limited than that for phytoalexins, but some suggestive demonstrations have appeared (e.g., Asada et al., 1976; Vance and Sherwood, 1976). The two mechanisms are closely related, since the monomeric unit(s) contributing to the inducibly formed structural polymer(s) are analogous to traditional low molecular weight phytoalexins. In light of this, it would not be surprising to find examples of simultaneous accumulation of classic phytoalexins and induced barriers, as the work of Goodliffe and Heale (1978), Heale and Sharman (1977), and Garcia-Arenal and Sagasta (1977) has suggested. The current concept is that hypersensitive resistance frequently, if not always, involves phytoalexin production. Therefore manipulating the natural HR of plants to effect disease control probably involves manipulating phytoalexins.

The classic association of host cell death with the HR has been challenged (Kiraly et al., 1972; Mayama et al., 1975), but more recent work appears to suggest a functional role. Rahe and Arnold (1975) reported that small freeze-injury sites on hypocotyls of *Phaseolus vulgaris* seedlings accumulated significant levels of phaseollin. Phytoalexin elicitors were presumably liberated from the dead or dying cells, and any phytoalexin produced by the dying cells or surrounding healthy ones must have preferentially accumulated in the dead cells. Hargreaves and Bailey (1978) bisected bean hypocotyls, killed one half, and then placed the two halves back into proximity before incubating. Again, phaseollin accumulated only in the dead half of the hypocotyl, although presumably it was synthesized only in the cells of the adjacent living segment. Coupling these findings with the evidence for endogenous phytoalexin turnover in living plant cells, might not the physiologic function of dead host cells in the HR be as a reservoir for the accumulation of phytoalexins? The dead cells would not per se be the cause of pathogen restriction as has been suggested, but they would be of importance as deposition sites for phytoalexins produced by the living host cells surrounding them.

Does the HR/phytoalexin system occur in other aspects of higher plant culture than disease defense? Limited evidence suggests that certain physiological disorders may in fact be the HR out of place. Tomatoes carrying the Cf_2 gene for leaf-mold resistance are prone to physiologic leaf spotting (Langford, 1948), but since the biochemistry of the normal HR to this disease is not understood, we cannot assess whether the autogenous necrosis is functionally the HR or not. Mace and Bell (1978) noted that certain cotton hybrids gave spontaneous necrotic stelar streaks and that these areas contained high levels of terpenoid compounds and flavonols, considered to function as phytoalexins in *Gossypium* spp. (Bell and Stipanovic, 1978). Keen and Taylor (1975) observed that ozone-damaged soybean leaves accumulated coumestrol, which also occurs during the normal HR to bacterial pathogens. This raised the possibility that ozone was at least in part initiating the HR. Further tests should be made of the idea that certain cases of physiologic disorder or pollutant damage are in fact the HR.

Other possible unrecognized occurrences of the HR are in graft incompatibility (Schneider et al., 1978) and certain pollen-stigma incompatibility reactions (Heslop-Harrison, 1975; Clarke et al., 1977; Lewis, 1980). These incompatibility reactions frequently involve necrosis of reacting tissue or deposition of induced structural barriers, and accordingly have similarities to the HR. Teasdale et al. (1974) noted chemical and cytological similarities between the incompatible responses of pea endocarp cells to plant pathogenic fungi and foreign pollen, animal tumor cells, and vegetative cells of *Phaseolus vulgaris*. Hodgkin and Lyon (1979) made the interesting observation that rishitin inhibited the germination of *Solanum* pollen. Several agents also interfere with self- and cross-incompatibility in certain plants, thus allowing successful crosses to be made (Bates and Deyoe, 1973; Bushnell, 1979). It is noteworthy that some of these (e.g., heat, translation inhibitors) would be expected to interfere with reactions involving inducible production of phytoalexins or structural barriers.

NEW DEVELOPMENTS IN PHYTOALEXIN RESEARCH

Several new phytoalexins have been discovered since the last review (Friend, 1979), and a few of them are reviewed here. Takasugi et al. (1978a,b) isolated new stilbene and benzofuran phytoalexins from mulberry shoots inoculated with fungi. Two of the compounds, oxyresveratrol and its 4-prenyl derivative (Figure 1), were only produced in xylem tissues, while the benzofurans were only obtained from fungus-challenged phloem and cortical tissues. These results are most interesting because they suggest tissue specificity for phytoalexin production and also constitute the first firm demonstration of phytoalexins from a perennial tree. Many trees are known to store preformed toxic chemicals in the heartwood, and indeed the stilbenes observed by the Japanese group had previously been reported to be preformed heartwood constituents in mulberry. It is therefore possible that perennial plants may utilize the same biosynthetic pathways for inducible production of phytoalexins in juvenile tissues as for their constitutive production in heartwood. The mulberry stilbenes are chemically interesting because they are similar to resveratrol and its prenylated derivative (Figure 1), isolated as phytoalexins from peanuts (*Arachis hypogaea*) by Ingham (1976b) and Keen and Ingham (1976). The benzofuran phytoalexins from mulberry (moracins) are also similar in structure to vignafuran (Figure 1), isolated from *Vigna unguiculata* leaves by Preston et al. (1975).

Another new and interesting group of phytoalexins is the viniferins (Figure 1), a class of fused stilbenes (Pryce and Langcake, 1977; Langcake and Pryce, 1977). These workers also reported that pterostilbene (Figure 1) functions as a phytoalexin in leaves of the same plant (Langcake et al., 1979). The list of isoflavonoid and acetylenic phytoalexins from the Leguminosae has been extended with reports for *Cajanus cajan* (Ingham, 1976a, 1979a), hyacinth bean (*Lablab niger*) (Ingham, 1977c), hare's foot clover (*Trifolium arvensae*) (Ingham and Dewick, 1977), winged bean (*Psophocarpus tetragonolobus*) (Preston, 1977), *Lens culinaris* (Robeson, 1978), *Glycyrrhiza glabra* (Ingham, 1977a), cowpea (*Vigna unguiculata*) (Preston, 1975), *Caragana* sp. (Ingham, 1979b), *Dalbergia sericea* (Ingham, 1979c), and from *Anthyllis, Lotus,* and *Tetragonolobus* spp. by Ingham (1977b).

Preliminary genetic experiments by Tegtmeier and Van Etten (1979) with *Fusarium solani* have supported the idea that phytoalexins may be involved with conferral of disease resistance and that the ability to tolerate host phytoalexins may be one factor required for pathogenicity by compatible pathogens.

Yoshikawa (1978) made the important discovery that biotic and abiotic factors elicited accumulation of glyceollin in soybean cotyledons by distinct mechanisms. Yoshikawa estimated the glyceollin biosynthetic rate by short-term feeding experiments with the labeled phenylalanine and estimated the rate of degradation when glyceollin was exogenously supplied to the elicitor-treated cotyledons. The key finding was that glyceollin biosynthesis was stimulated in wounded control cotyledons, but net accumulation did not occur because of degradation by a constitutive system. Although the tested β-glucan biotic elicitors stimulated net biosynthesis of glyceollin, the abiotic elicitors caused little or no stimulation of biosynthesis over the controls; instead they all inhibited or inactivated the degrading system. If Yoshikawa's observations are general in plants, several important consequences are at hand. First, phytoalexin elicitation by abiotic substances such as detergents and heavy metal salts, used as a detracting argument against the specificity of phytoalexin elicitation (Ward and Stoessl, 1976), would constitute artifacts relative to elicitation by biotic factors. Second, blocked phytoalexin turnover could possibly also be involved in the normal expression of hypersensitive resistance to pathogens, as indicated by more recent work (Yoshikawa et al., 1979) with soybean hypocotyls. Also Glazener and Van Etten (1978) reported that *Phaseolus vulgaris* suspension cells synthesized and degraded phaseollin simultaneously. They offered the plausible speculation that phytoalexin accumulation may be self-sustaining because the phytoalexin inactivates its degrading system, thus promoting further accumulation.

One way in which decreased phytoalexin turnover may occur during the HR is by host cell death. Some phytoalexins are rapidly converted to non-toxic metabolites by healthy plant cells. Stoessl et al. (1977a) noted the metabolism of capsidiol by pepper tissues and later identified the initial metabolites of capsidiol and rishitin produced by suspension culture cells of pepper and potato, respectively, as the 13-hydroxy derivatives (Ward et al., 1977). Tomiyama's group (Ishiguri et al., 1978) simultaneously obtained the same results with rishitin (the metabolite was called M-1) and also isolated a second rishitin metabolite, called M-2, in which the methylene double bond was reduced (Figure 2) (Murai et al., 1977). In both pepper and potato these primary hydroxylated metabolites completely lacked antifungal activity and were further metabolized to unidentified water-soluble compounds. Ishiguri et al. (1978) observed that freshly sliced potato tubers did not metabolize rishitin, but that aged ones did, thus possibly indicating that the degrading system was inducible rather than constitutive as concluded by Yoshikawa (1978) for soybean. Of additional interest is the observation by Masamune et al. (1978) that rishitin-M-1 was not transported from healthy to infected cells as was rishitin.

Ingham et al. (1980) recently isolated a hydroxylated adduct of glyceollin isomer III from soybean leaves challenged with incompatible *Pseudomonas* spp. or sodium iodoacetate (Figure 2). The hydroxylated product lacked antifungal activity, which is consistent with similar findings for a prenyl-hydroxylated derivative of kievitone (Kuhn et al., 1977). Loss of fungitoxicity by hydroxylation indicates that the nonpolar side chains must be important for the activity of certain isoflavonoid and sesquiterpene phytoalexins. This is also indicated by the report that the simple isoflavone genistein was almost inactive against fungi, but its 6-prenyl derivative, wighteone (Ingham et al., 1977), was much more active (Figure 3). The isoflavone hydroxygenistein has been isolated from several legumes but has little antifungal activity; the corresponding isoflavanone dalbergioidin has considerably a greater activity, however (Figure 3) (Ingham, 1977b). Although only one systematic synthetic study has been made (Carter et al., 1978), we are empirically assembling a structure–activity picture for phytoalexins.

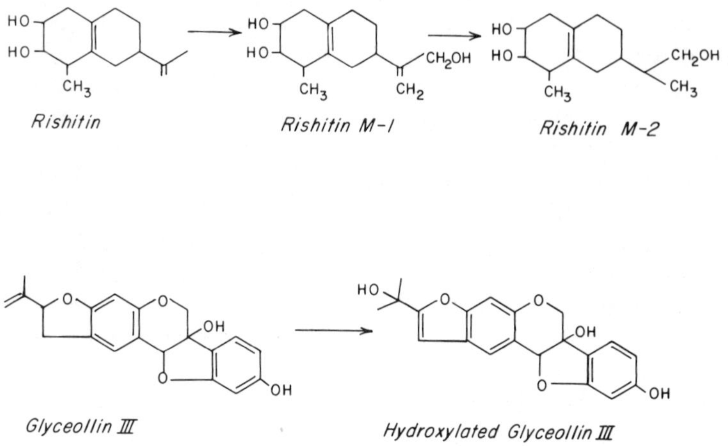
HO
HO
CH3
Rishitin
HO
HO
CH3
CH2
CH2OH
Rishitin M-1
HO
HO
CH3
CH2OH
CH3
Rishitin M-2
O
O
OH
O
OH
Glyceollin III
HO
O
O
OH
O
OH
Hydroxylated Glyceollin III

Figure 2 Higher plant metabolites of certain phytoalexins.

Progress has recently been made in the enzymology of phytoalexin biosynthesis. Following the pioneering research of Uritani's group with ipomeamarone (e.g., Oba et al., 1976) and Sitton and West (1975) on the complete biosynthesis of casbene in cell-free enzyme systems obtained from fungus-challenged castor bean seedlings, some information has now been obtained on the crucial terminal enzymes involved in biosynthesis of the legume phytoalexins. Gustine et al. (1978) found that fungus-challenged jackbean callus showed a 3–4-fold increase in the activity of O-methyltransferase concomitant with accumulation of medicarpin. Transferase activity capable of methylating the 4′-hydroxyl of isoflavonoids increased following challenge, but not methyltransferases accepting hydrox-

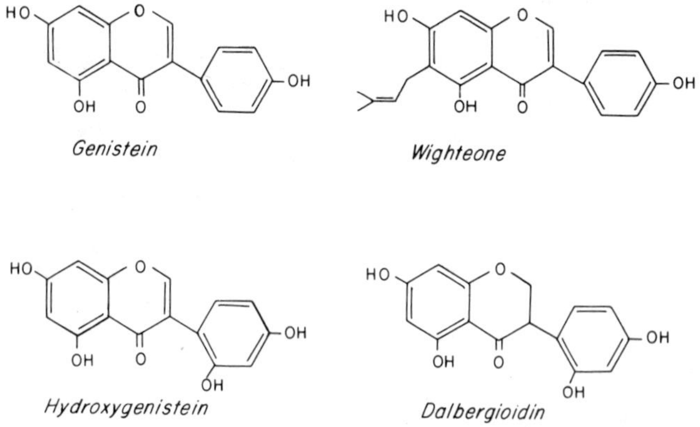
HO
O
OH
OH
O
Genistein
HO
O
OH
OH
O
Wighteone
HO
O
OH
OH
O
OH
Hydroxygenistein
HO
O
OH
OH
O
OH
Dalbergioidin

Figure 3 Isoflavonoids with little antifungal activity (left) and derivatives with much greater activity (right).

ylated cinnamic acids or flavones. Therefore, it appears that a selective increase occurs in only that enzyme catalyzing the methylation of isoflavonoids and that this may be important in regulating induced medicarpin biosynthesis.

Dewick used labeled precursors to show that *Trifolium pratense* and *Medicago sativa* biosynthesize medicarpin and maackiain via the chalcone, isoflavone/isoflavanone route previously suggested (Wong, 1975), but that a pterocarp-6a-ene intermediate is apparently not involved (Dewick, 1977; Dewick and Martin, 1979a,b). They also demonstrated that medicarpin and sativan are interconvertible in alfalfa, thus indicating that they share a common biosynthetic precursor. It will be of interest to see if the enzymes involved in biosynthesis of the pterocarpan nucleus can be obtained in cell-free systems.

Grisebach's group has made important recent progress in understanding the biosynthesis of prenylated isoflavonoid phytoalexins using cell-free extracts from *Lupinus albus*. They demonstrated production of wighteone (Figure 3) when genistein was supplied as substrate and of luteone when hydroxygenistein was supplied (Schröder et al., 1979). In further work, the same group obtained prenylation of 6a,3,9-trihydroxypterocarpan with cell-free extracts from elicitor-challenged soybean cotyledons (Zähringer et al., 1979). The major product obtained was prenylated at carbon-2 and is presumed to be the immediate biosynthetic precursor of glyceollin isomers II and III (Figure 4). No prenylation occurred with extracts from unchallenged cotyledons, thus indicating that the prenyl transferase must be induced following elicitor application. It would also be of great interest to investigate whether the enzyme(s) catalyzing formation of glyceollin isomers II and III

6a, 3, 9 trihydroxypterocarpan

2-prenyl derivative

Glyceollin II

Glyceollin III

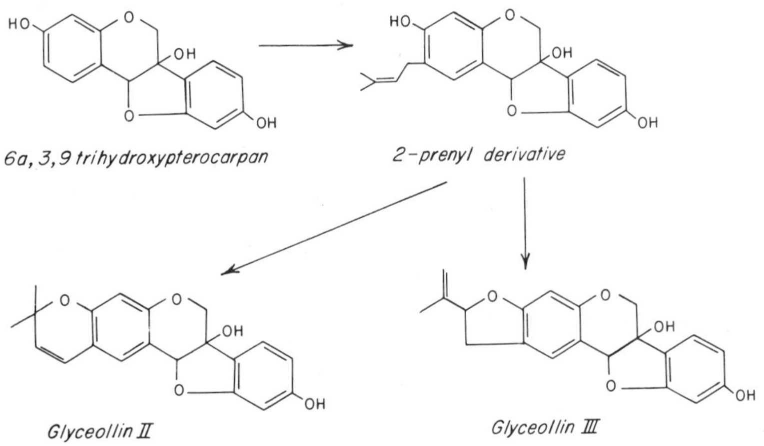

Figure 4 Prenylation of 6a,3,9-trihydroxypterocarpan demonstrated in cell-free system by Zähringer et al. (1979) and presumed conversion of the prenyl adduct to glyceollin isomers II or III.

from the 2-prenyl, trihydroxypterocarpan (Figure 4) could be recovered in cell-free system. The major impact of the work by Zähringer et al. (1979), however, is that it constitutes the first direct proof of the hypothesis that one or more enzymes in the terminal biosynthetic pathway to glyceollin is not present in healthy soybean tissues but is inducibly formed following elicitor challenge as a consequence of gene activation and *de novo* transcription and translation (see Keen and Bruegger, 1977).

Stoessl et al. (1977b) have elegantly used ^{13}C-NMR in biosynthetic studies with the sesquiterpene phytoalexins of the Solanaceae, but no cell-free systems catalyzing terminal biosynthetic steps have been reported.

Cyanide insensitive respiration may contribute to the accumulation of phytoalexins in certain plants. Haard (1977) first showed that elicitor-treated sweet potato root slices accumulated five times greater levels of ipomeamarone if ethylene, carbon monoxide, or cyanide were also supplied. He suggested that stimulated cyanide insensitive respiration might account for the increased phytoalexin levels. Henfling et al. (1978) confirmed the ethylene effect with irish potato tuber slices, and Alves et al. (1979) recently showed that the ethylene effect in potatoes was in turn stimulated by higher O_2 partial pressures. Since this is also similar to effects on cyanide insensitive respiration, they concluded that a possible function of this elusive respiratory system might precisely be in the biosynthesis of phytoalexins.

Little additional information has recently accumulated on the mode of action of phytoalexin toxicity, but it is becoming clear that pterocarpan and sesquiterpene phytoalexins are as toxic to plant cells as to microorganisms, even to the plant in which they accumulate (Glazener and Van Etten, 1978; Lyon and Mayo, 1978). Kaplan et al. (1980) found that three glyceollin isomers were potent inhibitors ($K_i = 2 \times 10^{-6}$ M) of the electron transport system in isolated soybean and table beet mitochondria. The precise site of inhibition was not determined, but the compounds did not act as uncouplers. These observations and the well-known effects of the related compound rotenone on the electron transport system raise the possibility that glyceollin toxicity might result in part by an internal mechanism of action. The fact that phytoalexin-resistant mutants have not been reported in any organism likely means that the mechanism of action of phytoalexins is pleiotropic, and accordingly there are probably several sites of action in addition to the frequently reported permeability effects (Van Etten and Pueppke, 1976). For instance, Ravise and Kirkiacharian (1976) noted that certain isoflavanones inhibited lytic enzymes produced by *Phytophthora parasitica*. Beretz et al. (1978) also reported that several flavonoids, most notably flavonols, increased cAMP levels in cells, and inhibited cAMP phosphodiesterase. The active compounds were also effective as muscle relaxants, a finding related to the inhibitory effects of certain isoflavonoids on nematode motility (Kaplan et al., 1980). It would be of interest to see if isoflavonoid and other phytoalexins had similar effects on cellular cAMP levels.

PHYTOALEXIN ELICITORS

A comprehensive treatment will not be attempted, since the topic will be discussed in a forthcoming review and few developments of significance have occurred since the reviews of Albersheim and Anderson-Prouty (1975), Callow (1977), Keen and Bruegger (1977), and Albersheim and Valent (1978). The latter reviews cite several examples of fungal metabolites such as branched β-1,3-glucans that are efficient phytoalexin elicitors but exhibit no race or cultivar specificity. Since none of these have been demonstrated to be surface-borne or otherwise detected by plant cells during pathogenesis, their physiological role is questionable. Phytoalexin elicitors have also been obtained from plants (Hargreaves and Selby, 1978; Hargreaves and Bailey, 1978) but their physiological relevance is also unclear.

As suggested earlier, the work of Yoshikawa (1978) and Zähringer et al. (1979) has indicated that biotic but not abiotic elicitors cause the accumulation of glyceollin in soybean tissues by direct or indirect induction of certain enzymes in the biosynthetic pathway. Since considerable evidence suggests that this induction is more pronounced in incompatible than compatible combinations of soybean and *Phytophthora megasperma* var. *sojae* races (see Keen and Bruegger, 1977), incompatible fungus races must contain metabolites that are recognized as more efficient elicitors by the plant than those of compatible races. We have presented evidence indicating existence of the predicted specific glyceollin elicitors from various *P. megasperma* var. *sojae* races (see Keen and Bruegger, 1977) and that they may be surface glycoproteins (Keen, 1978). Ziegler and Albersheim (1977) also discovered that the extracellular glycoprotein enzyme, invertase, exhibits structural variation in the glycosyl portion among three races of *Phytophthora megasperma* var. *sojae*. This may be very significant, since Smith and Ballou (1974) showed that structural differences in the glycosyl portion of the external invertase from *Saccharomyces cerevisiae* strains were also faithfully reflected in the corresponding structure of the surface glycoprotein complex. If *P. megasperma* var. *sojae* operates similarly, then Ziegler and Albersheim's observations portend structural uniqueness in the surface glycoproteins of the various fungus races.

Albersheim et al. (1978) and Wade and Albersheim (1979) suggested that similar if not identical fungal glycoproteins to those discussed above from *P. megasperma* var. *sojae* function not as phytoalexin elicitors but by some other mechanism to protect incompatible soybean hypocotyls from fungus growth. It is impossible to rule out the possibility, however, that glyceollin was in fact specifically elicited in their work but was not detected due to insensitivity of techniques or other reasons.

Stekoll and West (1978) have presented the most convincing evidence that glycoproteins in fact function as phytoalexin elicitors. A glycoprotein has been purified from cultures of *Rhizopus stolonifer* that is an efficient elicitor of casbene synthesis in castor bean seedlings. Integrity of the protein part of the molecule is necessary for activity (West, personal communication). Anderson (1979) has also detected specific phytoalexin elicitors in culture fluids of two *Colletotrichum lindemuthianum* races. The active fractions appeared to contain glycoproteins but the elicitors have not yet been purified and characterized. Although remaining a most appealing hypothesis, proof of the role of surface glycoproteins in recognition for gene-for-gene plant–fungus parasite systems is not yet available.

Bruegger and Keen (1979) solubilized factors from the outer membrane complex of *Pseudomonas glycinea* cells with sodium dodecyl sulfate that functioned as race-specific glyceollin elicitors in soybean cotyledons. With one exception, extracts from all incompatible races elicited higher levels of glyceollin in cotyledons of two soybean cultivars than did extracts from compatible races. Extracts from the nonpathogen *P. pisi* gave high glyceollin elicitor activity similar to incompatible *P. glycinea* races. Although the bacterial elicitors have not been purified and characterized, the observations suggest the possible association of specific elicitors with the outer membranes of *P. glycinea* and *P. pisi*.

In contrast to earlier suggestions that specific elicitors of potato sesquiterpenoids may occur in incompatible *Phytophthora infestans* races (Metlitskii et al., 1973; Tomiyama et al., 1974), Kuc et al., (1979) have proposed that all *P. infestans* races contain common elicitors of rishitin and related terpenoids in potato tubers and that race specificity is based on the production of race-specific suppressor molecules (blockers) that interact with the various dominant resistance genes in potato. The blockers have been partially characterized as relatively low molecular weight glucans (Doke et al., 1979).

LESSONS FROM PATHOGEN ESCAPE MECHANISMS

It is useful to compare the ways in which successful plant and higher animal parasites escape or circumvent the respective inducible host defense mechanisms. Bloom (1979) has summarized several of the ways in which successful animal parasites evade the immune systems. These are as follows:

1. *Antigenic variation,* a scheme very similar to our concept with virulent gene-for-gene plant parasite races; African trypanosomes have developed the ability to alter, presumably by mutation and selection, the antigenic nature of their surface glycoprotein. At approximately weekly intervals, the population of trypanosomes in the blood decreases due to immunologic recognition of the current surface antigen; this, however, leads to selection of a new antigenic type of parasite that is not immediately recognized by the immune system and thereby multiples until it is in turn recognized about one week later. This cyclical state is, with the exception

of a shorter frequency, exactly the situation experienced with gene-for-gene plant–parasite systems.

2. *Antigenic mimicry,* in which the parasite physically binds antigens of the host onto its surface so that the host no longer recognizes the parasite as foreign. A most devious version of this scheme cited by Bloom is that of certain murine parasites that bind precisely the mouse H-2 histocompatibility (MHC) antigens. This epitomizes mimicry of self!
3. *Coexist with the immune system,* for instance by existing in macrophages or by evading killing by macrophages; in some parasites this is taken to the extreme that the parasite requires the immune system for its survival!
4. *Suppress expression of the immune system;* one way of doing this is to generally stimulate the immune system to make many different antibodies so that the specific response to the parasite will be much less. Another is to generally impair the immune reponse so that the patient becomes more susceptible to many diseases.

The known ways in which plant parasites escape phytoalexin defense mechanisms are equally interesting. At present, these break down into the following categories:

a. *Passive failure to elicit production.* This corresponds to (1) above and is the case thus far observed in gene-for-gene systems, where all studied compatible interactions appear to result from failure of the host to respond defensively (Keen and Bruegger, 1977); this mechanism can be distinguished from (b) below because equal mixtures of compatible and incompatible races give completely incompatible responses [viz., the *Pseudomonas glycinea*–soybean system (Keen and Kennedy, 1974) and the *Xanthomonas vesicatoria*–pepper system (Stall et al., 1974)].
b. *Active suppression of phytoalexin production.* This mechanism is analogous to (4) above and has been demonstrated by Oku, Ouchi, and co-workers (Shiraishi et al., 1978) to account for infection of pea leaves by *Mycosphaerella pinodes.* Two fractions, believed to be low molecular weight peptides, were isolated from germination fluids of the fungus and shown to suppress pisatin production by pea leaves in response to high molecular weight elicitors from the fungus. The peptide suppressors also blocked phytoalexin production and thereby allowed colonization of pea leaves by the normally nonpathogenic fungus *Stemphyllium sarcinaeforme* (Shiraishi et al., 1978). Suppression of phytoalexin production may have wider significance among plant pathogens, and study of its underlying mechanisms should be worthwhile. For instance, host-specific toxins produced by *Helminthosporium* and *Alternaria* species may be working in precisely this way. Certain toxin insensitive genotypes of oats respond to *Helminthosporium victoriae* infection with a classical hypersensitive reaction followed by cessation of fungus growth (see Scheffer and Pringle, 1967). Toxin sensitive genotypes, however, do not undergo

an HR and the fungus colonizes the tissue. It is illuminating from the point of the HR suppression idea to note that exogenous victorin applied to toxin sensitive but not resistant oat tissues rendered them susceptible to colonization by not only *H. victoriae* but also by the nonpathogen *H. carbonum* (Yoder and Scheffer, 1969).

c. *Chemical degradation of phytoalexins.* This appears to be commonly employed by facultative fungus parasites (e.g., Friend, 1979; Kuhn et al., 1977; VanEtten and Pueppke, 1976), but will not be considered further here.

d. *Tolerance of phytoalexins.* In these cases, analogous to (3) above, the pathogen does not chemically degrade the phytoalexin, but is able to grow in the presence of levels of the phytoalexin that would inhibit related organisms. When further studied, this will probably turn up some interesting molecular mechanisms. *Aphanomyces eutiches* is sensitive to pisatin *in vitro* and elicits high levels of pisatin *in vivo,* yet is still able to grow in the host (Pueppke and VanEtten, 1976). Although this observation was taken as a detracting argument for the whole concept of phytoalexin defense (Pueppke, 1978), it is more plausible to assume that the fungus has an unknown mechanism to escape the phytoalexin, analogous to some of the better explored mechanisms described by Bloom (1979) for animal parasites. Notably, pathogens that tolerate the plant's phytoalexins, either *in vivo* or *in vitro,* are often members of that troublesome group for which genetic resistance has not been found or is rare in the host species. One example is the *Aphanomyces* disease of peas and another is certain *Fusarium solani formae speciales* (Van Etten and Stein, 1978). There is no known genetic resistance in soybean (or in many other cultivated plants) to *Meloidogyne javanica,* but genes are available that confer hypersensitive resistance to the related nematode, *M. incognita*. The resistance to *M. incognita* appears to be mediated by glyceollin production and these nematodes are sensitive to glyceollin *in vitro* (Kaplan et al., 1980). However, *M. javanica* was tolerant to high concentrations of glyceollin *in vitro,* which perhaps explains the lack of effective resistance genes. There are probably many other such examples of phytoalexin tolerance among successful plant pathogens.

It would be hoped that understanding the escape mechanisms used by successful parasites would aid in devising control methods. This has not happened thus far with successful plant or animal parasites (Bloom, 1979), but the next section presents a current wish-list for plant parasites.

FUTURE PROSPECTS AND DIRECTIONS

Although there is obvious need for additional research on the HR/phytoalexin system in higher plants, I am unable to outline any new rationale based on it that is likely to be quickly usable for disease control in the field. Immunization, the classic approach to control of animal diseases, is successful because of the specificity and memory aspects of the MHC–immunoglobulin systems. Plants have a corresponding recognition system with high specificity, but, with the possible exception of systemic protection to pathogens, they totally lack the memory capacity. Without this, plant pathologists can only exploit the HR/phytoalexin system for disease control by attempting to modulate the informational and expressive systems. Some speculative rationales for this are listed in Table 2. The historic cardinal plant disease control measure, resistance breeding, has capitalized on the high specificity for recognition of the pathogen gained by incorporating hypersensitive resistance genes into the plant that the pathogen had not seen recently. Of course, new virulent races frequently appeared and this has been claimed by some to doom breeding for monogenic resistance. I disagree. Why have not plant breeders–pathologists simply "managed the major genes" better? Knowing from experience the probable useful life of a newly released gene, why not sequentially and routinely release near-isogenic lines with first one gene for a predetermined time, then completely withdraw it for another gene, then another, etc., and eventually start over with the first gene? This simply says, "let stabilizing selection work for us." A second aspect that needs further investigation is the search for what Van der Plank calls "strong genes"—those that are not easily overcome by virulent races. As Van der Plank states (1975, p. 155), the Sr6/Sr9d gene combination in wheat successfully controlled wheat stem rust in the northern United States–Canada wheat belt for several years; another apparent

Table 2. Possible Rationales for Effecting Disease Control by Manipulating the HR/Phytoalexin System

1. Manage the major genes better
2. Search more diligently for "strong" resistance genes or combinations of genes (Van der Plank, 1975; Robinson, 1976)
3. Launch a major effort to more fully understand the biochemistry of recognition and expression of the HR
4. Attempt to find usable "sensitizers" or potentiators of the phytoalexin response (Cartwright et al., 1977)
5. Breed plants for the proper blend of phytoalexins (Bell and Stipanovic, 1978)
6. Breed plants for finite constitutive levels of phytoalexins (Hadwiger et al., 1976)
7. Attempt to breed plants that superproduce phytoalexins in response to biotic elicitors (Avazkhodzhaev et al., 1976)
8. Attempt to apply phytoalexins to plants to effect disease control
9. Apply phytoalexin elicitors to plants (Albersheim quoted in Maugh, 1976)

"strong" gene is the "T" gene in barley against *Puccinia graminis* f. sp. *tritici* (Sellam and Wilcoxson, 1976). Why not look for other examples?

Additional effort is needed to investigate more intensively the mechanisms underlying the HR (Table 2, no. 3). Whereas immunologists have launched major basic research efforts into the MHC/immunoglobulin systems, research on HR/phytoalexin systems plods along at far less than 1% this level of funding.

I have already advanced numbers 4 (Cartwright et al., 1977) and 5 (Table 2) in Keen and Bruegger (1977) and will not discuss them in detail here except to mention two promising aspects of number 4. One possibility is to consider exposing infected plants to ethylene, carbon monoxide, or similar agents capable of stimulating cyanide insensitive respiration. The rationale is that the anticipated higher rate of phytoalexin production might effect disease control. The normal occurrence of phytoalexin turnover in healthy plant tissues and Yoshikawa's demonstration of inhibited turnover by certain chemicals also presents the possibility that chemicals might be developed that will block phytoalexin turnover at the infection site of normally compatible pathogens, thereby leading to a resistant reaction.

Recommendation 6 (Table 2) says that if a plant can be developed with a significant constitutive level of its particular phytoalexin(s), but not so high that autogenous toxicity results, then it may be better able to contain a normally compatible pathogen due to the "head start." Hadwiger et al. (1976) obtained an interesting mutant of peas that appeared to behave in this way, but subsequent papers have not appeared to establish its agronomic and plant pathologic utility. A related objective is number 7 (Table 2). If one could, by breeding or cell culture techniques, select plant lines that produced phytoalexins faster and to higher levels than the wild types in response to pathogens or biotic elicitors, these might exhibit resistance to pathogens that were compatible on the wild-type genome. This appears to have been done by Avazkhodzhaev et al. (1976) and should be attempted with other plants. Keen and Bruegger (1977) discussed rationales 8 and 9 (Table 2); although it is possible that one or both of these approaches may work, they are fraught with practical and theoretical problems.

The immediate future is unlikely to see realization of any of the schemes listed in Table 2 except perhaps numbers 1 and 2. Neither long-range basic research to understand the HR nor shorter term research to test feasibility of any of the other objectives in Table 2 has been done on a significant scale. Bluntly stated, none of them is likely to surface unless the prevailing low level of financial support for such research is increased and the required biochemical talent marshalled. We are lucky in plant pathology that genetic resistance is so widely effective. Unlike our counterparts in human medicine, we appear unwilling or unable to pursue more forcefully the basic research that will be required to control those plant diseases in which resistance is not currently successful.

ACKNOWLEDGMENT

The author's research was supported by NSF grant PCM7724346.

REFERENCES

Albersheim, P., and A. J. Anderson-Prouty. (1975) Carbohydrates, proteins, cell surfaces, and the biochemistry of pathogenesis. *Annu. Rev. Plant Physiol.* **26**:31–52.

Albersheim, P., and B. S. Valent. (1978) Host-pathogen interactions in plants. Plants, when exposed to oligosaccharides of fungal origin, defend themselves by accumulating antibiotics. *J. Cell Biol.* **78**:627–643.

Albersheim, P., B. S. Valent, M. Wade, M. Hahn, and K. Cline. (1978) On the molecular basis of disease resistance in plants. *Philip Morris Sci. Symp., 3rd,* Richmond Va.

Alves, L. M., E. G. Heisler, J. C. Kissinger, J. M. Patterson, and E. B. Kalan. (1979) Effects of controlled atmospheres on production of sesquiterpenoid stress metabolites by white potato tuber. Possible involvement of cyanide-resistant respiration. *Plant Physiol.* **63**:359–362.

Anderson, A. J. (1979) Differential initiation of resistance in bean by culture filtrate preparations from two races of *Colletotrichum lindemuthianum.* (Abstr.) *Phytopathology* **69**:913.

Asada, Y., T. Ohguchi, and I. Matsumoto. (1976) Biosynthesis of lignin in Japanese radish root infected by downy mildew fungus. In *Biochemistry and Cytology of Plant–Parasite Interaction* (K. Tomiyama et al., eds.). Elsevier, Amsterdam. Pp. 200–212.

Avazkhodzhaev, M. Kh., A. E. Egamberdiev, I. Kh. Gailbullaev, and I. Zh. Zhumaniyazoo. (1976) Nature of wilt resistance in new cotton chemical mutants. *Uzb. Biol. Zh.* **1976**:60–62. (Russian original not read; abstr. in *Chem. Abstr.* **85**:106698).

Bates, L. S., and C. W. Deyoe. (1973) Wide hybridization and cereal improvement. *Econ. Bot.* **27**:401–412.

Bell, A. A., and J. T. Presley. (1969) Temperature effects upon resistance and phytoalexin synthesis in cotton inoculated with *Verticillium albo-atrum. Phytopathology* **59**:1141–1146.

Bell, A. A., and R. D. Stipanovic. (1978) Biochemistry of disease and pest resistance in cotton. *Mycopathologia* **65**:91–106.

Beretz, A., R. Anton, and J. C. Stoclet. (1978) Flavonoid compounds are potent inhibitors of cyclic AMP phosphodiesterase. *Experientia* **34**:1054–1055.

Bloom, B. R. (1979) Games parasites play: How parasites evade immune surveillance. *Nature* **279**:21–26.

Bridge, M. A., and W. L. Klarman. (1973) Soybean phytoalexin, hydroxyphaseollin, induced by ultraviolet irradiation. *Phytopathology* **63**:606–609.

Bruegger, B. B., and N. T. Keen. (1979) Specific elicitors of glyceollin accumulation in the *Pseudomonas glycinea*–soybean host–parasite system. *Physiol. Plant Pathol.* **15**:43–51.

Bushnell, W. R. (1979) The nature of basic compatibility: Comparisons between pistil–pollen and host–parasite interaction. In *Recognition and Specificity in Plant-Parasite Interactions* (J. M. Daly and I. Uritani, eds.). Japan Science Society Press, Tokyo. Pp. 211–227.

Callow, J. A. (1977) Recognition, resistance and the role of plant lectins in host–parasite interactions. *Adv. Bot. Res.* **4**:1–49.

Carter, G. A., K. Chamberlain, and R. L. Wain. 1978. Investigations on fungicides. XX. The fungitoxicity of analogues of the phytoalexin 2(2′-methoxy-4′-hydroxyphenyl)-6-methoxybenzofuran (vignafuran). *Ann. Appl. Biol.* **88**:57–64.

Cartwright, D., P. Langcake, R. J. Pryce, and D. P. Leworthy. (1977) Chemical activation of host defense mechanisms as a basis for crop protection. *Nature* **267**:511–513.

Chamberlain, D. W., and J. W. Gerdemann. (1966) Heat-induced susceptibility of soybeans to *Phytophthora megasperma* var. *sojae, Phytophthora cactorum,* and *Helminthosporium sativum. Phytopathology* **56**:70–73.

Chamberlain, D. W., and J. D. Paxton. (1968) Protection of soybean plants by phytoalexin. *Phytopathology* **58**:1349–1350.

Clarke, A. E., S. Harrison, R. B. Knox, J. Raff, P. Smith, and J. J. Marchalonis. (1977) Common antigens and male–female recognition in plants. *Nature* **265**:161–163.

Cruickshank, I. A. M. (1963) Phytoalexins. *Annu. Rev. Phytopathol.* **1**:351–374.

Deverall, B. J. (1977) *Defense Mechanisms of Plants.* Cambridge University Press, Cambridge. P. 54.

Dewick, P. M. (1977) Biosynthesis of pterocarpan phytoalexins in *Trifolium pratense. Phytochemistry* **16**:93–97.

Dewick, P. M., and M. Martin. (1979a) Biosynthesis of pterocarpan and isoflavan phytoalexins in *Medicago sativa:* The biochemical interconversion of pterocarpans and 2′-hydroxyisoflavans. *Phytochemistry* **18**:591–596.

Dewick, P. M., and M. Martin. (1979b) Biosynthesis of pterocarpan, isoflavan, and coumestan metabolites of *Medicago sativa:* Chalcone, isoflavone, and isoflavanone precursors. *Phytochemistry* **18**:597–602.

Doke, N., N. A. Garas, and J. Kuc. (1979) Partial characterization and aspects of the mode of action of a hypersensitivity-inhibiting factor (HIF) isolated from *Phytophthora infestans. Physiol. Plant Pathol.* **15**:127–140.

Ellingboe, A. H. (1975) Horizontal resistance: An artifact of experimental procedure? *Aust. Plant Pathol. Soc. Newsl.* **4**:44–46.

Evans, R. L., and D. J. Pluck. (1978) Phytoalexins produced in barley in response to the halo spot fungus, *Selenophoma donacis. Ann. Appl. Biol.* **89**:332–336.

Friend, J. (1979) Phenolic substances and plant disease. *Recent Adv. Phytochem.* **12**:557–588.

Garcia-Arenal, F., and E. M. Sagasta. (1977) Callose deposition and phytoalexin accumulation in *Botrytis cinerea* infected bean (*Phaseolus vulgaris*). *Plant Sci. Lett.* **10**:305–312.

Glazener, J. A., and H. D. VanEtten. (1978) Phytotoxicity of phaseollin to, and alteration of phaseollin by, cell suspension cultures of *Phaseolus vulgaris*. *Phytopathology* **68**:111–117.

Goodliffe, J. P., and J. B. Heale. (1978) The role of 6-methoxy mellein in the resistance and susceptibility of carrot root tissue to the cold-storage pathogen *Botrytis cinerea*. *Physiol. Plant Pathol.* **12**:27–43.

Gustine, D. L., R. T. Sherwood, and C. P. Vance. (1978) Regulation of phytoalexin synthesis in jackbean callus cultures. Stimulation of phenylalanine ammonia-lyase and O-methyltransferase. *Plant Physiol.* **61**:226–230.

Haard, N. F. (1977) Potentiation of wound induced formation of ipomeamarone by cyanide insensitive respiration in sweet potato (*Ipomoea batatas*) root slices. *Z. Pflanzenphysiol.* **81**:364–368.

Hadwiger, L. A., C. Sander, J. Eddyvean, and J. Ralston. (1976) Sodium azide-induced mutants of peas that accumulate pisatin. *Phytopathology* **66**:629–630.

Hargreaves, J. A., and J. A. Bailey. (1978) Phytoalexin production by hypocotyls of *Phaseolus vulgaris* in response to constitutive metabolites released by damaged bean cells. *Physiol. Plant Pathol.* **13**:89–100.

Hargreaves, J. A., and C. Selby. (1978) Phytoalexin formation in cell suspensions of *Phaseolus vulgaris* in response to an extract of bean hypocotyls. *Phytochemistry* **17**:1099–1102.

Heale, J. B., and S. Sharman. (1977) Induced resistance to *Botrytis cinerea* in root slices and tissue cultures of carrot (*Daucus carota* L.). *Physiol. Plant Pathol.* **10**:51–61.

Henfling, J. W. D. M., N. Lisker, and J. Kuc. (1978) Effect of ethylene on phytuberin and phytuberol accumulation in potato tuber slices. *Phytopathology* **68**:857–862.

Heslop-Harrison, J. (1975) Incompatibility and the pollen-stigma interaction. *Annu. Rev. Plant Physiol.* **26**:403–425.

Hodgkin, T., and G. D. Lyon. (1979) Inhibition of *Solanum* pollen germination *in vitro* by the phytoalexin rishitin. *Ann. Bot.* **44**:253–255.

Ingham, J. L. (1976a) Induced isoflavonoids from fungus-infected stems of pigeon pea (*Cajanus cajan*). *Z. Naturforsch., Teil C* **31**:504–508.

Ingham, J. L. (1976b) 3,5,4′-Trihydroxystilbene as a phytoalexin from groundnuts (*Arachis hypogaea*). *Phytochemistry* **15**:1791–1793.

Ingham, J. L. (1977a) An isoflavan phytoalexin from leaves of *Glycyrrhiza glabra*. *Phytochemistry* **16**:1457–1458.

Ingham, J. L. (1977b) Isoflavan phytoalexins from *Anthyllis, Lotus,* and *Tetragonolobus*. *Phytochemistry* **16**:1279–1282.

Ingham, J. L. (1977c) Phytoalexins of hyacinth bean (*Lablab niger*). *Z. Naturforsch., Teil C* **32**:1018–1020.

Ingham, J. L. (1979a) A revised structure for cajanol. *Z. Naturforsch., Teil C* **34**:159–161.

Ingham, J. L. (1979b) Phytoalexin production by species of the genus *Caragana*. *Z. Naturforsch., Teil C* **34:**293–295.

Ingham, J. L. (1979c) Isoflavonoid phytoalexins from leaflets of *Dalbergia sericea*. *Z. Naturforsch., Teil C* **34:**630–631.

Ingham, J. L., and P. M. Dewick. (1977) Isoflavonoid phytoalexins from leaves of *Trifolium arvense*. *Z. Naturforsch., Teil C* **32**:446–448.

Ingham, J. L., N. T. Keen, and T. Hymowitz. (1977) A new isoflavone phytoalexin from fungus-inoculated stems of *Glycine wightii*. *Phytochemistry* **16**:1943–1946.

Ingham, J. L., N. T. Keen, and R. Lyne. (1980) New isoflavonoids from hypersensitive soybean leaves. *Phytochemistry* (in press).

Ishiguri, Y., K. Tomiyama, N. Doke, A. Murai, N. Katsui, F. Yagihashi, and T. Masamune. (1978) Induction of rishitin-metabolizing activity in potato tuber tissue disks by wounding and identification of rishitin metabolites. *Phytopathology* **68**:720–725.

Kaplan, D. T., N. T. Keen, and I. J. Thomason. (1980) Studies on the mode of action of glyceollin in soybean incompatibility to the root knot nematode, *Meloidogyne incognita*. *Physiol. Plant. Pathol.* **16**:319–325.

Keen, N. T. (1978) Surface glycoproteins of *Phytophthora megasperma* var. *sojae* function as race specific glyceollin elicitors in soybeans. (Abstr.) *Phytopathology News* **12**:221.

Keen, N. T., and B. Bruegger. (1977) Phytoalexins and chemicals that elicit their production in plants. In *Host Plant Resistance to Pests* (P. Hedin, ed.). *Am. Chem. Soc. Symp. Ser. 62*. Pp. 1–26.

Keen, N. T., and J. L. Ingham. (1976) New stilbene phytoalexins from American cultivars of *Arachis hypogaea*. *Phytochemistry* **15**:1794–1795.

Keen, N. T., and B. W. Kennedy. (1974) Hydroxyphaseollin and related isoflavanoids in the hypersensitive resistant response of soybeans against *Pseudomonas glycinea*. *Physiol. Plant Pathol.* **4**:173–185.

Keen, N. T., and L. J. Littlefield. (1979) The possible association of phytoalexins with resistance gene expression in flax to *Melampsora lini*. *Physiol. Plant Pathol.* **14**:265–280.

Keen, N. T., and O. C. Taylor. (1975) Ozone injury in soybeans. Isoflavonoid accumulation is related to necrosis. *Plant Physiol.* **55**:731–733.

Kiraly, Z., B. Barna, and T. Ersek. (1972) Hypersensitivity as a consequence, not the cause, of plant resistance to infection. *Nature* **239**:456–458.

Klarman, W. L., and J. W. Gerdemann. (1963) Induced susceptibility in soybean plants genetically resistant to *Phytophthora sojae*. *Phytopathology* **53**:863–864.

Kuc, J., J. Henfling, N. Garas, and N. Doke. (1979) Control of terpenoid metabolism in the potato–*Phytophthora infestans* interaction. *J. Food Protec.* **42**:508–511.

Kuhn, P. J., D. A. Smith, and D. F. Ewing. (1977) 5,7,2′,4′-Tetrahydroxy-8-(3″,hydroxy-3″-methyl-butyl) isoflavonone, a metabolite of kievitone produced by *Fusarium solani* f. sp. *phaseoli*. *Phytochemistry* **16**:296–297.

Langcake, P., and R. J. Pryce. (1977) A new class of phytoalexins from grapevines. *Experientia* **33**:151–152.

Langcake, P., C. A. Cornford, and R. J. Pryce (1979) Identification of pterostilbene as a phytoalexin from *Vitis vinifera* leaves. *Phytochemistry* **18**:1025–1027.

Langford, A. N. (1948) Autogenous necrosis in tomatoes immune from *Cladosporium fulvum* Cooke. *Can. J. Res.* **C26**:35–64.

Lewis, D. H. (1980) Are there inter-relations between the metabolic role of boron, synthesis of phenolic phytoalexins and the germination of pollen? *New Phytol.* **84**:261–270.

Lyon, G. D., and M. A. Mayo. (1978) The phytoalexin rishitin affects the viability of isolated plant protoplasts. *Phytopathol. Z.* **92**:298–304.

Mace, M. E., and A. A. Bell. (1978) Association of induced terpenoid aldehydes and flavonols with genetic lethal reactions in the stele of a *Gossypium* hybrid. *Am. Phytopathol. Soc. Meeting,* Tucson, AZ. Abstr. 443.

Masamune, T., A. Murai, Y. Takahashi, F. Yagihashi, N. Katsui, N. Doke, and K. Tomiyama. (1978) Isolation of rishitin-M-1 from diseased potato tuber tissue slices. *Chem. Lett.* 1207–1208.

Maugh, T. H. (1976) Plant biochemistry: Two new ways to fight pests. *Science* **192**:874–876.

Mayama, S., J. M. Daly, D. W. Rehfeld, and C. R. Daly. (1975) Hypersensitive response of near-isogenic wheat carrying the temperature-sensitive Sr6 allele for resistance to stem rust. *Physiol. Plant Pathol.* **7**:35–47.

Metlitskii, L. V., Yu T. D'yakov, and O. L. Ozeretskovskaya. (1973) Double induction—A new hypothesis for the immunity of plants to blight and similar diseases. *Dokl. Akad. Nauk SSSR* **213**:209–212.

Murai, A., N. Katsui, F. Yagihashi, T. Masamune, Y. Ishiguri, and K. Tomiyama. (1977) The structure of rishitin M-1 and M-2, metabolites of rishitin in healthy potato tuber tissues. *J. Chem. Soc., Chem. Commun.* **1977**:670–671.

Murch, R. S., and J. D. Paxton. (1977) Glyceollin concentrations in *Phytophthora* resistant soybean: Light influences. *Proc. Am. Phytopathol. Soc.* **4**:135–136.

Nelson, R. R. (1978) Genetics of horizontal resistance to plant diseases. *Annu. Rev. Phytopathol.* **16**:359–378.

Oba, K., H. Tatematsu, K. Yamashita, and I. Uritani. (1976) Induction of furanoterpene production and formation of the enzyme system from mevalonate to isopentenyl pyrophosphate in sweet potato root tissue injured by *Ceratocystis fimbriata* and by toxic chemicals. *Plant Physiol.* **58**:51–56.

Paxton, J. D., and D. W. Chamberlain. (1967) Acquired local resistance of soybean plants to *Phytophthora* spp. *Phytopathology* **57**:352–353.

Preston, N. W. (1975) 2′-O-methylphaseollidinisoflavan from infected tissue of *Vigna unguiculata*. *Phytochemistry* **14**:1131–1132.

Preston, N. W. (1977) Induced pterocarpans of *Psophocarpus tetragonolobus*. *Phytochemistry* **16**:2044–2045.

Preston, N. W., K. Chamberlain, and R. A. Skipp. (1975) A 2-arylbenzofuran phytoalexin from cowpea (*Vigna unguiculata*). *Phytochemistry* **14**:1843–1844.

Pryce, R. J., and P. Langcake. (1977) β-viniferin: An antifungal resveratrol trimer from grapevines. *Phytochemistry* **16**:1452–1454.

Pueppke, S. G. (1978) Some intriguing aspects of the putative role of isoflavonoid phytoalexins in plant disease. *Mycopathologia* **65**:115–119.

Pueppke, S. G., and H. D. VanEtten. (1976) The relation between pisatin and the development of *Aphanomyces euteiches* in diseased *Pisum sativum*. *Phytopathology* **66**:1174–1185.

Rahe, J. E., and R. M. Arnold. (1975) Injury-related phaseollin accumulation in *Phaseolus vulgaris* and its implications with regard to host–parasite interaction. *Can. J. Bot.* **53**:921–928.

Ravise, A., and B. S. Kirkiacharian. (1976) Influence de la structure de composés phénoliques sur inhibition du *Phytophthora parasitica* et d'enzymes participant aux processes parasitaires. *Phytophathol. Z.* **85**:74–85.

Robeson, D. J. (1978) Furanoacetylene and isoflavonoid phytoalexins in *Lens culinaris*. *Phytochemistry* **17**:807–808.

Robinson, R. A. (1976) *Plant Pathosystems. Adv. Ser. Agric. Sci. 3*. Springer-Verlag, Berlin.

Scheffer, R. P., and R. B. Pringle. (1967) Pathogen-produced determinants of disease and their effects on host plants. In *The Dynamic Role of Molecular Constituents in Plant–Parasite Interaction* (C. J. Mirocha and I. Uritani, eds.). American Phytopathological Society, St. Paul, Minn. Pp. 217–236.

Schneider, H., R. G. Platt, W. P. Bitters, and R. M. Burns. (1978) Diseases and incompatibilities that cause decline in lemons. *Citrograph* **63**:219–221.

Schröder, G., U. Zähringer, W. Heller, J. Ebel, and H. Grisebach. (1979) Biosynthesis of antifungal isoflavonoids in *Lupinus albus*. Enzymatic prenylation of genistein and 2′hydroxygenistein. *Arch. Biochem. Biophys.* **194**:635–636.

Sellam, M. A., and R. D. Wilcoxson. (1976) Development of *Puccinia graminis* f. sp. *tritici* on resistant and susceptible barley cultivars. *Phytopathology* **66**:667–668.

Shiraishi, T., H. Oku, M. Yamashita, and S. Ouchi. (1978) Elicitor and suppressor of pisatin induction in spore germination fluid of pea pathogen, *Mycosphaerella pinodes*. *Ann. Phytopathol. Soc. Jpn* **44**:659–665.

Sitton, D., and C. A. West. (1975) Casbene: An anti-fungal diterpene produced in cell-free extracts of *Ricinus communis* seedlings. *Phytochemistry* **14**:1921–1925.

Smith, W. L., and C. E. Ballou. (1974) Immunochemical characterization of the mannan component of the external invertase (β-fructofuranosidase) of *Saccharomyces cerevisiae*. *Biochemistry* **13**:355–361.

Stall, R. E., J. A. Bartz, and A. A. Cook. (1974) Decreased hypersensitivity to *Xanthomonads* in pepper after inoculations with virulent cells of *Xanthomonas vesicatoria*. *Phytopathology* **64**:731–735.

Stekoll, M., and C. A. West. (1978) Purification and properties of an elicitor of castor bean phytoalexin from culture filtrates of the fungus *Rhizopus stolonifer*. *Plant Physiol.* **61**:38–45.

Stoessl, A., J. R. Robinson, G. L. Rock, and E. W. B. Ward. (1977a) Metabolism of capsidiol by sweet pepper tissue: Some possible implications for phytoalexin studies. *Phytopathology* **67**:64–66.

Stoessl, A., E. W. B. Ward, and J. B. Stothers. (1977b) Biosynthetic relationships of sesquiterpenoidal stress compounds from the Solanaceae. In *Host Plant Resistance to Pests* (P. Hedin, ed.). *Am. Chem. Soc. Symp. Ser. 62*. Pp. 61–77.

Takasugi, M., L. Munoz, T. Masamune, A. Shirata, and K. Takahashi. (1978a) Stilbene phytoalexins from diseased mulberry. *Chem. Lett.* **1978**:1241–1242.

Takasugi, M., S. Nagao, S. Ueno, T. Masamune, A. Shirata, and K. Takahashi. (1978b) Moracins C and D, new phytoalexins from diseased mulberry. *Chem. Lett.* **1978**:1239–1240.

Tani, T. (1979) Personal communication. Laboratory of Plant Pathology, Faculty of Agriculture. Kagawa University, Kagawa, Japan.

Teasdale, J., D. Daniels, W. C. Davis, R. Eddy, and L. A. Hadwiger. (1974) Physiological and cytological similarities between disease resistance and cellular incompatibility responses. *Plant Physiol.* **54**:690–695.

Tegtmeier, K. J., and H. D. VanEtten. (1979) Genetic analysis of sexuality, phytoalexin sensitivity and virulence of *Nectria haematococca* MP VI (*Fusarium solani*). (Abstr.) *Phytopathology* **69**:1047.

Tomiyama, K., H. S. Lee, and N. Doke. (1974) Effect of hyphal homogenate of *Phytophthora infestans* on the potato tuber protoplasts. *Ann. Phytopathol. Soc. Jpn* **40**:70–72.

Vance, C. P., and R. T. Sherwood. (1976) Regulation of lignin formation in reed canarygrass in relation to disease resistance. *Plant Physiol.* **57**:915–919.

Van der Plank, J. E. (1975) *Principles of Plant Infection.* Academic Press, NY.

VanEtten, H. D., and S. G. Pueppke. (1976) Isoflavonoid phytoalexins. In *Biochemical Aspects of Plant–Parasite Relationships* (J. Friend and D. R. Threlfall, eds.). Academic Press, London. Pp. 239–289.

VanEtten, H. D., and J. I. Stein. (1978) Differential response of *Fusarium solani* isolates to pisatin and phaseollin. *Phytopathology* **68**:1276–1283.

Wade, M., and P. Albersheim. (1979) Race-specific molecules that protect soybeans from *Phytophthora megasperma* var. *sojae*. *Proc. Natl. Acad. Sci. USA* **76**:4433–4437.

Ward, E. W. B., and A. Stoessl. (1976) On the question of "elicitors" or "inducers" in incompatible interactions between plants and fungal pathogens. *Phytopathology* **66**:940–941.

Ward, E. W. B., A. Stoessl, and J. B. Stothers. (1977) Metabolism of the sesquiterpenoid phytoalexins capsidiol and rishitin to their 13-hydroxy derivatives by plant cells, *Phytochemistry* **16**:2024–2025.

Wong, E. (1975) The isoflavonoids. In *The Flavonoids,* part 2 (J. B. Harborne et al., es.). Academic Press, New York. Pp. 743–800.

Yoder, O. C., and R. P. Scheffer. (1969) Role of toxin in early interactions of *Helminthosporium victoriae* with susceptible and resistant oat tissue. *Phytopathology* **59**:1954–1959.

Yoshikawa, M. (1978) Diverse modes of action of biotic and abiotic phytoalexin elicitors. *Nature* **275**:546–547.

Yoshikawa, M., K. Yamauchi, and H. Masago. (1978) Glyceollin: Its role in restricting fungal growth in resistant soybean hypocotyls infected with *Phytophthora megasperma* var. *sojae*. *Physiol. Plant Pathol.* **12**:73–82.

Yoshikawa, M., K. Yamauchi, and H. Masago. (1979) Biosynthesis and biodegration of glyceollin by soybean hypocotyls infected with *Phytophthora megasperma* var. *sojae*. *Physiol. Plant Pathol.* **14**:157–169.

Zähringer, U., J. Ebel, L. J. Mulheirn, R. L. Lyne, and H. Grisebach. (1979) Induction of phytoalexin synthesis in soybean. Dimethylallylpyrophosphate: trihydroxypterocarpan dimethylallyl transferase from elicitor-induced cotyledons. *FEBS Lett.* **101**:90–92.

Zeigler, E., and P. Albersheim. (1977) Host-pathogen interactions. XIII. Extracellular invertases secreted by three races of a plant pathogen are glycoproteins which possess different carbohydrate structures. *Plant Physiol.* **59**:1104–1110.

NONPHYTOALEXIN HOST RESPONSES IN VASCULAR DISEASES OF PLANTS

Alberto Matta

Vascular parasites as a group are characterized by the localization and complexity of their aggressive action. They can partly escape the reaction of the host being self-limiting to the rather inert structures of the conductive system. On the basis of their ability to colonize healthy extravascular tissues, most of the vascular parasites ought to be considered weak parasites. They make up for this weakness with that peculiar adaptability to the environment of the xylem vessels not present in other organisms.

In a few vascular diseases the parasite grows in the vessels and in the annexed parenchyma cells at the same time. An extravascular phase in the roots is in every case a necessary prerequisite for infection unless direct access to the vessels is afforded by wounds. Thus even in the most typical vascular diseases three phases can generally be distinguished. The first occurs in the extravascular tissues of the roots, and has been defined as determinative (Talboys, 1958a) because of the influence it can exert on the success of the infection. It is followed by a phase of progressive colonization of the vessels, responsible for symptom expression. A final degenerative phase occurs in which the parasite escapes from the vessels and the moribund surrounding tissues.

The plant responds locally, at the point of entry of the pathogen and in the invaded vessels, or systemically.

Pathogenesis of vascular infection is complicated by the interference of toxic, lytic, nutritional, and hormonal aspects. With vascular plant diseases a uniquely wide range of morphological and physiological responses are exhibited by the host. I have restricted my attention to the morphological responses and a few physiological reactions which seem to be more strictly related.

MORPHOLOGICAL RESPONSES

Unless wounds nullify any extravascular barrier (which is quite common under normal culture conditions), the first visible responses to the vascular pathogens are found in the root cortical tissues.

In some cases penetration occurs without significant morphological reactions (Khadr and Snyder, 1967; Al-Shukri, 1968), but normally reactions follow in the epidermis, in the cortical layers and/or in the endodermis.

Piliferous, epidermal, and cortical cells generally react to penetration by producing wall thickenings which can be localized at the point of contact with the hyphal peg or extend all over the wall (Talboys, 1958b; Griffiths and Lim, 1964). Such thickenings, which are mostly due to lignin deposition, can even be detected a few cells farther from the point of invasion (Griffiths and Isaac, 1966). The root hairs are less reactive, but they respond to *Verticillium dahliae* by the deposition of electron-dense material (Griffiths, 1971).

Protrusions of the host cell wall at the point of entry of the fungal peg are known as lignitubers and have been observed in many different cases. Externally, lignitubers are composed of lignin-like substance, but their inner layer is more probably cellulosic (Griffiths and Isaac, 1966). The formation of lignitubers apparently is identical in such unrelated species as pea, sunflower, and tomato plants (Griffiths, 1973). It starts with the extrusion of vesicles in the paramural spaces. The vesicles aggregate and coalesce, giving rise to a pad of lamellar material, which appears as a distinct structure deposited on the inner side of the host cell wall (Griffiths, 1971, 1973).

Deposition of suberin in the endodermal cell walls can be another reaction to fungal invasion in the roots of some plants (Talboys, 1958a) but not in others (Selman and Buckley, 1959; Khadr and Snyder, 1967).

In the vessels the presence of the pathogen incites a complex series of morphological alterations, which can markedly differ in various vascular diseases. These alterations can be distinguished as active or passive in relation to the participation of the host, but such a distinction cannot always be easily done. It can be presumed that erosion of the rather inert secondary wall structures is passively endured or that tyloses are an active response of the plant, but we do not know whether the collapse of the vessels wall (Chambers and Corden, 1963) is caused by pressure of the surrounding hyperplastic cells, weakening of the wall structures, or increased tension of the liquid column. Gel occlusions (Crandall and Baker, 1950; Beckman et al., 1962) may be metabolic products of the pathogen, an effect of the action of enzymes of the pathogen on cell wall constituents, or reaction products of the plant.

The pectic nature of the gelatinous plugs formed in the vessels has been suggested for different vascular wilt diseases (Waggoner and Dimond, 1955; Ludwig, 1952; Gagnon, 1967a; Beckman and Zaroogian, 1967) but substances of different nature and origin could contribute to the formation of this type of occlusion.

There is controversy concerning whether or not plugging material is derived from the lytic action of the enzymes of the parasite. Swelling of the primary wall and degradation of secondary thickenings suggest a conspicuous participation of wall-degrading enzymes in tomato plants infected with *Corynebacterium michiganense* (Wallis, 1977). The disruption of the vessel wall allows the passage of the cells of this bacterium into the intercellular spaces of the xylem parenchyma and a subsequent maceration of the surrounding tissue. In spite of this, there is no evidence that plugging material persists in the vessels. It is easy to infer that gel deposition is prevented by the strong lytic activity of the parasite.

The participation of pectic or cellulolytic enzymes of the parasite in the induction of vascular gels also has to be rejected because abnormal thickenings of the wall are more commonly detected instead of erosion or disruption (Catesson et al., 1972). In *Phialophora cinerescens*-infected carnation stems and in *Ceratocystis ulmi*-infected elm twigs the pit membranes persist and are reinforced by deposition of gums (polysaccharide and phenolic material) (Catesson et al., 1976), which could have a waterproofing effect on the vessel wall (Corden and Chambers, 1966).

Two alternative possibilities remain, that gels are derived from deposition of viscous substances produced by the parasite or from secretion of the xylem parenchyma cells. The first one is supported by the finding that the multiplication of *Pseudomonas solanacearum* in the vessels is accompanied by accumulation of large amounts of viscous material, apparently a bacterial extracellular polysaccharide (Husain and Kelman, 1958; Wallis and Truter, 1978). In support of the second alternative, swelling of perforation plates in vessels is considered to be a host reaction, which is responsible for the formation of gels in banana roots (Beckman et al., 1962). Also it has been clearly shown that the gums in infected vessels of carnation stems and elm twigs are actively secreted by xylem parenchyma cells in the paramural spaces and are extruded into the vessels through the pits (Catesson et al., 1976).

Tyloses may be associated with or produced instead of gums. The vessels can be plugged by a single large tylose or by many small tyloses, each bulging from a different pit. Their walls thicken, lignify, and brown or degenerate, releasing gums (Graniti and Matta, 1969). Pathological tyloses do not substantially differ from normal tyloses. Tylose formation is a hypertrophic phenomenon but is functionally comparable to gummosis. The prevalence of one or the other seems to depend on the size of vessels and pits (Chataway, 1949; Struckmeyer et al., 1954). According to Beckman et al. (1974) there is an inverse relationship between the occurrence of tyloses and vascular browning.

Tyloses and gumming are the most evident and constant histological responses of plants to vascular invasion, but other morphological features can occur less constantly. These are hyperplastic parenchyma tissues that can cause deformation of the vessels (Linford, 1931), rings of dedifferentiated cells that multiply actively around infected bundles (Chambers and Corden, 1963), demarcation tissues surrounding infected xylem sectors (Hellmers, 1958; Garibaldi, 1969), etc. New vascular bundles or new xylem tissues, which compensate for the dysfunctioned vessels, can be the result of increased cambial activity in many different cases (Talboys, 1958b; Schoeneweiss, 1959; Moreau, 1963). The repeated occurrence of these processes in carnation plants can modify the gross plant morphology with the formation on the base of the stem of tumor-like structures on which adventitious roots will be initiated (Moreau, 1963).

HORMONAL PHYSIOLOGICAL RESPONSES

Tyloses, activation of cambial activity, and dedifferentiation of parenchyma cells with production of new vessels all contribute along with external symptoms such as epinasty, stunting, premature leaf abscission, and adventitious roots formation to suggest hormonal imbalances in vascular diseases.

The relation between external and internal symptoms and hormonal imbalance has been shown in various ways. One is the reproduction of symptoms following treatment with growth regulators. Epinasty, stunting, stem swelling, adventitious roots, are obtained in tomato plants following treatments with auxin or auxin-like substances (Davis and Dimond, 1953; Pegg and Selman, 1959; Matta, 1962). Tyloses and gum formation were induced by the same type of growth regulators in other plants (Smalley, 1962; Sorokin et al., 1962; Matta and Gentile, 1964; Mace and Solit, 1966; Bottallico and Graniti, 1974). Also exogenous ethylene can induce various symptoms of vascular wilt such as tyloses (Bottalico and Graniti, 1974; Pegg, 1976), epinasty (Dimond and Waggoner, 1953), chlorosis, and foliar abscission. Chlorosis and foliar abscission are not always induced (Gentile and Matta, 1975), but this depends on the dose and the exposure time, the age, and the physiological state of the plant.

Another proof of the participation of hormonal substances in pathogenesis of vascular wilt is their increase preceding or associated with the expression of disease symptoms. Marked increases of β-indoleacetic acid (IAA) content have been detected in *Verticillium* (Pegg and Selman, 1959) and *Fusarium* (Matta and Gentile, 1964) infected tomato plants, in bacterial wilt of tobacco (Sequeira and Kelman, 1962) and in *Verticillium* wilt of cotton (Wiese and De Vay, 1970). In tomato plants, *Fusarium* wilt induces increases of the auxin content in the aerial parts, but not in the roots, and in the stem more than in the leaves. Changes of the auxin gradients in the plant could account for premature leaf abscission.

Increased rate of ethylene evolution is another common consequence of vascular infections (Dimond and Waggoner, 1953; Wiese and DeVay, 1970; Talboys, 1972, Gentile and Matta, 1975; Pegg and Cronshaw, 1976) and it is believed that ethylene can account for most of the symptoms of hormonal imbalance exhibited by infected plants (Pegg, 1976). It is well known that IAA is a growth promoter at low concentrations and a growth inhibitor at higher concentrations and that its inhibitory effect coincides in many tissues with enhanced ethylene production (Galston and Davies, 1969). It is not surprising, therefore, that the two hormones can have similar effects.

The condition determined by the increase of IAA content in the stem of infected plants might be compared with that occurring in plants treated with auxin-like herbicides. Ethylene is one of the factors involved even if it has no herbicidal properties by itself. The herbicide 2,4-D suppresses normal cell division and elongation at growing points, but induces increases in the weight of the stem axis. Because this aberrant growth of the stem occurs at the expense of other plant parts, the leaves age prematurely and collapse (Hanson and Slife, 1969). The hyperauxiny in the stem of infected plants may be responsible for the establishment of a dominant metabolic sink, which induces senescence and physiological dysfunction in the leaves. In such a situation ethylene manifests its toxic effects by the induction of symptoms such as yellowing and leaf abscission.

Evolution of ethylene from diseased tissues may result from damage. Its production can be induced by different phytotoxic products of vascular pathogens (Wilson et al., 1972; Mussell, 1973; Cronshaw and Pegg, 1976; Ferraris and Matta, 1977). Pectolytic enzymes of the hydrolase or lyase type deserve particular attention because they may release glucose oxidase from the plant cell walls, leading to increased production of hydrogen peroxide, which is a limiting factor in the biosynthesis of ethylene from methionine (Lund and Mapson, 1970).

Although many vascular pathogens are able to produce IAA as well as ethylene, it is generally believed that the increase of these compounds depends mainly on the activity of the host plant. Unfortunately the evidence to this regard is mostly indirect. Only Sequeira (1965), on the basis of differences in the pathways followed by the host and the parasite in the synthesis of IAA, had the opportunity to prove that during the initial and intermediate stages of the bacterial wilt of tobacco, most of the IAA was contributed by the plant and not by the parasite.

In tobacco infection by *P. solanacearum* induces marked decreases in IAA oxidase activity, but at the same time there are increases in scopoletin content (Sequeira, 1964). This *o*-dihydroxyphenol stimulates (at low concentrations) and inhibits (at high concentrations) IAA oxidase (Imbert and Wilson, 1970), and is then well suited to play a major role in the regulation of IAA levels. Similar functions can be ascribed to other *o*-dihydroxyphenols and it is believed that their accumulation can generally account for pathological hyperauxiny.

Such a mechanism of hyperauxiny has not been confirmed in the *Fusarium* wilt of tomato. Stem slices and partially purified extracts of *Fusarium* infected tomato stems showed increased IAA oxidase activity (Gentile and Matta, 1973). *Fusarium* wilt of tomato is associated with a strong production of IAA from tryptophan by stem slices *in vitro* (Matta and Gentile, 1965). There is some evidence that in this disease, as with *Fusarium* wilt of banana (Mace and Solit, 1966), IAA synthesis takes place by the quinone-mediated oxidative deamination of tryptophan proposed by Gordon and Paleg (1961). This reaction is not the usual pathway of auxin biogenesis in normal tissues, but may play quite an important role as a mechanism of auxin formation in pathological conditions. Indeed, vascular infections generally increase the availability of tryptophan, *o*-dihydroxyphenols, and polyphenol oxidase (Sequeira and Kelman, 1962; Mace and Solit, 1966; Gentile, 1967; Pegg and Sequeira, 1968; Matta et al., 1969; Matta and Gentile, 1968), which are required for a phenol–phenolase mediated system of IAA biogenesis.

Whether IAA accumulation occurs through inhibition of IAA oxidation or through a rather unusual pathway of enhanced biosynthesis, an activated metabolism of the aromatic compounds with accumulation of phenols is likely to be required in both case. It is generally accepted that the increase of phenols and of their oxidation products is a key response in the vascular diseases of plants.

NONPHYTOALEXIN RESPONSES AND RESISTANCE

Disease generally results from an interaction between host and pathogen. Interaction implies a response and healing effort by the host. The host responses are, at the same time, part of the disease syndrome and expression of defense mechanisms. Thus even responses traditionally ascribed to resistance, such as phytoalexin production, are part of the disease syndrome.

The capacity of any of the many different host reactions to act as an effective defense system depends on their readiness to give an adequate response to the pathogen. In vascular diseases this is evident in many different instances. Lignituber formation can slow down the invasion of the root cortex. They are induced in resistant and susceptible varieties by virulent or avirulent fungi, but marked differences have been detected in the speed of their formation, which is higher in incompatible combinations (Talboys, 1958a, 1964; Griffiths and Isaac, 1966).

The production of gums or tyloses in the vessels increases with disease severity. Thus they are formed in higher quantities in compatible than in incompatible combinations. Here again "the speed with which they are formed when the host is invaded" (Beckman, 1964) seems to regulate resistance (Beckman and Halmos, 1962; Beckman, 1966; Elgersma, 1973).

Other factors could act concurrently with the vessel-occluding reactions. The effectiveness of the vessel-sealing structures in arresting or retarding the systemic invasion by the parasite could be conditioned by the more or less packed structure of the xylem. An unquestionable negative correlation between conductivity for air and water of the wood and resistance to *C. ulmi* has been shown with a wide range of elm material (Elgersma, 1970). High elm susceptibility seems to be associated with the large size not only of the vessels but also of groups of continguous vessels not intermingled with parenchyma cells (McNabb et al., 1970). The higher resistance of latewood as compared to springwood and of plants grown under conditions favorable for the production of small vessels confirm the importance of xylem structure in delaying movement of *C. ulmi* (McNabb et al., 1970). Whether or not the vertical or lateral distribution of spores and mycelia can be hindered under the sole effect of morphological preinfectional features is unknown. Preinfectional mechanical barriers could hardly account for the different behavior in the same host of morphologically similar races of a pathogen. In this case the defensive role of such structures can be expressed only in conjunction with active host reactions.

The sealing effect in the vessels which delays the advance of the parasite may seriously impair water transport. Responses such as formation of adventitious roots and the production of new xylem could play an evident role in alleviating the dysfunction of the vessels.

Differences in phenol content in compatible and incompatible combinations suggest that resistance to vascular diseases is characterized by a marked postinfectional increase in phenols within the first days after inoculation (Okasha et al., 1968; Matta et al., 1969; Conway and MacHardy, 1974). Such increases appear to be particularly significant considering the localization of phenols in the tissues surrounding the vessels (Gagnon, 1967b) or in phenol-storing cells (Beckman and Mueller, 1970). Apart from the still uncertain validity of their direct antibiotic effect, the phenols can participate in the defense of the plant in a number of different ways. They can be promoters of hyperauxiny and indirectly responsible for the stimulation of tyloses and other hypertrophic and hyperplastic reactions, take part in lignification and suberization processes, infuse the gels or the vessel walls rendering them less accessible to attack by the enzymes of the parasite, etc.

Resistance against *Fusarium* wilt of tomato is increased by exogenous phenols or phenolic precursors (Raj and Mahadevan, 1970; Chet et al., 1978; Carrasco et al., 1978). The oxidation products of polyphenols are effective inhibitors of polygalacturonases. The natural increase of these substances after the infection could be a deterrent to the pectolytic activity of vascular parasites (Patil and Dimond, 1967).

Suberization, lignituber formation, tyloses, and phenolic accumulation share the common property of being more rapidly incited in resistant than in susceptible hosts. Because such reactions are highly nonspecific, and are even incited by saprophytes and by mechanical of chemical stimuli, their induction by avirulent microorganisms is not likely to depend on a specific recognition by the host. It seems more likely that the pathogen has the means for delaying the reaction of the host. There is some evidence for this: the formation of lignitubers induced by *V. albo-atrum* in tomato roots is suppressed by a high inoculum potential of the same fungus (Selman and Buckley, 1959); the rapid sealing-off of vascular infections by gels or tyloses is impeded by virulent forms of *F. oxysporum* (Beckman, 1966) and the initial accumulation of phenols induced by avirulent forms of *F. oxysporum* in tomato is suppressed in presence of the virulent form (Matta et al., 1970). Thus induced susceptibility, which turns off metabolic sequences leading to the buildup of defense barriers, seems to fit vascular diseases better than induced resistance.

There is no proof that incompatible host–parasite combinations are also characterized by rapid accumulation of growth regulators. I have failed on different occasions to detect increases of IAA content in resistant *Fusarium*-infected tomato plants (unpublished). Neither does ethylene significantly increase in incompatible *Fusarium*/tomato or *Verticillium*/tomato combinations (Gentile and Matta, 1975; Pegg and Cronshaw, 1976). Failure to detect accumulation of IAA or ethylene soon after infection does not eliminate the possibility that rapid increases occur at localized sites of invasion.

SOME IMPLICATIONS IN DISEASE MANAGEMENT

A decisive contribution to the everyday struggle against vascular diseases hardly can be expected from this incomplete and unilateral view of host responses. However, the ability to express morphological and physiological reactions that affect colonization, although innate and genetically determined, can be enhanced in different ways. There are many vascular diseases for which sources of resistance are not known or, if known, cannot be exploited for practical reasons. In such cases the possibilities of increasing the potential reactivity of the host becomes of paramount importance.

There are a number of environmental and cultural conditions that affect plant predisposition to diseases in general (Yarwood, 1976) and to vascular diseases specifically. Low soil moisture levels increase resistance against *Fusarium* wilt of tomato and *Verticillium* wilt of eggplant. In hardened tomato plants growth of the parasite is reduced (Matta and Garibaldi, 1972). High potassium and low nitrogen fertilization reduce the severity of most wilt diseases (Meyer, 1972). Unfortunately little is known about the relation between host nutrition and magnitude of the defensive host responses.

Are there other ways to manage vascular diseases based on the enhancement of the plant reactivity? The relationships between growth regulators and morphological responses in the xylem suggest that these substances might induce resistance against vascular parasites. With an empirical approach it was shown many years ago that auxin-like regulators can delay symptom expression of different diseases such as *Fusarium* wilt of tomato (Davis and Dimond, 1953), Dutch elm disease (Brener and Beckman, 1968), and *Phialophora* wilt of carnation (Matta et al., 1966). In *Phialophora* wilt of carnation, the protective effect was obtained under normal cultural conditions and was compatible with the commercial production of flowers. Auxin-like substances modify the structure of the vessels and enhance the formation of tyloses after inoculation (Brener and Beckman, 1968). The effect of ethylene is more controversial. It has been reported to enhance (Retig, 1974; Pegg, 1976) and to decrease (Gentile and Matta, 1975) resistance against vascular fungi. The paradoxical role of ethylene as a resistance inducer and a toxin synergist has been discussed by Pegg (1976).

Growth retardants also might be successfully applied against vascular diseases (Buchenauer and Erwin, 1976). They have morphological effects similar in part to those induced by auxin-like growth regulators. The mechanism of their protective action is quite intriguing, but has to be completely elucidated.

It is surprising that a strategy of vascular disease control based on the use of growth regulators has not been investigated more thoroughly. The development of systemic chemicals which can stimulate the plant to form antifungal or antibacterial factors would be a good subject for future research.

Predisposition towards vascular parasites also can be affected by previous inoculation with avirulent fungi or bacteria. This effect, known as preimmunity, cross-protection, or induced resistance, has been considered in more general terms by Kuć (this volume). The validity of this phenomenon has been shown many times with vascular diseases and exploited with success even in commercial crops (Matta, 1978). Avirulent invaders can induce resistance by stimulating production of defense barriers preceding the infection by the pathogen such as phenols and/or more specific antimicrobial substances, tyloses and gums in the vessels, new xylem and lateral roots. Induced resistance could also result in a state of increased sensitivity, enabling the tissues to react more promptly in case of infection by the virulent pathogen.

The extent and speed of the host reaction are influenced by the level of the inoculum potential. A high inoculum potential can suppress or delay the host responses. Modification of the level of inoculum in the soil could be another means of enhancing the effectiveness of defense mechanisms. The adjustment of soil fertility, with the aim of starving the pathogen, seems to be a correct and innovative approach to the problem. Woltz and Jones (1973) have shown that the multiplication of a vascular pathogen in the soil and the resulting disease level can be reduced by avoiding the presence of readily available organic matter, by reducing soil application fertilizers in favor of foliar sprays, and by avoiding unnecessary application of micronutrients to the soil.

REFERENCES

Al-Shukri, M. M. (1968) Mode of penetration of *Fusarium* and *Verticillium* species into intact radicle of cotton in culture. *Biológia (Bratislava)* **23**:819–824. In *Rev. Appl. Mycol.* **48**:211.

Beckman, C. H. (1964) Host responses to vascular infection. *Annu. Rev. Phytopathol.* **2**:231–252.

Beckman, C. H. (1966) Cell irritability and localization of vascular infections in plants. *Phytopathology* **56**:821–824.

Beckman, C. H., and S. Halmos. (1962) Relation of vascular occluding reactions in banana roots to pathogenicity of root-invading fungi. *Phytopathology* **52**:893–897.

Beckman, C. H., and W. C. Mueller. (1970) Distribution of phenols in specialized cells of banana roots. *Phytopathology* 60:79–82.

Beckman, C. H., and G. E. Zaroogian. (1967) Origin and composition of vascular gel in infected banana roots. *Phytopathology* **57**:11–13.

Beckman, C. H., S. Halmos, and M. E. Mace. (1962) The interaction of host, pathogen, and soil temperature in relation to susceptibility to *Fusarium* wilt of bananas. *Phytopathology* **52**:134–140.

Beckman, C. H., W. C. Mueller, and M. E. Mace. (1974) The stabilization of artificial and natural cell wall membranes by phenolic infusion and its relation to wilt disease resistance. *Phytopathology* **64**:1214–1220.

Bottalico, A., and A. Graniti. (1974) Fusicoccina, acido indolil-3-acetico e formazione di tille. *Phytopathol. Mediterr.* **13**:163–168.

Brener, W. D., and C. H. Beckman. (1968) A mechanism of enhanced resistance to *Ceratocystis ulmi* in American elms treated with sodium trichlorophenylacetate. *Phytopathology* **58**:555–561.

Buchenauer, H., and D. C. Erwin. (1976) Effect of the plant growth retardant Pydanon on *Verticillium* wilt of cotton and tomato. *Phytopathology* **66**:1140–1143.

Carrasco, A., A. M. Boudet, and G. Marigo. (1978) Enhanced resistance of tomato plants to *Fusarium* by controlled stimulation of their natural phenolic production. *Physiol. Plant Pathol.* **12**:225–232.

Catesson, A. M., Y. Czaninski, M. Péresse, and M. Moreau. (1972) Modification des parois vasculaires de l'oeillet infecté par le *Phialophora cinerescens. C. R. Acad. Sci. (Paris)* **275**:827–829.

Catesson, A. M., Y. Czaninski, M. Péresse, and M. Moreau. (1976) Sécrétion intravasculaire de substances gommeuses par les cellules associées aux vaisseaux en réaction à une attaque parasitaire. *Soc. Bot. Fr., Coll. Secrét. Végét.* **123**:93–107.

Chambers, H. L., and M. E. Corden. (1963) Semeiography of *Fusarium* wilt of tomato. *Phytopathology* **53**:1006–1010.

Chattaway, M. M. (1949) The development of tyloses and secretion of gum in heartwood formation. *Aust. J. Sci. Res., Ser. B* **2**:227–240.

Chet, I., D. Havkin, and J. Katan. (1978) The role of catechol in inhibition of *Fusarium* wilt. *Phytopathol. Z.* **91**:60–66.

Conway, W. S., and W. E. MacHardy. (1974) Phenol accumulation in resistant and susceptible tomato plants infected by *Fusarium oxysporum* f. sp. *lycopersici*, race 1 or race 2. *Proc. Am. Phytopathol. Soc.* **1**:134–135.

Corden, M. E., and H. L. Chambers. (1966) Vascular dysfunction in *Fusarium* wilt of tomato. *Am. J. Bot.* **53**:284–287.

Crandall, B. S., and W. L. Baker. (1950) The wilt disease of American persimmon caused by *Cephalosporium diospyri*. *Phytopathology* **40**:307–325.

Cronshaw, D. K., and G. F. Pegg. (1976) Ethylene as a toxin synergist in *Verticillium* wilt of tomato. *Physiol. Plant Pathol.* **9**:33–44.

Davis, D., and A. E. Dimond. (1953) Inducing disease resistance with plant growth-regulators. *Phytopathology* **43**:137–140.

Dimond, A. E., and P. E. Waggoner. (1953) The cause of epinastic symptoms in *Fusarium* wilt of tomatoes. *Phytopathology* **43**:663–669.

Elgersma, D. M. (1970) Length and diameter of xylem vessels as factors in resistance of elms to *Ceratocystis ulmi*. *Neth. J. Plant Pathol.* **76**:179–182.

Elgersma, D. M. (1973) Tylose formation in elms after inoculation with *Ceratocystis ulmi*, a possible resistance mechanism. *Neth. J. Plant Pathol.* **79**:218–220.

Ferraris, L., and A. Matta. (1977) Sintomi di fitotossicità, produzione di etilene e attività perossidasica indotti in recisi di pomodoro da enzimi pectolitici di *Fusarium oxysporum* f. sp. *lycopersici*. *Phytopathol. Mediterr.* **16**:87–95.

Gagnon, C. (1967a) Histochemical studies on the alteration of lignin and pectic substances in white elm, infected by *Ceratocystis ulmi*. *Can. J. Bot.* **45**:1619–1623.

Gagnon, C. (1967b) Polyphenols and discoloration in the elm disease investigated by histochemical techniques. *Can. J. Bot.* **45**:2119–2124.

Galston, A. W., and P. J. Davies. (1969) Hormonal regulation in higher plants. *Science* **163**:1288–1297.

Garibaldi, A. (1969) Personal communication. Istituto di Patologia vegetale dell'Università di Torino, Italy.

Gentile, I. (1967) Ulteriori ricerche sull'iperauxinia nella tracheofusariosi del pomodoro: Variazioni del contenuto in triptofano libero. *Phytopathol. Mediterr.* **6**:168–170.

Gentile, I., and A. Matta. (1973) Idoleacetic acid oxidase activity in relation to hyperauxiny in *Fusarium* wilt of tomato. *Phytopathol. Mediterr.* **12**:43–47.

Gentile, I., and A. Matta. (1975) Production of and some effect of ethylene in relation to *Fusarium* wilt of tomato. *Physiol. Plant Pathol.* **5**:27–37.

Gordon, S. A., and L. G. Paleg. (1961) Formation of auxin from tryptophan through action of polyphenols. *Plant Physiol.* **36**:838–845.

Graniti, A., and A. Matta. (1969) Indirizzi attuali degli studi sulla patogenesi delle malattie vascolari. *Ann. Phytopathol.* **1** (hors de série):77–120.

Griffiths, D. A. (1971) The development of lignitubers in roots after infection by *Verticillium dahliae*. *Can. J. Bot.* **17**:441–444.

Griffiths, D. A. (1973) An electron microscopic study of host reaction in roots following invasion by *Verticillium dahliae*. *Shokubutsu Byogai Kenkyu* **8**:147–154.

Griffiths, D. A., and I. Isaac. (1966) Host parasite relationships between tomato and pathogenic isolates of *Verticillium*. *Ann. Appl. Biol.* **58**:259–272.

Griffiths, D. A., and W. C. Lim. (1964) Mechanical resistance in root hairs to penetration by species of vascular wilt fungi. *Mycopathol. Mycol. Appl.* **24**:103–112.

Hanson, J. B., and F. W. Slife. (1969) Role of RNA metabolism in the action of auxin-herbicides. *Residue Rev.* **25**:59–67.

Hellmers, E. (1958) Four wilt diseases of perpetual-flowering carnations in Denmark. *Dan. Bot. Ark.* **18**:1–200.

Husain, A., and A. Kelman. (1958) Relation of slime production to mechanism of wilting and pathogenicity of *Pseudomonas solanacearum*. *Phytopathology* **48**:155–165.

Imbert, M. P., and L. A. Wilson. (1970) Stimulatory and inhibitory effects of scopoletin on IAA oxidase preparations from sweet potato. *Phytochemistry* **9**:1787–1794.

Khadr, A. S., and W. C. Snyder. (1967) Histology of early stage of penetration in *Fusarium* wilt of cotton. *Phytopathology* **57**:99.

Linford, M. B. (1931) Transpirational history as a key to the nature of wilting in the *Fusarium* wilt of peas. *Phytopathology* **21**:791–826.

Ludwig, R. A. (1952) Studies on the physiology of hadromycotic wilting in the tomato plant. *MacDonald Coll. Tech. Bull.* **20**:1–40.

Lund, B. M., and L. W. Mapson. (1970) Stimulation by *Erwinia corotovora* of the synthesis of ethylene in cauliflower tissues. *Biochem. J.* **119**:251–263.

Mace, M. E., and E. Solit. (1966) Interactions of 3-indoleacetic acid and 3-hydroxytyramine in *Fusarium* wilt of banana. *Phytopathology* **56**:244–247.

Matta, A. (1962) Endoterapia della fusariosi del pomodoro con sostanze ad azione regolatrice de crescita. *Riv. Patol. Veg.* **2**:234–244.

Matta, A. (1978) Esperienze di lotta biologica contro la verticilliosi del pomodoro mediante preinoculazione di funghi ipovirulenti. *Atti Giornate Fitopat., Acireale, Italy, March 8–10,* **1978**:325–331.

Matta, A., and A. Garibaldi. (1972) Influenza di diversi regimi indrici su alcune fitopatie da funghi del terreno. *Agric. Ital. (Pisa)* **72**:237–253.

Matta, A., and Gentile. (1964) Variazioni del contenuto in auxine indotte nel pomodoro dal *Fusarium oxysporum* f. *lycopersici*. *Riv. Patol. Veg.* **4**:208–237.

Matta, A., and I. Gentile. (1965) Sul meccanismo di accumulo dell'acido β-indolilacetico in piante di pomodoro infette da *Fusarium oxysporum* f. sp. *lycopersici*. *Phytopathol. Mediterr.* **4**:129–137.

Matta, A., and I. Gentile. (1968) The relation between polyphenoloxidase activity and ability to produce indoleacetic acid in *Fusarium*-infected tomato plants. *Neth. J. Plant Pathol. (Suppl.)* **74**:47–51.

Matta, A., A. Garibaldi, and M. Palenzona. (1966) Impiego dell'acido naftalenacetico contro le tracheomicosi del graofano. *Atti 1° Congr. Un. Fitopat. Mediterr.* **1**:218–222.

Matta, A., I. Gentile, and I. Giai. (1969) Accumulation of phenols in tomato plants infected by different forms of *Fusarium oxysporum*. *Phytopathology* **59**:512–513.

Matta, A., I. Gentile, and I. Giai. (1970) Effect of mixed inoculation with *Fusarium oxysporum* f. sp. *lycopersici* and *F. oxysporum* f. sp. *dianthi* on the phenol content of tomato plants. *Neth. J. Plant Pathol.* **76**:144–146.

McNabb, H. S., H. M. Heybroek, and W. L. MacDonald. (1970) Anatomical factors in resistance to Dutch elm disease. *Neth. J. Plant Pathol.* **76**:196–204.

Meyer, J. A. (1972) Soil-borne disease and fertilizer responses. *Institut International Potasse* **9**:449–460.

Moreau, M. (1963) Responses de l'oeillet aux attaques du *Fusarium roseum* et du *Phialophora cinerescens*. *Atti Conv. Internaz. Garofano, San Remo, Italy* **1963**:56–75.

Mussel, H. W. (1973) Endopolygalacturonase: Evidence for involvement in *Verticillium* wilt of cotton. *Phytopathology* **63**:62–70.

Okasha, K. A., K. Ryugo, S. Wilhelm, and R. S. Bringhurst. (1968) Inhibition of growth of *Verticillium albo-atrum* sporelings by tannins and polyphenols from infected crowns of *Verticillium*-resistant and -susceptible strawberry plants. *Phytopathology* **58**:1114–1117.

Patil, S. S., and A. E. Dimond. (1967) Inhibition of *Verticillium* polygalacturonase by oxidation products of polyphenols. *Phytopathology* **57**:492–496.

Pegg, G. F. (1976) The response of ethylene-treated tomato plants to infection by *Verticillium albo-atrum*. *Physiol. Plant Pathol.* **9**:215–226.

Pegg, G. F., and D. K. Cronshaw. (1976) Ethylene production in tomato plants infected with *Verticillium albo-atrum*. *Physiol. Plant Pathol.* **8**:279–295.

Pegg, G. F., and I. W. Selman. (1959) An analysis of the growth response of young tomato plants to infection by *Verticillium albo-atrum*. II. The production of growth substances. *Ann. Appl. Biol.* **47**:222–231.

Pegg, G. F., and L. Sequeira. (1968) Stimulation of aromatic biosynthesis in tobacco plants infected with *Pseudomonas solanacearum*. *Phytopathology* **58**:476–483.

Raj, S. A., and A. Mahadevan. (1970) Induction of wilt resistance in cotton by phenolic compounds. *Indian Phytopathol.* **23**:89–94.

Retig, N. (1974) Changes in peroxidase and polyphenoloxidase associated with natural and induced resistance of tomato to *Fusarium* wilt. *Physiol. Plant Pathol.* **4**:145–150.

Schoeneweiss, D. F. (1959) Xylem formation as a factor in oak wilt resistance. *Phytopathology* **49:335**–337.

Selman, I. W., and W. R. Buckley. (1959) Factors affecting the invasion of tomato roots by *V. albo-atrum*. *Trans. Br. Mycol. Soc.* **42**:227–234.

Sequeira, L. (1964) Inhibition of indoleacetic acid oxidase in tobacco plants infected by *Pseudomonas solanacearum*. *Phytopathology* **54**:1078–1083.

Sequeira, L. (1965) Origin of indoleacetic acid in tobacco plants infected by *Pseudomonas solanacearum*. *Phytopathology* **55**:1232–1236.

Sequeira, L., and A. Kelman. (1962) The accumulation of growth substances in plants infected by *Pseudomonas solanacearum*. *Phytopathology* **52**:439–448.

Smalley, E. B. (1962) Prevention and Dutch elm disease by treatments with 2,3,6-trichlorophenyl acetic acid. *Phytopathology* **52**:1090–1091.

Sorokin, H. P., S. N. Mathur, and K. V. Thimann. (1962) The effects of auxins and kinetin on xylem differentiation in the pea epicotyl. *Am. J. Bot.* **49**:444–454.

Struckmeyer, B. E., C. H. Beckman, J. E. Kuntz, and A. J. Riker. (1954) Plugging of vessels by tyloses and gums in wilting oaks. *Phytopathology* **44**:148–153.

Talboys, P. W. (1958a) Some mechanisms contributing to *Verticillium*-resistance in the hop root. *Trans. Br. Mycol. Soc.* **41**:227–241.

Talboys, P. W. (1958b) Association of tyloses and hyperplasia of the xylem with vascular invasion of the hop by *Verticillium albo-atrum*. *Trans. Br. Mycol. Soc.* **41**:249–260.

Talboys, P. W. (1964) A concept of the host-parasite relationship in *Verticillium* wilt diseases. *Nature* **202**:361–364.

Talboys, P. W. (1972) Resistance to vascular wilt fungi. *Proc. R. Soc. London, Ser. B* **181**:319–332.

Waggoner, P. E., and A. E. Dimond. (1955) Production and role of extracellular pectic enzymes of *Fusarium oxysporum* f. *lycopersici*. *Phytopathology* **45**:79–87.

Wallis, F. M. (1977) Ultrastructural histopathology of tomato plants infected with *Corynebacterium michiganense*. *Physiol. Plant Pathol.* **11**:333–342.

Wallis, F. M., and S. J. Truter. (1978) Histopathology of tomato plants infected with *Pseudomonas solanacearum,* with emphasis on ultrastructure. *Physiol. Plant Pathol.* **13**:307–317.

Wiese, M. V., and J. E. DeVay. (1970) Growth regulator changes in cotton associated with defoliation caused by *Verticillium albo-atrum*. *Plant Physiol.* **45**:304–309.

Wilson, D. M., B. Etherton, and R. Jagels. (1972) Evidence for ethylene injury in fusaric acid treated tomato. *Phytopathology* **62**:501.

Woltz, S. S., and J. P. Jones. (1973) Tomato *Fusarium* wilt control by adjustment in soil fertility: A systemic approach to pathogen starvation. *Bradenton AREC Res. Rep. GC 7*. 3 pp.

Yarwood, C. E. (1976) Modification of the host response: Predisposition. In *Encyclopedia of Plant Physiology,* Vol. 4 (R. Heitfuss and P. H. Williams, eds.). Springer-Verlag, New York. Pp. 703–718.

LACK OF CORRELATION BETWEEN FIELD AND LABORATORY TESTS FOR RESISTANCE WITH SPECIAL REFERENCE TO WHITE ROT OF ONIONS

James E. Rahe

Field resistance is usually shunned as a subject for physiological studies because of real and perceived difficulties associated with its detection or characterization in the laboratory or greenhouse environment. This avoidance derives in part from the quantitative nature of field resistance, but mainly from the fact that many types of field resistance do not readily lend themselves to analysis in terms of individual plants. These latter types of resistance are likely to be horizontal in nature and function to reduce the rate of epidemic increase of disease in genetically uniform host populations. From a practical viewpoint they are, in some situations, more valuable than quantitative vertical resistance. Robinson (1979) deals elegantly with the integrated function of vertical and horizontal resistance in wild and domesticated crop pathosystems, and gives procedures for exploiting horizontal field resistance by plant breeders.

Just as plant breeders have not yet achieved a balanced exploitation of vertical and horizontal resistance, with resulting "suboptimization of the crop pathosystem" (Robinson, 1979), so too has the study of plant disease resistance mechanisms been unbalanced, and a bias in knowledge of host–parasite interaction has been the result. The subject of this chapter is an examination of bases for noncorrelation between the expression of resistance in field and laboratory environments. Better understanding of causes for noncorrelation should serve to catalyze research effort on the physiological nature of the horizontal component of field resistance.

THE PROBLEM OF TERMINOLOGY

Roane (1973), in his review of trends in breeding for disease resistance in crops, credits Van der Plank (1963) for initiating a new era in resistance breeding by introduction of the terms "horizontal and vertical resistance." Although the concepts were already embodied in the older terms *qualitative* and *quantitative* resistance, the new terms served to intensify, in both principle and practice, awareness of the diversity of relationships between hosts and parasites.

Plant breeders are concerned with the range, level, durability (Johnson and Law, 1975; Johnson, 1979), and inheritance of resistance, and a frightening multiplicity of terms has been generated to deal with these concepts (Robinson, 1969). The term "field resistance" was described by Russell (1978) as "resistance which gives an effective control of a parasite under natural conditions in the field, but which is difficult to detect or to characterize in laboratory or greenhouse tests." Van der Plank's concept of field resistance is broader: "Resistance in the field is horizontal resistance, if measures are taken to exclude vertical resistance. Field resistance, mainly as horizontal resistance, was collected consciously or unconsciously down the ages, and was often strong." (Van der Plank, 1968).

While in this chapter we use the term "field resistance" in the broadest sense, incomplete or quantitative resistance is implicit because it is with quantitative resistance that noncorrelation between field and laboratory tests is most frequently encountered.

LACK OF CORRELATION BETWEEN FIELD AND LABORATORY TESTS FOR RESISTANCE DETECTED ON AN INDIVIDUAL PLANT BASIS

Noncorrelations will be considered in two categories. The first category involves resistance which is readily demonstrable on an individual plant basis. Noncorrelations within this category are due to critical variation between field and laboratory environments in one or more factors that directly affect individual units of host or parasite. The factors may be physical (e.g., temperature, humidity, water potential, pH, nutrient status, etc.) or biological (e.g., unrecognized pathogen biotypes, unrecognized effects of incidental organisms, antagonists, rhizosphere colonists, etc.). Many examples of noncorrelations due to environmental variation are well known, and the major physical (Colhoun, 1973; Cook and Papendick, 1972; Dimock, 1967; Jones, 1924; Walker, 1965; Yarwood, 1959) and biological (Atkinson et al., 1975; Klingner et al., 1971; Rovira, 1965; Schroth and Hildebrand, 1964; Sidhu and Webster, 1977) variables and their effects on host–pathogen interaction have been reviewed.

In theory, all cases of noncorrelation due to factors affecting host–parasite interaction on an individual plant basis can be resolved. The majority of such cases that are documented in the literature have been resolved, partly because there are barriers to publication of those that have not been resolved.

LACK OF CORRELATION BETWEEN FIELD AND LABORATORY TESTS FOR RESISTANCE DETECTED IN POPULATIONS OF PLANTS

While the more familiar types of resistance can be detected on an individual plant basis as a qualitative or quantitative effect, there are examples of field resistance that are readily detected only as quantitative effects in comparative populations of plants, at least under the constraints of our present level of understanding. These examples constitute the second category of noncorrelation between field and laboratory tests.

Two situations can give rise to resistance which is readily detected only as a quantitative effect in comparative populations of plants. The first is where the mechanism for primary (1°) infection is independent (−) of the resistance in question, while that for secondary (2°) infection is sensitive (+) to it. The second situation is the qualified reciprocal of the first, the qualification being that the difference in resistance to primary infection between the comparative populations is small. (Where the difference is large, the resistance is readily detected on an individual plant basis.) Small differences can be potentially useful when they are substantially amplified by resistance-independent secondary infection, which is possible in multiple-cycle diseases where the inoculum potential of the pathogen, as defined by Garrett (1956), is such that there is a low probability of primary infection occurring under field conditions on any individual plant.

For clarification, consider as respective examples of these two cases: (1) 20-fold and 30-fold amplifications, via resistance-sensitive secondary infections, of a common 2% level of resistance-independent primary infection in comparative host populations (1°−/2°+), and (2) a common 20-fold increase, via resistance-independent secondary infections, of 2% and 3% levels of resistance-sensitive primary infection in comparative populations (1°+/2°−). In both cases, the ultimate field losses are 40% and 60% in the resistant and susceptible populations, respectively. The occurrence of these mechanisms would not exclude the simultaneous occurrence of additional types of resistance in the same plant populations.

Resistance of the 1°−/2°+ type is more readily detected than is the reciprocal type. It may be encountered in multiple-cycle diseases where the resistance is expressed as a longer incubation period, a reduced rate of inoculum production, or reduced amounts of inoculum produced per infection site. Some types of slow rusting (Brown and Shipton, 1964; Martin et al., 1977) and some types of horizontal resistance in potato (Lapwood 1961; Russell, 1978) provide examples of this situation.

Resistance of the 1°+/2°− type is difficult to detect in cases where the difference in resistance to primary infection is small. Small differences are readily obscured by variation of environmental factors. Its detection is further complicated in the case of root diseases with long and variable incubation periods and indistinct foliar symptoms where non-destructive methods for detecting infection are lacking, and criteria for discrimination between primary and secondary infections are vague. Detection of 1°+/2°−resistance in the field is facilitated by large plant populations and uniform conditions in the test plots. Statistical analysis is essential.

WHITE ROT OF *ALLIUM* SPP.

Does 1°+/2°−resistance exist? Utkhede and Rahe (1978) reported resistance in *Allium cepa* to *Sclerotium cepivorum*, the causal agent of white rot, which appears to be of this type. We have subsequently established, based on five separate field trials over 3 years, that the field resistance of cultivars such as Ailsa Craig, Wolska, and Festival is reproducible in the muck soils of coastal British Columbia. We also obtained a statistically significant correlation between percent infection in the field and laboratory trials using muck soil, but the correlation was not close enough to permit effective laboratory screening for resistance. We tested 3-week-old onion seedlings for resistance by placing them in vials containing mycelium of *S. cepivorum* and found no evidence of resistance in any of the 64 cultivars tested, including ones with significant field resistance (Utkhede and Rahe, unpublished information). Haley (1978) compared the cultivars Ailsa Craig, Dako, and Autumn Spice grown in autoclaved sand–cornmeal mix inoculated with *S. cepivorum* at the University of Wisconsin (Ailsa Craig and Dako were the most resistant cultivars in our tials, and Autumn Spice was one of the more susceptible). She writes "In these tests the three onions either reacted the same or Autumn Spice had the greatest proportion of live plants after one month."

These results are not contradictory. Rather, they indicate that there is no resistance to *S. cepivorum* in the cultivars tested when the attack is by mycelium of the fungus growing from a nutrient base in the absence of soil fungistasis (Baker and Cook, 1974). The resistance observed in the field trials, and to a lesser extent in field soil in pots in the greenhouse, must therefore be expressed prior to establishment of a pathogen mycelium–host tissue interface; this most likely occurs via differential stimulation of sclerotial germination by resistant and susceptible cultivars.

Primary infection is sensitive to the resistance mechanism, whereas secondary infection is not. Secondary spread from infected to healthy onion plants is analogous to attack from a strong nutrient base in the absence of fungistasis, and is independent of the resistance mechanism in this case.

There is evidence that even nonhosts lack resistance to *S. cepivorum* at the mycelium–tissue interface. Young and Allen (1969) showed that nonhost seedlings were attacked when inoculated with a malt extract agar disc bearing mycelium of *S. cepivorum*. Earlier, Coley-Smith (1959) showed that mycelial inoculum of *S. cepivorum* did not attack plants in genera other than *Allium* when growing in soil. We tested sclerotia and mycelium of *S. cepivorum* in separate trials on 26 plant species representing 19 genera growing in muck soil and obtained results similar to those of Coley-Smith (Utkhede and Rahe, unpublished data). Thus, the specificity of *S. cepivorum* for *Allium* spp. is not due to tissue resistance of nonhost species, but to soil fungistasis (Allen and Young, 1968; Coley-Smith et al., 1967) acting in conjunction with specific stimulation of sclerotial germination by *Allium* spp. (Coley-Smith and Holt, 1966).

Recognition of the likelihood of a 1°+/2°− mechanism of resistance in *A. cepa* to *S. cepivorum* has implications beyond providing an explanation for noncorrelation between laboratory and field tests. If it can be established that field resistance of onion cultivars is due to qualitative or quantitative differences in production of sclerotial germination stimulants, it is likely that a rapid and effective laboratory screening technique can be developed based on direct measurement of these substances.

The importance of secondary spread of *S. cepivorum* in increasing disease loss is evident from the typical occurrence of irregular infection foci (groups of diseased plants) from evenly distributed inoculum in both laboratory (Scott, 1956b) and field (Ryan and Kavanagh, 1977) tests. Scott (1956a) and Ryan and Kavanagh (1977) have established that *S. cepivorum* has a negligible capacity to spread through soil as mycelium, hence it can be concluded that secondary infection requires virtual plant-to-plant contact. Clearly, interplant spacing is a critical variable in screening for 1°+/2°−resistance, particularly where the criteria for discrimination between primary and secondary infections are vague or subjective.

In theory, some control of white rot is possible via increasing interplant spacing. Whether this could be practical depends on the degree of disease increase via secondary infection. Ryan and Kavanagh (1977) provided diagrammatic representations of entire test plots showing the spatial distribution of diseased and healthy plants. From their 1972 field trial data it can be estimated (from the ratio of infected plants to infection foci) that disease increase via secondary infection was 2.5- to 4-fold. At a 2.5-fold level of disease increase it is doubtful whether increasing interplant spacing would provide a significant increase in the yield of healthy onions.

Under other conditions, secondary increase might occur at a level higher than that which was estimated from the data of Ryan and Kavanagh (1977). White rot is influenced by temperature, soil moisture, pH, inoculum density, and other factors (Adams and Papavizas, 1971; Utkhede et al., 1978; Walker, 1924), but it is not known whether these factors affect primary or secondary infection or both. Recognition of a likely 1°+/2°− mechanism of resistance clearly underscores the need for reevaluation of the effects of these environmental variables.

CONCLUSIONS

1. Noncorrelations between field and laboratory tests for resistance limit our capacity to screen for resistance, and may provide clues to aid in identification of previously unrecognized variables affecting host–parasite interaction.
2. Resistance which is readily detectable only in comparative populations of plants may occur in situations where the mechanism for primary infection is independent of the resistance in question, while that for secondary infection is sensitive to it (1°−/2°+), and vice versa (1°+/2°−).
3. Analysis of appropriate diseases in terms of 1°−/2°+ or 1°+/2°−resistance mechanisms can serve to focus experimental attention on critical aspects of host–parasite interaction. Improved techniques for resistance testing or screening, and better understanding of environmental influences and their possible exploitation via modified cultural practice may result.

REFERENCES

Adams, P. B., and G. C. Papavizas. (1971) Effect of inoculum density of *Sclerotium cepivorum* and some soil environmental factors on disease severity. *Phytopathology* **61**:1253–1256.

Allen, J. D., and J. M. Young. (1968) Soil fungistasis and *Sclerotium cepivorum* Berk. *Plant Soil* **39**:479–480.

Atkinson, T. G., J. L. Neal, Jr., and R. Larson. (1975) Genetic control of the rhizosphere microflora of wheat. In *Biology and Control of Soil-Borne Plant Pathogens* (G. W. Bruehl, ed.). American Phytopathological Society, St. Paul, Minn. Pp. 116–122.

Baker, K. F., and R. J. Cook. (1974) *Biological Control of Plant Pathogens*. W. H. Freeman, San Francisco, Calif. 433 pp.

Brown, J. F., and W. A. Shipton. (1964) Relationship of penetration to infection type when seedling wheat leaves are inoculated with *Puccinia graminis tritici*. *Phytopathology* **54**:89–91.

Coley-Smith, J. R. (1959) Studies on the biology of *Sclerotium cepivorum* Berk. III. Host range: Persistence and viability of sclerotia. *Ann. Appl. Biol.* **47**:511–518.

Coley-Smith, J. R., and R. W. Holt. (1966) The effect of species of *Allium* on germination in soil of sclerotia of *Sclerotium cepivorum* Berk. *Ann. Appl. Biol.* **58**:273–278.

Coley-Smith, J. R., J. E. King, D. J. Dickinson, and R. W. Holt. (1967) Germination of sclerotia of *Sclerotium cepivorum* Berk. under aseptic conditions. *Ann. Appl. Biol.* **60**:109–115.

Colhoun, J. (1973) Effects of environmental factors on plant disease. *Annu. Rev. Phytopathol.* **11**:343–364.

Cook, R. J., and R. I. Papendick. (1972) Influence of water potential of soils and plants on root diseases. *Annu. Rev. Phytopathol.* **10**:349–374.

Dimock, A. W. (1967) Controlled environment in relation to plant disease research. *Annu. Rev. Phytopathol.* **5**:265–284.

Garrett, S. D. (1956) *Biology of Root-Infecting Fungi.* Cambridge University Press, Cambridge. 293 pp.

Haley, A. B. (1978) Personal communication. 112 Whitecliff Dr., Vallejo, Calif. 94590.

Jones, L. R. (1924) The relation of environment to disease in plants. *Am. J. Bot.* **11**:601–609.

Johnson, R. (1979) The concept of durable resistance. *Phytopathology* **69**:198–199.

Johnson, R., and C. N. Law. (1975) Genetic control of durable resistance to yellow rust (*Puccinia striiformis*) in the wheat cultivar Hybride de Bersée. *Ann. Appl. Biol.* **81**:385–391.

Klingner, A. E., D. C. Hildebrand, and S. Wilhelm. (1971) Occurrence of *Erwinia carotovora* in the rhizosphere of cotton plants which escape *Verticillium wilt. Plant Soil* **34**:215–218.

Lapwood, D. H. (1961) Potato haulm resistance to *Phytophthora infestans.* II. Lesion production and sporulation. *Ann. Appl. Biol.* **49**:316–330.

Martin, C. D., L. J. Littlefield, and J. D. Miller. (1977) Development of *Puccinia graminis* f. sp. *tritici* in seedling plants of slow-rusting wheats. *Trans. Br. Mycol. Soc.* **68**:161–166.

Roane, C. W. (1973) Trends in breeding for disease resistance in crops. *Annu. Rev. Phytopathol.* **11**:463–486.

Robinson, R. A. (1969) Disease resistance terminology. *Rev. Appl. Mycol.* **48**:593–606.

Robinson, R. A. (1979) Permanent and impermanent resistance to crop parasites: A reexamination of the pathosystem concept with special reference to rice blast. *Z. Pflanzenzuecht.* **83**:1–39.

Rovira, A. D. (1965) Interactions between plant roots and soil microorganisms. *Annu. Rev. Microbiol.* **19**:241–266.

Russell, G. E. (1978) *Plant Breeding for Pest and Disease Resistance.* Butterworths, Toronto, Canada. 485 pp.

Ryan, E. W., and T. Kavanagh. (1977) White rot of onion (*Sclerotium cepivorum*). 3. Epidemiology. *Ir. J. Agric. Res.* **16**:57–63.

Schroth, M. N., and D. C. Hildebrand. (1964) Influence of plant exudates on root-infecting fungi. *Annu. Rev. Phytopathol.* **2**:101–132.

Scott, M. R. (1956a) Studies of the biology of *Sclerotium cepivorum* Berk. I. Growth of the mycelium in soil. *Ann. Appl. Biol.* **44**:576–583.

Scott, M. R. (1956b) Studies of the biology of *Sclerotium cepivorum* Berk. II. The spread of white rot from plant to plant. *Ann. Appl. Biol.* **44**:584–589.

Sidhu, G. S., and J. M. Webster. (1977) Genetics of simple and complex host-parasite interactions. In *Induced Mutations Against Plant Diseases.* Proceedings of the International Atomic Energy Agency, Vienna. 580 pp.

Utkhede, R. S., and J. E. Rahe. (1978) Screening commerical onion cultivars for resistance to white rot. *Phytopathology* **68**:1080–1083.

Utkhede, R. S., J. E. Rahe, and D. J. Ormrod. (1978) Occurrence of *Sclerotium cepivorum* sclerotia in commercial onion farm soils in relation to disease development. *Plant Dis. Rep.* **62**:1030–1034.

Van der Plank, J. E. (1963) *Plant Diseases: Epidemics and Control.* Academic Press, New York. 349 pp.

Van der Plank, J. E. (1968) *Disease Resistance in Plants.* Academic Press, New York. 206 pp.

Walker, J. C. (1924) White rot of *Allium* in Europe and America. *Phytopathology* **14**:315–322.

Walker, J. C. (1955) Use of environmental factors in screening for disease resistance. *Annu. Rev. Phytopathol.* **3**:197–208.

Yarwood, C. E. (1959) Microclimate and infection. In *Plant Pathology, Problems and Progress, 1900–1958.* University of Wisconsin Press, Madison, Wis. Pp. 548–556.

Young, J. M., and J. D. Allen. (1969) In vitro infection and pathogenesis of plants from several genera by the specific soilborne pathogen *Sclerotium cepivorum.* *Plant Dis. Rep.* **53**:821–823.

NONHOST RESISTANCE

Michèle C. Heath

Each plant species is host to only a few of the thousands of microorganisms known to cause diseases of higher plants; thus each plant species is resistant to infection by the majority of plant pathogens. (We define plant pathogens as those organisms known to infect successfully at least one species of higher plant.) Typically all cultivars of the nonhost plant show resistance which is effective against all races of the pathogen. Commonly a nonhost plant develops no visible symptoms of infection, a condition often called immunity. However, this term is extremely difficult to define, and it should probably be avoided. Although one also may encounter some difficulty in defining a nonhost plant, nonetheless it is generally recognized that nonhost resistance is not only highly effective but also of great durability. The mere fact that host (species)–parasite specificity exists, and in most cases has not noticeably changed over recorded history, suggests that such resistance is not easily overcome. Thus, theoretically, exploitation of nonhost resistance could provide an effective and durable control of crop diseases.

To evaluate the potential of such an approach to plant disease management, however, one needs a detailed knowledge of (1) the nature of nonhost resistance, and (2) the mechanism that makes such resistance inoperative or ineffective against plant pathogens for which the plant species is a host. Such knowledge also answers the question of why certain plant species only are susceptible to infection by a particular plant pathogen. Thus one is essentially asking for an understanding of the basis of host (species)–parasite specificity and it is probably true to say at the present time, there is no system where this is known with any certainty.

POSSIBLE DETERMINANTS OF NONHOST RESISTANCE

Viruses

Perhaps the situation where we have the least information is in the case of infection of a nonhost species by a plant virus (see review by Atabekov, 1975). Conceivably, nonhost plants (i.e., those in which the virus will not multiply) may (1) lack receptor sites necessary for the entry of the virus into the cell, (2) in some way prevent the uncoating of the viral genome, and/or (3) be unable to successfully translate and transcribe the genome once released. Unequivocal evidence for any one of these suggestions is lacking. Specific interference between a virus and its homologous coat protein suggests that host plants contain specific receptor sites for certain viruses (e.g., Novikov and Atabekov, 1970), but whether these are absent from nonhosts is unknown. Gaard and De Zoeten (1979) have recently suggested that attachment by tobacco rattle virus to the cell wall, and its subsequent uncoating, is not host-specific; however, this may not be true for all viruses, although attempts to extend the host range of a virus by changing its protein coat, or by using free RNA for infection, have not been successful (Atabekov, 1975).

In light of this current lack of knowledge, it is almost impossible to predict whether future research will uncover information useful in the control of viral plant diseases. Certainly any direct interference with processes other than entry into the cell seems unlikely in the near future since this involves tampering with the vital metabolic machinery of the cell. Plant hybridization or exploitation of cross protection (see chapter by Kuć in this volume) seem to hold more immediate promise for the control of viral diseases.

Bacteria and Fungi

By far the greatest amount of information concerning specificity-determining mechanisms comes from work with bacterial and fungal diseases. The following is a brief description of some of the factors for which there is at least some evidence for a role in nonhost resistance.

Tropic Responses During Preinfection Phases of Disease Development

The most convincing evidence that tropic responses may be involved in host–pathogen specificity comes from studies of rust fungi where, on nonhosts, germ tubes from uredospores are often less efficient at locating, and/or recognizing, the stomata through which the fungus enters the tissue. Several reports have related this phenomenon to differences between host and nonhost plants in surface topography of the leaf (Heath, 1974, 1977a; Wynn, 1976). For further information on this topic see the chapter by Wynn and Staples in this volume.

Nutrients, Metabolites, and Morphogenic Factors

There is only one convincing report in the literature that pathogenesis depends on the presence of essential nutrients in the host, which are presumably absent from nonhosts. The infection of wheat and other plants by *Fusarium graminearum* correlates well with the presence in these plants of choline and betaine (Strange et al., 1974). However, such a simple explanation cannot explain the host range of all pathogens, since, for example, *Pseudomonas pisi* can grow perfectly well in intercellular fluid removed from leaf tissue of the nonhost tobacco (Klement, 1965). Similarly axenic cultures of the normally biotrophic (i.e., deriving nutrition from living tissue) rust fungus *Cronartium ribicola* can be supported by molecules diffusing from tissue cultures of the nonhost *Pseudotsuga menziesii* (Harvey and Grasham, 1971). Nevertheless, it is conceivable that certain factors may not be *made available* to the pathogen in nonhost plants, particularly if the pathogen is located primarily in the intercellular spaces. Young and Paton (1972) have suggested that pathogenic bacteria in their host stimulate the release of nutrients, whereas saprophytic bacteria do not. Whether lack of nutrient release could be a basis for nonhost resistance remains to be determined.

Related to the requirement of certain nutrients for pathogenesis is the suggestion that exudates from host plants may be more effective than those from nonhosts in counteracting soil fungistasis towards *Sclerotium cepivorum* and *Thielaviopsis basicola* (Mitchell, 1976). In a similar vein, Grambow and Riedel (1977) have suggested that the induction of infection structures by *Puccinia graminis* depends on the presence of volatile and nonvolatile factors which may be absent from nonhost species.

Immobilization and Envelopment of Bacteria

Currently, one of the earliest detectable responses of nonhost plants to some plant pathogenic bacteria is the apparent immobilization and envelopment of the bacteria by wall-like materials (Goodman et al., 1976). This response seems similar, if not identical, to the response of the host to avirulent strains of the pathogen and is described by Sequeira in this volume.

Physical Barriers to Fungal Growth

Although there is a report linking nonhost resistance to cuticle thickness (Hashioka, 1938), conclusive evidence that this is a causal factor in such resistance is lacking. Similarly, although there are reports of nonhost resistance being correlated with the development of papilla-like deposits (Aist, 1976; Ride and Pearce, 1979; Sherwood and Vance, 1976; Vance and Sherwood, 1976, 1977), wall thickenings (Leath and Rowell, 1969), silicon deposits (Heath, 1979a), and lignification (Ride, 1975) at the site of infection, it is extremely difficult to prove that these are acting merely as physical barriers. Even the fact that susceptibility is increased by treatments which inhibit the formation of these structures (Heath, 1979b; Leath and Rowell, 1969; Vance and Sherwood, 1976) does not preclude the possibility that other concurrent biochemical phenomena are more important for resistance. For example, Johnston and Sproston (1965) attributed the lack of fungal penetration of isolated *Ginkgo biloba* cuticles to the presence inhibitory compounds in the cuticle rather than the latter's physical properties.

Inhibitory Compounds

In addition to the work on *Ginkgo biloba* described above, there have been several other reports that preformed inhibitory compounds may play a role in nonhost resistance, particularly toward fungal plant pathogens. Kovács and Szeöke (1956) reported that concentrated leaf washings from certain nonhosts prevented spore germination in three species of plant pathogens, but they did not determine whether spore germination was a significant feature of resistance in any of the fungus–plant combinations examined. However, such lack of spore germination does seem to be a feature of other nonhost–plant pathogen interactions (e.g., Hashioka, 1938; Heath, 1974; Johnson et al., 1978; Kuć, 1957; Raggi, 1964; Sempio and Barbieri, 1966; Staub et al., 1974), although the underlying mechanisms have not been elucidated. Preformed compounds that are toxic to several plant pathogens that do not normally attack potatoes have been reported to occur in potato peel (Clark et al., 1959; Kuć, 1957), and Greathouse and Rigler (1941) have concluded the host range of *Phymatotrichum omnivorum* is related to the presence of toxic alkaloids in nonhost species.

Inhibition of fungal growth, presumably due to toxic factors, has been reported for several species of the biotrophic rust fungi growing in nonhost plants (Gibson, 1904; Heath, 1974, 1977a, 1979b; Leath and Rowell, 1966), but the latter authors (Leath and Rowell, 1970) concluded that a postinfectionally formed, rather than a preformed, inhibitor was involved. Postinfectionally formed antifungal compounds (phytoalexins) have been found in almost every nonhost plant species challenged with a variety of living and nonliving agents (Kuć, 1972; see also chapter by Keen in this volume), and many studies have suggested that such phytoalexins accumulate at the right time and in high enough concentrations to be responsible for nonhost resistance (e.g., Heath and Higgins, 1973; Kuć, 1972; Shiraishi et al., 1977; Ward, 1976). Treatments that increase susceptibility of the nonhost tissue also result in reduced levels of phytoalexin accumulation (e.g., Chamberlain and Gerdemann, 1966; Jerome and Müller, 1958; Teasdale et al., 1974), but such correlations do not conclusively prove that phytoalexins are causal agents of nonhost resistance. Indeed there are several reports that suggest that phytoalexin production cannot be the prime determinant in the restriction of fungal growth. Teasdale et al. (1974), in a careful study of the growth of *Fusarium solani* f. sp. *phaseoli* on pea endocarp tissue, concluded that germ tube growth is inhibited before significant quantities of phytoalexin have accumulated. Similarly, nonhost red clover is resistant to *Stemphylium botryosum* in spite of low levels of phytoalexin accumulation presumably brought about by the ability of the fungus to convert them to other compounds (Duczek and Higgins, 1976). It should also be mentioned that conclusions as to the efficacy of phytoalexin accumulation *in vivo* are based on *in vitro* studies of toxicity and there is evidence that the latter may bear little relationship to the sensitivity of the fungus to phytoalexins in the tissue (Pueppke and Van Etten, 1976). For further information on the role of phytoalexins as determinants of host and cultivar specificity, the reader is referred to the excellent review by Kuć (1976).

Phytoalexins are usually fairly complex organic molecules (Kuć, 1972) but a postinfectionally released simpler toxin, hydrogen cyanide, has been reported to be formed in birdsfoot trefoil during infection by fungi nonpathogenic towards this plant; the eight nonpathogens tested proved to be highly sensitive to this compound *in vitro* (Millar and Higgins, 1970).

Hypersensitive Cell Death

Many interactions of nonhost plants with plant pathogenic bacteria and fungi result in the rapid necrosis of one or more plant cells (e.g., Elliston et al., 1976; Heath, 1972, 1974; Johnson et al., 1978; Klement et al., 1964; Sempio and Barbieri, 1966; Staub et al., 1974; Ward, 1976). For bacteria, this hypersensitive cell death occurs after bacterial immobilization (see above) and therefore may be a secondary phenomenon unrelated to the determination of specificity. However, the situation is not nearly so clear in the case of fungal pathogens. For biotrophic fungi, cell death may be a means of resistance in itself merely by restricting the accessibility of the organism to components only available from living cells (Heath, 1976), although there is little evidence to support such a hypothesis. However, cell death per se cannot be a sufficient reason to inhibit the growth of nonbiotrophic fungi which often cause cell death, and yet continue growing, in their host plant. Thus in cases of nonhost resistance toward this latter group of fungi, cell death must be accompanied by other events, such as the production of phytoalexins (Hargreaves and Bailey, 1978), which restrict fungal growth. Whether cell death is always a vital accompaniment to these events is unknown and therefore the role of hypersensitive cell death in nonhost resistance is as controversial and poorly understood as its role in cultivar resistance (Heath, 1976; Ingram, 1978).

Receptor Sites

For those situations where pathogenesis seems to be associated with the presence of receptor sites, such as binding of the crown gall bacterium *Agrobacterium tumefaciens* to its host (Lippincott and Lippincott, 1976) or the interaction of host-selective toxins with various cell components (see Yoder in this volume), it is conceivable that nonhost plants lack such receptor sites. So far there is little evidence for or against such a hypothesis except for the observation that germ tube walls of *Uromyces phaseoli* var. *typica* bind to sections of host, but not nonhost, tissue (Mendgen, 1978).

Degradative Enzyme Activity in the Higher Plant

In the nonhost broad bean infected by *Uromyces phaseoli* var. *vignae* Heath (1974) has demonstrated that the first-formed haustorial mother cell often becomes detached from the nonhost cell wall through the breakage of the outer layer of the fungal wall. It is possible that this phenomenon is the result of the secretion of hydrolytic enzymes by the nonhost cell, and in support of such a hypothesis the activity of such enzymes has been detected in a number of investigations involving fungal pathogens in their host plants (Albersheim and Valent, 1974; Netzer et al., 1979; Pegg, 1977). Whether such degradation is a common or critical feature of nonhost resistance remains to be determined.

Comments and Conclusions Concerning Mechanisms of Nonhost Resistance

One fact that emerges from the above-described information is that nonhost resistance most commonly involves, or is accompanied by, active responses of the higher plant. Several studies have shown that treatment of tissue with heat shock, chloroform, or inhibitors of the synthesis of RNA or protein increases the susceptibility of the tissue to subsequent infection by bacteria or fungi for which the plant is usually a nonhost (e.g., Chamberlain, 1972; Chamberlain and Gerdemann, 1966; Heath, 1977b, 1979b; Leath and Rowell, 1969; Lyon and Wood, 1977; Tani et al., 1976; Vance and Sherwood, 1976), as well as abolishing the responses normally associated with nonhost resistance. However, it should be pointed out that these results can also be interpreted in terms of a "susceptibility factor," which is released by the above treatments and whose absence in untreated tissue is the real reason for nonhost resistance. Heath (1980) could find no evidence for such a factor being released after heat-shock treatment of French bean, but such a possibility has been little studied in other situations.

Another point which emerges from the literature is that there is no single mechanism that can account for nonhost resistance in all plants. However, there is a certain amount of evidence that there is not a high degree of specificity in the responses of nonhosts to infection. The fungal compounds which elicit phytoalexins do so in more than one nonhost species (Cline et al., 1978), and many plant pathogens elicit apparently similar responses in the same nonhost (e.g., Chamberlain and Gerdemann, 1966; Heath, 1977a; Klement et al., 1964; Kuć, 1972; Sherwood and Vance, 1976). Teasdale et al. (1974) have suggested that the response in peas is similar to that elicited by a number of foreign biological agents such as pollen and mouse tumor cells and may be part of a general cellular incompatibility response. However, there *are* examples where nonhost plants respond differently to different fungi (Heath, 1977a, 1979b; 1980; Kuć, 1976), and there is much additional evidence that nonhost plants possess more than one mechanism of resistance. Several investigators (Heath, 1974; Johnson et al., 1978) have demonstrated that different responses can be expressed at successive stages of infection by rust and powdery mildew fungi. Potatoes are seemingly protected against invasion by *Helminthosporium carbonum* by the toxic phenolic compounds in potato peel (Clark et al., 1959), but this fungus can also elicit the increase of toxic steroid alkaloids when placed directly on the tuber (Locci and Kuć, 1967). Even after the abolition of the normal haustorium-inhibiting response of a number of nonhosts to rust fungi, the tissue still does not become fully susceptible to infection (Heath, 1977b, 1979b). Similarly both phytoalexin production and wall thickenings appear to play a role in the resistance of corn to *Puccinia graminis* (Leath and Rowell, 1966, 1969, 1970).

Presumably a successful pathogen of any plant species must have evolved mechanisms to overcome all these potential barriers to infection (Heath, 1974). In determining the durability of nonhost resistance if exploited for disease control, it is important to know how pathogens have managed to establish this "basic compatibility" (Ellingboe and Gabriel, 1977) between themselves and their host plants.

HOW BACTERIA AND FUNGI OVERCOME NONHOST RESISTANCE

Pathogenic bacteria are not bound to host cell walls and do not elicit hypersensitive cell death (Goodman et al., 1976; Klement et al., 1964), and it has been suggested that this absence of response in the host is due to modifications in the structure of the lipopolysaccharide of the bacterium (see chapter by Sequeira in this volume). For fungi, since there is so little conclusive evidence as to the critical features of nonhost resistance, there obviously is equally little evidence as to why these resistance mechanisms are inoperative or ineffective toward pathogens in the host plant. If resistance is governed by preformed antifungal factors, presumably pathogens are those fungi that can degrade, are tolerant of, or are not affected by these factors. There is relatively little evidence for any of these hypotheses but Kuć (1957) has reported that fungi that attack potatoes are less sensitive to the toxic phenolic compounds present in the potato peel than fungi for which potato is a nonhost.

In the case of "induced" defense mechanisms, theoretically the pathogen in its susceptible host may not elicit them, may be tolerant of them, or may actively prevent their expression. All three mechanisms have been suggested in the literature. *Phytophthora megasperma* var. *sojae* has been suggested to specifically *not* elicit phytoalexins in susceptible host cultivars of soybean (Keen, 1978); *Colletotrichum lindemuthianum* produces in culture a protein which specifically inhibits the host glucanase, which can degrade the pathogen's cell wall (Albersheim and Valent, 1978); the pathogenicity of

Stemphylium botryosum in alfalfa, and *Botrytis cinerea* and *Fusarium oxysporum* f. *vasinfectum* in pepper, has been suggested to be related to their ability to degrade phytoalexins (Higgins and Millar, 1969; Ward and Stoessl, 1972); *Helminthosporium avenae* may be able to degrade the papillae formed in reed canarygrass (Vance and Sherwood, 1976); the growth of *Aphanomyces euteiches* in pea and *Alternaria alternata* in pepper has been correlated with their low sensitivity to phytoalexins *in vivo* (Pueppke and Van Etten, 1976; Ward et al., 1973); the insensitivity of *Stemphylium loti* to hydrogen cyanide has been postulated as a contributing factor toward its pathogenicity to birdsfoot trefoil (Millar and Higgins, 1970). Elicitation of a phytoalexin, and its concomitant suppression, has been suggested to be important in the pathogenesis of pea by *Mycosphaerella pinodes* (Oku et al., 1977; Shiraishi et al., 1978) and potato by *Phytophthora infestans* (Doke et al., 1979; Garas et al., 1979), and in the latter case hypersensitive cell death is also inhibited by this pathogen. Such "induced susceptibility" (Daly, 1972; Ward and Stoessl, 1976), is also supported by several reports of increased susceptibility of nonhost plants following infection of the plant with a fungus for which it is a host (Heath, 1980; Ouchi et al., 1976a,b; Shiraishi et al., 1978; Yarwood, 1977). Low molecular weight compounds have been extracted from the spore germination fluids of *M. pinodes* (Oku et al., 1977; Shiraishi et al., 1978) and from French-bean leaves infected with *Uromyces phaseoli* var. *typica* (Heath, 1980), which increase the susceptibility of these higher plants to pathogens for which they are normally nonhosts.

POTENTIAL EXPLOITATION OF NONHOST RESISTANCE FOR CROP PROTECTION AGAINST BACTERIA AND FUNGI

Assuming that some of the described features of nonhosts, and the acttivities of the pathogen in the host, are true determinants of host–pathogen specificity, the theoretical possibilities for man to interfere with the process of pathogenesis seem endless. For example, surface features of the host could be modified through breeding programs to promote incorrect topic responses (see chapter by Wynn and Staples in this volume) and plant breeding could also be used to introduce toxic secondary metabolites into the host. Specific chemicals could be applied to block the pathogen's inhibitors of resistance or raise the levels of phytoalexins in the tissue. However, if one tries to be realistic, it seems clear that many of these approaches are unlikely to be useful in practice. The feasibility of using externally applied chemicals to interfere with biochemical pathways of host or pathogen cannot be determined until these pathways have been elucidated. Nevertheless, the development of any procedure aimed to specifically attack the metabolism of the pathogen seems unlikely to be economically worthwhile unless the metabolic process concerned is shared by more than one pathogenic organism. Chemicals which specifically elicit nonhost resistance may be more useful, particularly if they evoke responses not normally elicited during successful infection, but for this approach to be worth pursuing, the crop must be of sufficient economic importance and the elicited response must provide protection against a wide array of pathogens.

At present, current information is too sparse to determine which responses should be looked at with this in mind. Phytoalexin production may not be the best choice, since so many pathogens can detoxify these compounds, even those from nonhost species (e.g., Bailey et al., 1977; Duczek and Higgins, 1976). In addition, raising the levels of these or other antifungal substances in crop plants would be unwise until their toxicity to humans and other animals has been evaluated. Similar hazards may accompany the use of externally applied phytoalexins as fungicides (Ward et al., 1975) since there is no a priori reason why phytoalexins should provide fewer environmental or health problems than do currently available pesticides, apart from the possibly mixed blessing that the host plant may be able to metabolize the phytoalexin (Stoessl et al., 1977) instead of accumulating it. Of more interest in the context of eliciting nonhost resistance is the report that a particular fungicide may work through stimulating the host's natural defense mechanisms during infection by the pathogen (Cartwright et al., 1977). Whether typical nonhost resistance is evoked in this case is uncertain, but the report does suggest that this is an approach worth investigating further. However, our lack of knowledge of resistance mechanisms at present means that the search for such chemicals will have to be made empirically.

Compared with the development of control measures involving externally applied chemicals, introduction of nonhost resistance into host plants though hybridization between hosts and nonhosts would seem to be the simplest solution to the problem. Interspecific and intergeneric crosses are often difficult to make but it is possible that techniques such as protoplast fusion (see chapter by Earle in this volume) will make such hybrids easier to initiate than is now the case, although whether these hybrids can be propagated, other than vegetatively, is uncertain (Knott and Dvořák, 1976). Nevertheless, it should be pointed out that such crosses are already used fairly extensively in the search for disease resistance and have been the source of many of the major genes for resistance found in currently used crop cultivars (e.g., Knott and Dvořák, 1976; Jones and Pickering, 1978). Such resistance has commonly been overcome eventually by the pathogen, raising the question of whether these genes are indeed the ones involved in nonhost resistance of the donor plant or whether breeders inadvertently have selected for something else. Perhaps plant breeding could better exploit such hybridization if the selection of progeny was based on specific traits or biochemical pathways (if we knew which were important) rather than the lack of visible infection after inoculation with the pathogen.

AREAS OF FUTURE RESEARCH

From the above discussion, it may seem that the exploitation of nonhost resistance does not hold great promise for disease management in the near future. However, I believe that this conclusion is primarily the result of our current ignorance of what is really involved when a nonhost plant is attacked by a plant pathogen. In particular, there seem to me to be two major areas where research is badly needed before nonhost resistance can be properly evaluated as a means of disease control in crop plants.

The Identification of the Determinants of Host–Parasite Specificity

The need for such information is obvious but it is equally obvious that producing such information is not going to be an easy task since it has baffled plant pathologists for many years. However, relatively few scientists have investigated nonhost resistance rather than cultivar resistance, and what is especially needed is a multidiscipline approach, involving cooperation between cytologists, geneticists, and biochemists, so that the most critical events of a given nonhost–pathogen interaction can be pinpointed. Too often in the past only one aspect of particular interest, such as phytoalexin production, has been examined in detail for any one system.

The Comparison of Host and Nonhost Resistance

As yet it is completely unknown whether there are any fundamental differences between the resistance shown by nonhost plants and that shown by resistant cultivars of the host. Certainly there are no clear criteria that distinguish the two in terms of morphology or biochemistry. For example, responses of nonhosts to bacteria seem similar to those elicited in the host by avirulent strains of the pathogen (Sequeira, 1978) and most of the above described interactions between plant pathogenic fungi and nonhost plants have some counterpart in interactions involving pathogens in resistant host cultivars. Moreover, several examples exist where fungal growth in nonhost plants exceeds that in highly resistant cultivars of the host (e.g., Raggi, 1964). Nevertheless there are some important differences between the two types of resistance, particularly in terms of durability and genetical basis. In contrast to cultivar resistance, nonhost resistance is unlikely to be governed by specific gene-for-gene interactions (Ellingboe and Gabriel, 1977), since it is inconceivable that each plant possesses a specific gene for resistance towards every pathogen for which it is a nonhost. As mentioned earlier, it seems more likely that each plant species has a battery of potential resistance mechanisms, many probably nonspecifically elicited by some fungal or bacterial activity, of which one or more is certain to be active against most potential plant pathogens. It therefore becomes important to know whether the durability of nonhost resistance lies solely with the "stacking" of potential mechanisms for defense or whether these defense mechanisms themselves are fundamentally different from those elicited in cultivars of the host plant. Only with this knowledge can the plant breeder know what to select for when trying to introduce nonhost resistance into host plants. Often the current practice of adding more genes for a similar type of resistance (usually expressed as hypersensitive cell death) does not result in durable resistance of the cultivar: perhaps if nonhost resistance does depend on its multiplicity of defense mechanisms, the important feature is that these are of various different types and are expressed at many stages during infection.

Cultivar resistance has been suggested to be superimposed on the basic compatibility established between a pathogen and its host (e.g., Ellingboe and Gabriel, 1977). Presumably such compatibility must be related to the pathogen's ability to overcome the host's "nonhost" systems of resistance, and the elucidation of the genetical basis of basic compatibility is as important to our understanding of host–parasite specificity as the determination of the biochemical processes involved. As a preliminary step towards the former goal, Ellingboe and Gabriel (1977) recently reported the isolation of conditional mutants of *Colletotrichum lindemuthianum*, whose mutated genes are apparently vital for successful parasitism. Further research in this area is badly needed.

One practical problem which is likely to arise in any comparative study of nonhost and cultivar resistance is the definition of a nonhost plant. Superficially, nonhosts seem easy to identify and, indeed, apparently indisputable nonhosts have been used for most of the investigations described in this paper. However, there are cases where a wide enough search has provided isolates of a fungus that are pathogenic on what was previously thought to be a nonhost species (e.g., Jones and Pickering, 1978); this obviously is a serious problem for the plant breeder if nonhost and cultivar resistance prove to be fundamentally different. There are also examples in the literature of symptomless hosts, which would be recorded as nonhosts if not examined microscopically (Horner, 1954; Lacy and Horner, 1966). In addition there is the case of a plant which is a "nonhost" to certain strains or races of a pathogen but the host for another. Is resistance in this case true nonhost resistance? Most researchers seem to assume that this is so, but for the rust fungi, for example, light-microscope studies of the growth of various *forma speciales* of *Puccinia graminis* in "inappropriate" cereals (Ogle and Brown, 1971; Stakman, 1915) show that this more closely resembles growth in resistant host cultivars than the typical growth of rust fungi in less disputable nonhosts (Gibson, 1904; Heath, 1977a; Leath and Rowell, 1966). Perhaps this is a problem restricted to fungi that attack cereals, since it is a strain of another cereal pathogen, *Ophiobolus graminis,* which has overcome the "nonhost" resistance of oats by producing an enzyme which detoxifies the otherwise inhibitory glucoside, avenacin (Turner, 1961); nonhost resistance in other situations is rarely so easily overcome. Thus, until the different types of resistance have been clearly defined (or shown that they truly do not differ), it is probably unwise to assume that the resistance of a plant species to only certain strains of a pathogen is an example of true nonhost resistance; moreover, special care should be taken to ensure that all studies of the biochemical and genetical basis of nonhost immunity use plants that are indisputable nonhosts to the pathogens chosen.

One final cautionary note which should be added at this point is that in spite of the nonspecific nature of nonhost immunity to plant pathogens, there is some evidence that interactions between pathogen and nonhost may vary between different cultivars of the nonhost species (Heath, 1974, 1980); up to now, most published reports on nonhost resistance are based on investigations of only one plant cultivar and these results may not be representative of the species as a whole.

REFERENCES

Aist, J. R. (1976) Papillae and related wound plugs of plant cells. *Annu. Rev. Phytopathol.* **14**:145–163.

Albersheim, P., and B. S. Valent. (1978) Host–pathogen interactions in plants when exposed to oligosaccharides of fungal origins, defend themselves by accumulating antibiotics. *J. Cell Biol.* **78**:627–643.

Atabekov, J. G. (1975) Host specificity of plant viruses. *Annu. Rev. Phytopathol.* **13**:127–145.

Bailey, J. A., R. S. Burden, A. Mynett, and C. Brown. (1977) Metabolism of phaseollin by *Septoria nodorum* and other non-pathogens of *Phaseolus vulgaris*. *Phytochemistry* **16**:1541–1544.

Cartwright, D., P. Langcake, R. J. Pryce, D. P. Leworthy, and J. P. Ride. (1977) Chemical activation of host defence mechanisms as a basis for crop protection. *Nature* **267**:511–513.

Chamberlain, D. W. (1972) Heat-induced susceptibility to nonpathogens and cross-protection against *Phytophthora megasperma* var. *sojae* in soybean. *Phytopathology* **62**:645–646.

Chamberlain, D. W., and J. W. Gerdemann. (1966) Heat-induced susceptibility of soybeans to *Phytophthora megasperma* var. *sojae*, *Phytophthora cactorum*, and *Helminthosporium sativum*. *Phytopathology* **56**:70–73.

Clark, R. S., J. Kuć, R. E. Henze, and F. W. Quackenbush. (1959) The nature and fungitoxicity of an amino acid addition product of chlorogenic acid. *Phytopathology* **49**:594–597.

Cline, K., M. Wade, and P. Albersheim. (1978) Host–pathogen interactions. XV. Fungal glucans which elicit phytoalexin accumulation in soybean also elicit the accumulation of phytoalexins in other plants. *Plant Physiol.* **62**:918–921.

Daly, J. M. (1972) The use of near-isogenic lines in biochemical studies of the resistance of wheat to stem rust. *Phytopathology* **62**:392–400.

Doke, N., N. A. Garas, and J. Kuć. (1979) Partial characterization and aspects of the mode of action of a hypersensitivity-inhibiting factor (HIF) isolated from *Phytophthora infestans*. *Physiol. Plant Pathol.* **15**:127–140.

Duczek, L. J., and V. J. Higgins. (1976) The role of medicarpin and maakiain in the response of red clover leaves to *Helminthosporium carbonum, Stemphylium botryosum,* and *S. sarcinaeforme*. *Can. J. Bot.* **54**:2609–2619.

Ellingboe, A. H., and D. W. Gabriel. (1977) Induced conditional mutants for studying host/pathogen interactions. In *Induced Mutations Against Plant Diseases*. International Atomic Agency, Vienna. Pp. 35–44.

Elliston, J., J. Kuć, and E. B. Williams. (1976) A comparative study of the development of compatible, incompatible, and induced incompatible interactions between *Colletotrichum* spp. and *Phaseolus vulgaris*. *Phytopathol. Z.* **87**:289–303.

Gaard, G., and G. A. De Zoeten. (1979) Plant virus uncoating as a result of virus–cell wall interactions. *Virology* **96**:21–31.

Garas, N. A., N. Doke, and J. Kuć. (1979) Suppression of the hypersensitive reaction in potato tubers by mycelial components from *Phytophthora infestans*. *Physiol. Plant Pathol.* **15**:117–126.

Gibson, C. M. (1904) Notes on infection experiments with various Uredineae. *New Phytol.* **3**:184–191.

Goodman, R. N., P.-Y. Huang, and J. A. White. (1976) Ultrastructural evidence for immobilization of an incompatible bacterium, *Pseudomonas pisi,* in tobacco leaf tissue. *Phytopathology* **66**:754–764.

Grambow, H. J., and S. Riedel. (1977) The effect of morphogenically active factors from host and nonhost plants on the *in vitro* differentiation of infection structures of *Puccinia graminis* f. sp. *tritici*. *Physiol. Plant Pathol.* **11**:213–224.

Greathouse, G. A., and N. E. Rigler. (1941) Alkaloids from *Zephyranthes texana, Cooperia pedunculata* and other amaryllidaceae and their toxicity to *Phymatotrichum omnivorum*. *Am. J. Bot.* **28**:702–704.

Hargreaves, J. A., and J. A. Bailey. (1978) Phytoalexin production by hypocotyls of *Phaseolus vulgaris* in response to constitutive metabolites released by damaged bean cells. *Physiol. Plant Pathol.* **13**:89–100.

Harvey, A. E., and J. L. Grasham. (1971) Production of the nutritional requirements for growth of *Cronartium ribicola* by a nonhost species. *Can. J. Bot.* **49**:1517–1519.

Hashioka, Y. (1938) The mode of infection by *Sphaerotheca fuliginea* (Schlecht.) Poll. in susceptible, resistant and immune plants. *Natural History Soc. Taiwan (Formosa), Trans.* **28**:47–60.

Heath, M. C. (1972) Ultrastructure of host and nonhost reactions to cowpea rust. *Phytopathology* **62**:27–38.

Heath, M. C. (1974) Light and electron microscope studies of the interactions of host and non-host plants with cowpea rust–*Uromyces phaseoli* var. *vignae*. *Physiol. Plant Pathol.* **4**:403–414.

Heath, M. C. (1976) Hypersensitivity, the cause or the consequence of rust resistance? *Phytopathology* **66**:935–936.

Heath, M. C. (1977a) A comparative study of non-host interactions with rust fungi. *Physiol. Plant Pathol.* **10**:73–88.

Heath, M. C. (1977b) The effect of heat treatment on nonhost interactions with cowpea rust. (Abstr.) *Proc. Am. Phytopathol. Soc.* **4**:169.

Heath, M. C. (1979a) Partial characterization of the electron-opaque deposits formed in the non-host plant, French bean, after cowpea rust infection. *Physiol. Plant Pathol.* **15**:141–148.

Heath, M. C. (1979b) Effects of heat shock, actinomycin D, cycloheximide and blasticidin S on nonhost interactions with rust fungi. *Physiol. Plant Pathol.* **15**:211–218.

Heath, M. C. (1980) Effects of infection by compatible species or injection of tissue extracts on the susceptibility of nonhost plants to rust fungi. *Phytopathology* **70**:356–360.

Heath, M. C., and V. J. Higgins. (1973) *In vitro* and *in vivo* conversion of phaseollin and pisatin by an alfalfa pathogen *Stemphylium botryosum*. *Physiol. Plant Pathol.* **3**:107–120.

Higgins, V. J., and R. L. Millar. (1969) Comparative abililties of *Stemphylium botryosum* and *Helminthosporium turcicum* to induce and degrade a phytoalexin from alfalfa. *Phytopathology* **59**:1493–1499.

Horner, C. E. (1954) Pathogenicity of *Verticillium* isolates to peppermint. *Phytopathology* **44**:239–242.

Ingram, D. S. (1978) Cell death and resistance to biotrophs. *Ann. Appl. Biol.* **89**:291–295.

Jerome, S. M. R., and K. O. Müller. (1958) Studies on phytoalexins. II. Influence of temperature on resistance of *Phaseolus vulgaris* towards *Sclerotinia fructicola* with reference to phytoalexin output. *Aust. J. Biol. Sci.* **11**:301–314.

Johnson, L. E. B., R. J. Zeyen, and W. R. Bushnell. (1978) Defense patterns in inappropriate higher plant species to two powdery mildew fungi. (Abstr.) *Phytopathol. News* **12**:89.

Johnston, H. W., and T. Sproston, Jr. (1965) The inhibition of fungus infection pegs in *Ginkgo biloba*. *Phytopathology* **55**:225–227.

Jones, I. T., and R. A. Pickering. (1978) The mildew resistance of *Hordeum bulbosum* and its transference into *H. vulgare* genotypes. *Ann. Appl. Biol.* **88**:295–298.

Keen, N. T. (1978) Surface glycoproteins of *Phytophthora megasperma* var. *sojae* function as race specific glyceollin elicitors in soybeans. (Abstr.) *Phytopathol. News* **12**:221.

Klement, Z. (1965) Method of obtaining fluid from intercellular spaces of foliage and the fluid's merit as substrate for phytobacterial pathogens. *Phytopathology* **55**:1033–1034.

Klement, Z., G. L. Farkas, and L. Lovrekovich. (1964) Hypersensitive reaction induced by phytopathogenic bacteria in the tobacco leaf. *Phytopathology* **54**:474–477.

Knott, D. R., and J. Dvořák. (1976) Alien germ plasm as a source of resistance to disease. *Annu. Rev. Phytopathol.* **14**:211–235.

Kovács, A., and E. Szeöke. (1956) Die phytopathologische Bedeutung der kutikulären Exkretion. *Phytopathol. Z.* **27**:335–349.

Kuć, J. (1957) A biochemical study of the resistance of potato tuber tissue to attack by various fungi. *Phytopathology* **47**:676–680.

Kuć, J. (1972) Phytoalexins. *Annu. Rev. Phytopathol.* **10**:207–232.

Kuć, J. (1976) Phytoalexins and the specificity of plant–parasite interaction. *In* Specificity in Plant Diseases. (R. K. S. Wood and A. Graniti, eds.). Plenum Press, New York and London. Pp. 253–268.

Lacy, M. L., and C. E. Horner. (1966) Behaviour of *Verticillium dahliae* in the rhizosphere and on roots of plants susceptible, resistant, and immune to wilt. *Phytopathology* **56**:427–430.

Leath, K. T., and J. B. Rowell. (1966) Histological study of the resistance of *Zea mays* to *Puccinia graminis*. *Phytopathology* **56**:1305–1309.

Leath, K. T., and J. B. Rowell. (1969) Thickening of corn mesophyll cell walls in response to invasion by *Puccinia graminis*. *Phytopathology* **59**:1654–1656.

Leath, K. T., and J. B. Rowell. (1970) Nutritional and inhibitory factors in the resistance of *Zea mays* to *Puccinia graminis*. *Phytopathology* **60**:1097–1100.

Lippincott, J. A., and B. B. Lippincott. (1976) Morphogenic determinants as exemplified by the crown-gall disease. In *Encyclopedia of Plant Physiology*, vol. 4. *Physiological Plant Pathology*. (R. Heitefuss and P. H. Williams, eds.). Springer-Verlag, Berlin, Heidelberg, New York. Pp. 356–388.

Locci, R., and J. Kuć. (1967) Steroid alkaloids as compounds produced by potato tubers under stress. *Phytopathology* **57**:1272–1273.

Lyon, F., and R. K. S. Wood. (1977) Alteration of response of bean leaves to compatible and incompatible bacteria. *Ann. Bot.* **41**:359–367.

Mendgen, K. (1978) Attachment of bean rust cell wall material to host and non-host plant tissue. *Arch. Microbiol.* **119**:113–117.

Millar, R. L., and V. J. Higgins. (1970) Association of cyanide with infection of birdsfoot trefoil by *Stemphylium loti*. *Phytopathology* **60**:104–110.

Mitchell, J. E. (1976) The effects of roots on the activity of soil-borne plant pathogens. In *Encyclopedia of Plant Physiology*, Vol. 4. *Physiological Plant Pathology* (R. Heitefuss and P. H. Williams, eds.). Springer-Verlag, Berlin, Heidelberg, New York. Pp. 104–128.

Netzer, D., G. Kritzman, and I. Chet. (1979) β-(1.3) glucanase activity and quantity of fungus in relation to *Fusarium* wilt in resistance and susceptible near-isogenic lines of muskmelon. *Physiol. Plant Pathol.* **14**:47–55.

Novikov, V. K., and J. G. Atabekov. (1970) A study of the mechanisms controlling the host range of plant viruses. I. Virus-specific receptors of *Chenopodium amaranticolor*. *Virology* **41**:101–107.

Ogle, H. J., and J. F. Brown. (1971) Quantitative studies of the post-penetration phase of infection by *Puccinia graminis tritici*. *Ann. Appl. Biol.* **67**:309–319.

Oku, H., T. Shiraishi, and S. Ouchi. (1977) Suppression of induction of phytoalexin, pisatin. *Naturwissenschaften* **64**:643.

Ouchi, S., C. Hibino, and H. Oku. (1976a) Effect of earlier inoculation on the establishment of a subsequent fungus as demonstrated in powdery mildew of barley by a triple inoculation procedure. *Physiol. Plant Pathol.* **9**:25–32.

Ouchi, S., H. Oku, and C. Hibino. (1976b) Localization of induced resistance and susceptibility in barley leaves inoculated with the powdery mildew fungus. *Phytopathology* **66**:901–905.

Pegg, G. F. (1977) Glucanohydrolases of higher plants: a possible defence mechanism against parasitic fungi. *In* Cell Wall Biochemistry Related to Specificity in Host-Plant Pathogen Interactions (B. Solheim and J. Raa, eds.). Columbia University Press, New York. Pp. 305–345.

Pueppke, S. G., and H. D. Van Etten. (1976) The relation between pisatin and the development of *Aphanomyces euteiches* in diseased *Pisum sativum*. *Phytopathology* **66:**1174–1185.

Raggi, V. (1964) Affinità sistematica e affinità biologica nel caso di parassiti obbligati specializzati. *Phytopathol. Mediterr.* **3**:135–155.

Ride, J. P. (1975) Lignification in wounded wheat leaves in response to fungi and its possible rôle in resistance. *Physiol. Plant Pathol.* **5**:125–134.

Ride, J. P., and R. B. Pearce. (1979) Lignification and papilla formation at sites of attempted penetration of wheat leaves by non-pathogenic fungi. *Physiol. Plant Pathol.* **15**:79–92.

Sempio, C., and G. Barbieri. (1966) Aumento respiratorio prodotto da *Uromyces appendiculatus* in specie refrattarie. *Phytopathol. Z.* **57**:145–158.

Sequeira, L. (1978) Lectins and their role in host–pathogen specificity. *Annu. Rev. Phytopathol.* **16**:453–481.

Sherwood, R. T., and C. P. Vance. (1976) Histochemistry of papillae formed in reed canarygrass leaves in response to noninfecting pathogenic fungi. *Phytopathology* **66**:503–510.

Shiraishi, T., H. Oku, S. Ouchi, and Y. Tsuji. (1977) Local accumulation of pisatin in tissues of pea seedlings infected by powdery mildew fungi. *Phytopathol. Z.* **88**:131–135.

Shiraishi, T., H. Oku, M. Yamashita, and S. Ouchi. (1978) Elicitor and suppressor of pisatin induction in spore germination fluid of pea pathogen, *Mycosphaerella pinodes*. *Ann. Phytopathol. Soc. Jpn* **44**:659–665.

Stakman, E. C. (1915) Relation between *Puccinia graminis* and plants highly resistant to its attack. *J. Agric. Res.* **4**:193–200.

Staub, T., H. Dahmen, and F. J. Schwinn. (1974) Light- and scanning electron microscopy of cucumber and barley powdery mildew on host and nonhost plants. *Phytopathology* **64**:364–372.

Stoessl, A., J. R. Robinson, G. L. Rock, and E. W. B. Ward. (1977) Metabolism of capsidiol by sweet pepper tissue: Some possible implications for phytoalexin studies. *Phytopathology* **67**:64–66.

Strange, R. N., J. R. Majer, and H. Smith. (1974) The isolation and identification of choline and betaine as the two major components in anthers and wheat germ that stimulate *Fusarium graminearum in vitro*. *Physiol. Plant Pathol.* **4**:277–290.

Tani, T., H. Yamamoto, G. Kadota, and N. Naito. (1976) Development of rust fungi in oat leaves treated with blasticidin S, a protein synthesis inhibitor. *Tech. Bull. Fac. Agric. Kagawa Univ.* **27** (Ser. 59):95–103.

Teasdale, J., D. Daniels, W. C. Davis, R. Eddy, Jr., and L. A. Hadwiger. (1974)

Physiological and cytological similarities between disease resistance and cellular incompatibility responses. *Plant Physiol.* **54**:690–695.

Turner, E. M. C. (1961) An enzymic basis for pathogenic specificity in *Ophiobolus graminis*. *J. Exp. Bot.* **34**:169–175.

Vance, C. P., and R. T. Sherwood. (1976) Cycloheximide treatments implicate papilla formation in resistance of reed canarygrass to fungi. *Phytopathology* **66**:498–502.

Vance, C. P., and R. T. Sherwood. (1977) Lignified papilla formation as a mechanism for protection in reed canarygrass. *Physiol. Plant Pathol.* **10**:247–256.

Ward, E. W. B. (1976) Capsidiol production in pepper leaves in incompatible interactions with fungi. *Phytopathology* **66**:175–176.

Ward, E. W. B., and A. Stoessl. (1972) Postinfectional inhibitors from plants. III. Detoxification of capsidiol, an antifungal compound from peppers. *Phytopathology* **62**:1186–1187.

Ward, E. B. W., and A. Stoessl. (1976) On the question of 'elicitors' or 'inducers' in incompatible interactions between plants and fungal pathogens. *Phytopathology* **66**:940–941.

Ward, E. W. B., C. H. Unwin, and A. Stoessl. (1973) Postinfectional inhibitors from plants. VII. Tolerance of capsidiol by fungal pathogens of pepper fruit. *Can. J. Bot.* **51**:2327–2332.

Ward, E. W. B., C. H. Unwin, and A. Stoessl. (1975) Experimental control of late blight of tomatoes with capsidiol, the phytoalexin from peppers. *Phytopathology* **65**:168–169.

Wynn, W. K. (1976) Appressorium formation over stomates by the bean rust fungus; response to a surface contact stimulus. *Phytopathology* **66**:136–146.

Yarwood, C. E. (1977) *Pseudoperonospora cubensis* in rust-infected bean. *Phytopathology* **67**:1021–1022.

Young, J. M., and A. M. Paton. (1972) Development of pathogenic and saprophytic bacterial populations in plant tissue. In *Proceedings of the Third International Conference on Plant Pathogenic Bacteria* (H. P. Maas Geesteranus, ed.). University of Toronto Press, Toronto, Canada. Pp. 77–80.

III

NEW DIRECTIONS IN DEVELOPMENT OF PLANTS RESISTANT TO DISEASE

CONCEPTUAL AND PRACTICAL CONSIDERATIONS WHEN BREEDING FOR TOLERANCE OR RESISTANCE

Ivan W. Buddenhagen

Although many crop varieties are products of recent scientific breeding, many, surprisingly, are not. With all the emphasis on breeding new varieties with superior quality, adaptability, disease and insect resistance, it is a shock to realize that millions of acres of many crops are varieties, or land-races, or clones selected by ancient men or women in prehistory, or at least before agricultural science was developed. This is so for nearly all the bananas and plantains of the world, for pineapples, for cassava, for the *Dioscorea* yams, for most of the rice and cowpeas in West Africa, and for most of the maize and beans in Latin America. It is still true for several million acres of rice in Asia, and much of the potato crop of the Andes in South America. Even in the United States, the most important potato variety, representing some 40% of the total acreage, is of a chance seedling selection made 100 years ago. Sorghum and millet in tropical Africa are largely old land-races, as is much of the forage grass acreage of the world. These old or ancient selections were made by people who lived (and often still live) intimately with their crops on a day-to-day and year-to-year basis in environments that have changed with time, and which they could not manipulate.

We are still in the gathering stage for most of our forest supplies and our seafood, and only at an "improved-cultural-practice-stage" with ancient selections for more of our food crop area than we usually realize. Part of the antiquity can be ascribed to insufficient effort in varietal development in the tropics and elsewhere, but part is due to other reasons, one of which is the continued stability and quality of ancient types.

Can our breeding programs and research investment decisions profit from asking why ancient cultivars persist and are successful? I think so; however, first, a few general comments about concepts, terms, and levels of approach. Let us assume that our research exists because of the belief that diseases cause too much crop loss, and that, with enough knowledge, these losses can be reduced. Let us further assume that crop loss through disease is a function of interaction of crop population genotype, physical environment, and biological environment over time. (Yield is a function of the same factors.) Thus any knowledge gained to reduce losses must affect population/environmental interactions over time to favor yield and inhibit disease development or effects of disease.

I have attempted to analyze the titles in this volume in relation to the above conceptual base of definitions and interactions. Most titles are concerned with the *individual plant, with its nature or response at the biochemical, physiological, or microstructural level.* Mechanisms affecting infection (attraction, recognition, induction, immunity) are to be elaborated, or physiological changes due to cohabitation of cells or tissues are to be described. One paper deals purely with pathogen physiology. Only this chapter and that by Robinson deal with *potential future host-pathogen population/environment interaction,* that is, *breeding.*

These comments emphasize the contrast between research to understand the *mechanisms* of single-plant host–parasite interactions at the molecular or physiological level and the requirement of reduction of crop loss by disease through influencing interaction of host–pathogen *populations.* Exactly how do we foresee this physiological knowledge altering such population interaction in thousands of farmers' fields each season? There are various hypothetical ways such knowledge could be used. However, as a field-oriented breeder/pathologist, I am quite aware of how difficult it is even to influence positively future crop/pathogen *population* interaction by what would appear to be the easiest method—altering varietal population resistance.

We should be aware of the continuing emphasis in crop breeding programs on "improving disease and insect resistance." This must mean that we have not yet achieved a desired level of resistance in spite of considerable past efforts. Why is this the case? Surely we have tried; there are numerous articles over the years on nature *of,* breeding *for,* and inheritance *of* resistance. There are numerous varieties released as "disease resistant," but few remain "resistant" for long. However, a dwarf variety always remains dwarf. Thus we must be dealing with different phenomena, and indeed we are. We are dealing with a very complex interaction of existing populations influencing the evolution of future populations, all influenced by variable environments.

It is the general failure to recognize and understand this dynamic complexity, plus the ambiguous use of imprecise terms applied to host–pathogen interaction, which are largely responsible for the paucity of real long-term success in resistance breeding.

In the case of host–parasite physiological research we should realize that the subject for the research is a plant, bred by plant breeders (usually), and that the kind of reaction the physiologist measures is a result of selection methods and criteria previously applied by the plant breeder. If the breeder had selected a different host genotype, the host–parasite system would not be exactly the same. Thus, the plant breeder is in control of the *answers* to be found by the students of host–parasite physiology. He also determines, inadvertently or not, the level of and durability of "resistance" of new varieties.

The question then becomes: Is the plant breeder utilizing the best approaches and methods for parent, progeny and environment selection to ensure maximum levels and durabilities of disease resistance? Often, probably not. Why not, and what can be done about it? The answers are complex and largely unresearched. This paper suggests key points to consider while examining these questions for individual breeding programs.

BREEDING AND BREEDING OBJECTIVES

Breeding crop or forest varieties is a practical activity involving four basic steps where resources and decisions affect the final outcome:

1. Genetic diversity of and selection of parents.
2. Selection of crossing methods (or changing genotype by methods other than by crossing).
3. Selection of progeny.
4. Selection of environments for steps 1–3.

Decisions made for each of these steps will influence the final disease/pest host–crop relationship. A practical way to examine, and then alter a breeding program–disease relationship is to examine each of these four points in turn for this relationship. To do this one must know how to obtain realistic pathogen challenge and development in relation to field crop loss. Then one must examine how one's plots, parents, progenies and environments where one is making decisions relate to the potential of both disease development and yield in farmers' fields. This sounds simple, but it is not.

The difficulty is due to the following problems:

1. *Epidemiological principles of disease development are not satisfied.* There is a gap between an instant measurement of disease level on a plant in a screening nursery and the ability of that plant genotype to generate its own epidemic. This gap will vary with the type of disease, the type of challenge given and with the environment (Clifford and Clothier, 1974; Habgood, 1972; Johnson and Taylor, 1976; Van der Plank, 1963, 1968; Zadoks and Schein, 1979).
2. *The potential for microevolutionary pathogen population shifts is not determined.* These shifts are the cause of "breakdown of resistance."

They are indirectly influenced by the coevolutionary background of the host and pathogen (Buddenhagen, 1977) and directly influenced by the nature of the resistance factor(s) introduced into the hosts' genetic background (Robinson, 1973, 1976).

3. *Environmental influences on the disease process are slighted.* The variable environments (physical and biological) under which the crop is grown, which affect field disease development and crop loss, are not fully represented in experimental plots where disease resistance is judged on a single instance.
4. *Single plants in segregating populations do not represent plants in homogeneous populations, especially in terms of disease development.* This nonrepresentational characteristic contributes cryptic error to predictability of disease reaction when such plants are grown later as a homogeneous crop variety.
5. *There is great ambiguity of definition and misleading use of terms for describing disease/host/environment relationships.* This is so fundamental that it will be discussed separately.

Effects of the first four problems in biasing judgment of disease predilection can be minimized by appropriate designs and methods. Each disease type and host crop, with their evolutionary backgrounds, and each existing breeding/selection methodology can be examined in relation to the principles governing disease development and future disease change, in order to maximize selection for yield stability.

The objectives of the breeder, however, include disease or insect resistance as only one (often small) part of varietal development. The degree to which a breeder (or an improvement team) can or will give time and effort to a concern for these complexities will influence varietal yield stability in the longer term. The tendency to oversimplify "disease resistance" within the overall objectives of varietal improvement is almost overwhelming. Too many instant decisions are necessary, and a qualitative static view of a "disease reaction" is the simplest approach. One deals with many thousands or even hundreds of thousands of plants for many nondisease characteristics. Each disease or pest is an entity. How do we integrate all into one program? Or, should we have separate programs of disease resistance, insect resistance, yield, quality? To place perspective: a very good potato breeding program in the United States grows progeny initially under four environmental conditions where decisions are made on 15 growth and quality characters up to and including harvest-time evaluation, 4 characters of storability (one of which is "rot"), 1 virus disease, 5 fungal diseases, 1 bacterial disease, 1 insect pest, 1 mite pest (Martin, 1978, 1979).

Thus, determining objectives and balancing efforts on various objectives are indeed fundamental, difficult and important problems facing those directly involved in crop improvement.

Objectives of high-yield potential, high quality, and disease resistance sound good but "disease resistance" or "incorporation of disease resistance into a high-yielding variety" are too pat as objectives. They look good in a textbook but operationally they do not explain what to look for or how to obtain it. I prefer the overall objectives of stable high farm yield and high quality—which must include all necessary aspects of parasite/crop/environment interaction. How to translate these two basic objectives into operational procedures that enable accurate selection for them is the key problem.

These objectives can be translated into developing varieties that will:

1. Yield high under (defined) variable physical environmental conditions.
2. Have necessary quality characteristics.
3. Decelerate pathogen/pest generation cycles.
4. Limit increase of each pathogen/pest generation cycle.
5. Limit deleterious effects of infection.
6. Stabilize pathogen/pest evolution.

Each of these six objectives is measurable and can be applied operationally to any breeding program. Only point 6 is questionable in this regard and operational decisions regarding it can, at present, be based on logical but only hypothetical assumptions.

Operationally, these complexities require simplification, especially where many negative biological factors (diseases and pests) are involved. The simplest conceptual and practical approach is to breed for "total ecosystem adaptation." This means, simply, to circumscribe the area for the new potential variety and breed and select for *yield* in that area at a few representative sites, utilizing maximum recombination and whatever negative pressures that occur naturally. When many undefined diseases or pests influence yield potential, this is the most direct route to a new, stable, higher-yielding variety. If one includes local quality preferences as part of the ecosystem, then quality is satisfied as well. This approach has been recommended recently for cassava in Latin America (Lozano et al., 1976).

THE PROBLEM OF DEFINITION OF TERMS

We recognize, but often only subconsciously, that disease is a process, operating within a set of ecological conditions. The ambiguity and imprecision of terms used to simplify or circumscribe this process are major sources of cryptic error in our attempts to manipulate genotype/disease relationships, and to communicate about this activity. Although many recent papers discuss specific resistance topics, and some attempt to clarify terms, a basic reference on usage remains that by Robinson (1969).

RESISTANCE / SUSCEPTIBILITY

In my view, the terms "resistant" and "susceptible" are sources of great misunderstanding, especially between breeders and pathologists or entomologists. The breeder often considers "resistant" as a character of a variety just as "dwarf" is a character of a variety. The enlightened pathologist may (or may not) realize that "resistant" is really used when the relative amount of resistance/susceptibility (the amount of disease) is at some arbitrary ratio in a particular environment at a particular time with the host challenged by a certain race or races of the pathogen. Moreover, "resistant" is used for a breeding line if no disease appears on it at all (immune or escape) or if it is cellularly supersusceptible and responds with a hypersensitive reaction. The latter may represent only a nonmatching host/pathotype challenge, on, actually, a quite susceptible (when matched) breeding line. No wonder confusion is rampant. The use of these terms without qualifying exactly what is meant should be avoided in order to reduce a major source of error in crop breeding methods and in communication.

Part of the problem is that the approach often used for breeding and for genetic studies, of transferring visual estimates of plant or plot disease levels into numbers and then grouping these numbers into categories of resistance/susceptibility (R, MR, MS, S), is based on many tacit assumptions regarding the disease process. These assumptions are usually neither met nor considered, and a breeder utilizes an "R-category" parent source much as he would a "dwarf-category" parent source, as if it is an absolute genetic character. He wishes to transfer the "R character" into an otherwise superior genotype. This biases the activity toward single genes with major effects and it biases methods of challenge to sort out progeny toward detecting only such major effects. Disease, as a complex epidemiological process, is obscured. Future disease, due to pathogen population shifts brought about by microevolutionary selection processes operating on the new variety, is ignored. Thus we have a gap between a nursery screen reaction and field epidemic potential and another gap in relation to pathogen shifts of the future.

TOLERANCE / TOLERANT

Tolerance is variously defined and redefined and used to mean different things (Posnette, 1969; Robinson, 1969; Schafer, 1971; Mussell, 1980). A widely accepted definition (Schafer, 1971) is basically "less yield reduction with equal amounts of disease." I consider this definition to be relatively nonutilizable in resistance breeding due to two basic problems: the first problem is that single plant selection in the important early segregating generations (for inbreeders) cannot be made on the basis of this "tolerance" since there is no way of knowing the plant's potential yield in the absence of the disease effect. One can only observe plants with more or with less disease or symptoms. The second basic problem arises later with genetically fixed lines, where such "tolerance" measurements could be made (although relatively less useful in breeding at this point): both yield and disease are interactions of genotype, physical environment and biological environment (including potential pathogens) over time. But the amount of disease (or degree of symptoms) is actually estimated at one or only a few times during the crop cycle. Thus the discrepancy between yield reduction and observed disease severity among varieties is really a combination of some level of this phenomenon (if any) mixed with some degree of nonmeasurement and nonintegration of disease levels throughout the growth of the crop. A careful examination of several papers in which the 'tolerance' is defined to mean "less yield reduction with equal amounts of disease" reveals that this interpretation is confounded with non-measured differences in disease buildup with time (Simons, 1966, 1969; Caldwell et al., 1958). (Simons recognized this problem and suggested a broader definition of "tolerance".) A further complexity in considering this type of 'tolerance' is the source/sink relationship of yield limitation. Equal disease on two varieties that differ in degree of source limitation to yield should cause differing effects on grain yield reduction that is a reflection of the source/sink balance differences—not of "tolerance" to disease per se, except as expressed in this sense.

Can tolerance be defined so that it is both an explicit logical term and one that can be used in a practical breeding program? I think so; by confining "tolerance" to an expression of host performance when infected by pathogens causing systemic disease. A plant is "tolerant" if it is less sick than its neighbors when infected similarly by a systemic pathogen. It is more, or it is less tolerant as measured by degree of continued growth and eventual yield when sick. If plants can be seen to be infected when segregating populations are equally challenged, then one needs only to select for plants with good yield and some symptoms. Less sick in the extreme is equal yield when infected as when not. Virologists use "latent infection" for a degree of tolerance so high that symptoms cannot be seen (Posnette, 1969). Selecting for such a high degree of tolerance is practical only where efficient indexing methods are available. In a breeding program, by avoiding the use of "tolerance" for a varietal or population reaction, one does not confound an individual plant reaction with epidemiological parameters influencing disease spread and buildup in populations. However, once developed, a population of plants all equally tolerant (usually a clone, an inbreeder or an F_1 hybrid) can validly be referred to as a "tolerant cultivar," if no judgment of predilection for disease incidence is inferred.

Tolerance, *sensu* Schafer, for spot or blight diseases, or root infection non-systemic diseases is theoretically measurable in fixed cultivars, but in practical breeding operations, other than for selection of parents, it is best considered either not to exist or to be part of "resistance," if it does exist.

TOLREMIC—A NEW TERM

Varieties of plants with some, but less, disease than others growing nearby under the same conditions are more tolremic—they have limited the development of disease in relation to plant growth rate more than others. They have limited the epidemiological *r* of disease development. They are "tolremic": they have tolerated and/or resisted the development of disease or its effects, on a population basis.

I introduce tolremic as a specific name to indicate what we are actually trying to select for in resistance breeding. The word is an adjective for a plant, a breeding line, variety, or a landrace-mixture of varieties. Its opposite is "intolremic." A tolremic plant or variety develops little disease where an intolremic variety develops more. The terms are relative. The terms can be applied to populations or to single plants for non-systemic diseases. For systemic diseases they can be applied only to populations. For single plants with systemic disease, "tolerance" is used instead.

The practicability (and conceptual validity) of these definitions for tolerant/intolerant and tolremic/intolremic are revealed if one examines what is seen, interpreted and decided about plants in the field in a breeding/selection program.

Tolerant / Tolremic Selection for Systemic Diseases

The environment or researcher should provide enough challenge to infect all plants equally in segregating generations. Symptoms appear. The plants least sick are "tolerant." Plants without symptoms are either escapes or are so tolerant they cannot be detected as sick. If desired, one can index to determine which and use the latter. Or, if too much trouble, one can throw all of these away and rely on the most tolerant, but detectably sick plants. At this point, one does not know if the tolerant plant will make a tolremic variety (one that will slow the spread of disease in a population). If the disease is caused by a vectored virus it probably will. This can be tested later. For a clonal perennial, "tolremicity" may be less important than tolerance.

One can select for any degree of tolerance and for its expression in whatever diverse environmental conditions that may be desired. It is impossible to go wrong on a static seedling reaction because it is not used in selecting for yield in an infected population. A late seedling stage may be challenged but performance over full crop development to yield is the basis of selection. The effectiveness of the tolerance that is accepted can be measured later, in pure lines, by agronomic comparison with protected or isolated plots which remain healthy. This "tolerance" may be operating through limitation of internal pathogen reproduction, limitation of internal systemic spread, or through low cellular and tissue reaction to the pathogen and its products.

These or other mechanisms can be ignored or researched later. After tolerant breeding lines are fixed and established, they may be compared for degree of "tolremicity" by establishing plots of suitable size, shape and isolation, depending on the nature of the pathogen (Thresh, 1976). Plants at one end (or the middle) of the plots are infected and *incidence* recorded against time and distance. The average lessening in yield of sick plants, times incidence of sick plants, is the total degree of "resistance" (valid usage of this misused term) of the variety. By the above definitions and description, for systemic diseases, one could have, theoretically, a tolerant variety which does not lessen disease spread and thus it would be intolremic. This distinction is important since it maintains initial selection on an individual plant reaction that can be honestly challenged and judged in segregating populations. This reaction is not confused with a population reaction of incidence or epidemic development, which cannot be measured easily for systemic diseases on individual plants in segregating progenies.

There are many successful examples (Russell, 1978). We have successfully bred maize for tolerance to maize streak virus, the most important maize virus in Africa (Soto and Buddenhagen, 1980). Methods were developed to create equal young plant challenge; high tolerance (good yield when sick) was found and easily incorporated into populations. When tested for tolremicity, incidence was reduced in these populations from 35 to below one percent. The one percent diseased contained no plants rating severe and only 7% rating moderate, contrasting with 93% moderate to severe for the sick plants in the intolerant population. Thus breeding for tolerance also gave high tolremicity and the combination has reduced the disease to insignificance where resistant populations are grown.

In the rare case of inability to find sufficient tolerance, a practical alternative is to select families which are tolremic (show low incidence with equal challenge), recross such families to accumulate genes for this characteristic, ignoring the intolerance of the few plants becoming sick. This method has been used recently for the leafhopper-transmitted curly top virus in tomatoes and success has been obtained recently after 45 years of failure using other approaches (Martin, 1979). In effect, this method has converted a systemic disease into a non-systemic one, on a population basis. Plants becoming systemically infected are intolerant but rare. A requirement for a threshold level may be the explantion of this rather illogical situation.

Conversion by breeding, of normally systemic diseases to a nonsystemic state may well be development of very high tolerance, with infection resulting in tissue localization of the previously systemic pathogen. This apparently occurs for *Fusarium* wilt resistance in tomatoes and bananas—and in the former case it is known to be governed by a single gene. Where one is dealing with the manipulation of a disease by changing it from a systemic to a nonsystemic one, the problem becomes more complex but it is amenable to analysis and solution.

As long as one is dealing only with tolerance (reduced individual plant systemicity, or effects of systemic infection on total performance) one selects for plants with least-reduced yield following early and equal challenge. The problem so visualized becomes a simple one of obtaining sufficient diversity in total germplasm for screening, screening accurately, and then recrossing the most tolerant individuals found and repeating the process in segregating generations.

Tolremic Selection for Nonsystemic Diseases

The situation is somewhat different, but the concepts and terms apply. Here one cannot select for "tolerance" on an individual plant basis in segregating populations. Also one does not want to select for seedling resistance in the classical sense because it may mean a nonmatching race reaction, a supersusceptible hypersensitive reaction, a monocyclic reaction to a polycyclic disease (Zadoks, 1971, 1972), an escape, etc. Since one must select plants on an individual basis, how does one combine "disease resistance" with other criteria? One selects for "tolremic plants." That is, one designs a plant layout and disease challenge in selection plots so that a small amount of inoculum infects each widely spaced plant in a segregating population. Disease then develops on each plant during its life at a rate which largely reflects its tolremic level. The use of hexagonal grid designs (Fasoulas, 1979) with the addition of an internal check in every hexagon which has a certain level of known yield potential and tolremicity, increases the precision of the method in terms of reduction of environmental error on both yield and disease. The use of the hexagonal grid design to minimize physical environment variability is most compatible with disease tolerance/tolremic selection in the F_2 generation for inbreeders. In the F_3 and subsequent generations where family esodemic (see Robinson, 1976) comparisons are desired, the single plant/plot nature of the hexagonal grid design departs from epidemiological requirements for tolremic selection. The ability to yield in the face of some disease, combined with the ability to preclude development of much disease during the crop cycle, become the bases (among other key criteria) for selection. All comparisons are relative since the environment is relative and variable. Tolerance is ignored since it cannot be judged easily. Hopefully, part of the yield reflects tolerance, but no matter if it doesn't. We want low r, low ability to perpetuate an esodemic—a self-generated epidemic. Any error in selection in this way will err in terms of belief in higher disease loss than will actually occur in farmers' fields because of microplot alloinfection from neighboring more susceptible (intolremic) plants, countered by belief in lower disease loss than will occur in farmers' fields because of the dilution effect of inoculum loss from widely spaced plants. It is hoped that these opposing factors will cancel out, but probably they will vary depending on the system.

Recognition of the need to select for tolerant/tolremic plants and for a tolremic variety should result in changing how resistance breeding and selection are conducted. If selecting for tolremic plants and varieties is accepted, we should no longer use methods that select for a "resistant reaction," a criterion that does not enable distinguishing plants that will slow the development of an epidemic. The focus becomes one of judging the ability of a plant to develop both large "damaged areas" and a large pathogen population of propagules suitable for self- and neighbor infection. This approach applies for insects, nematodes, root-infecting fungi or bacteria, pathogens causing stalk rots and cankers, and for the more classical and usually discussed leaf-spotting, rapid-evolving fungi.

The key then lies in how to design tests that mechanistically satisfy the concepts. What methods are both practical to carry out and conceptually valid? The answer is simple but nongeneral—one must design a system that fits one's pathosystem and ecosystem and genotype-quality-yield stability requirements. But the methods designed must be compatible with the basic concepts outlined. They must be practical in terms of one's budget and one's ability to supervise, manage, and follow the genetic material with an understanding of its environmental relationship.

For nonsystemic diseases one must satisfy:

1. Low-dose initial challenge so that buildup of disease on the plant is self-generating; a gradient challenge in plots F_3 onward.
2. Minimal contamination from overly susceptible neighbors.
3. Some disease on the plants selected (otherwise nonmatching susceptibility may be selected). This is especially needed for classical, race-involved fungal diseases. (For some exceptional pathosystems, nonvisible disease may work.)
4. Normal but variable growing conditions and normal or wider than normal spacing. The normal growing conditions should encompass the swings in environmental conditions which affect disease development that might be expected to occur over several years in the region intended for the new variety. (Slight modifications to favor epidemic development may be devised.)
5. Combined disease/yield performance and evaluation. As much as possible, one should keep total performance when lightly challenged as one operation, without separate disease/insect screens divorced from crop performance and normal conditions.

An Example

We (IITA, 1979) and others (Bidaux, 1978; Notteghem, 1977) have developed a method for selecting for horizontal resistance (tolremicity) to blast (*Pyricularia oryzae*) in rice, which satisfies these criteria. Breeding rice using this approach should quickly change rice blast from the most important world rice disease to one of insignificance. It should result in durable tolremicity and thus, varietal stability. The method is simple but it deviates considerably from standard rice breeding methodology. The key points are the following:

1. Parents must get some blast (to minimize vertical gene nonmatching escapes).
2. Segregating F_2 populations are grown under wide spacing in a hexagonal grid with a reference plant in every hexagon that is known *not* to be a vertical gene escape in the location. Even for rice intended for paddy (flooded), this population is grown under upland (dryland) conditions where blast develops better. Either no blast challenge is offered or an early challenge is offered to the reference plants only.
3. F_2 selection is based on vigor, yield, plant type, and grain quality. Yield will reflect blast tolremicity to some degree.
4. $F_{3\text{-}5}$ families are grown in wide-spaced long rows perpendicular to an inoculum spreader that is densely sown, composed of many rice varieties. This provides a gradient multiracial inoculum challenge in time and space. Families are chosen in which the disease proceeds slowly down the rows and slowly upward on the developing plants. Families without disease are not selected if they are probably vertical-gene nonmatching escapes. Individuals are selected within families on the basis of less disease than their immediate neighbors, combined with best vigor, plant type, yield, and quality.

This design and method attempts to simulate, in segregating generations, the requirements of epidemiological disease buildup that would occur in fields of a pure variety during its complete phenology. The keys are nonimmunity, a gradient challenge and isoinoculum production on families. Wide spacing in families eliminates allocompetition for vigor and growth, a major contributor to bias in selection, as well as reducing alloinfection bias.

CONCLUDING REMARKS

To return to the conceptual base of interaction that must underlie the design of any practical approach: one should consider the following topics in breeding and selection for improved, balanced, productive varieties:

1. The evolutionary and geographical character of the pathosystem (Buddenhagen, 1977; Cherrett and Sagar, 1977).
2. The degree of stability of the natural or old pathosystem in the existing technology and the level of host damage at normal balance in it (Buddenhagen, 1978).
3. The epidemiological characteristics of disease development, with special concern for environmental parameters affecting this development.
4. The annual cycle of pathogen (and host) populations, with emphasis on the source of initial new-crop inoculum.
5. The type of disease in regard to its systemic or nonsystemic nature.

6. The flexibility of pathogen evolution.
7. The total germplasm resource within the crop species, genus, etc. and the flexibility of reassembly.

It gives one pause to consider the *potential* of breeding new varieties in relation to their pests and pathogens. *At one extreme is the eradication of a major disease.* Can this be done? Possibly, but it will take vision and imagination, plus support for the type of heresy that once attempted to refute the obvious truth of the sun's revolution around the center of the universe—the earth. An impediment to support for this approach is the collective mentality that knows the oft-failure of endurance of resistance that seemed so complete and so simple in breeding nurseries and in the varieties they generated. This was, however, *temporary* disease eradication. The oft-failure can be understood now by critical analysis and logical evaluation of genetic, evolutionary, ecological, and epidemiological principles. But it seldom is.

But such a logical and analytical approach also offers a newly focused objective at the other extreme—*breeding for a certain low level of host/pathogen coexistence*—tolerance for systemic diseases and tolremicity for both systemic and nonsystemic diseases. I believe that we will then obtain new superior varieties that are stable, environmentally balanced and economically useful. Maybe we should test the approaches at both extremes and try them with vigor!

REFERENCES

Bidaux, J. M. (1978) Screening for horizontal resistance to rice blast (*Pyricularia oryzae*) in Africa. In *Rice in Africa* (I. W. Buddenhagen and G. J. Persley, eds.). Academic Press, New York. Pp. 159–174.

Buddenhagen, I. W. (1977) Resistance and vulnerability of tropical crops in relation to their evolution and breeding. *Ann. NY Acad. Sci.* **287**:309–326.

Buddenhagen, I. W. (1978) Varietal improvement in relation to stability and environmental balance in peasant tropical agriculture. Presented at the Third International Congress of Plant Pathology, Munich, August 1978.

Caldwell, R. M., J. F. Schafer, L. E. Compton, and F. L. Patterson. (1958) Tolerance to cereal leaf rusts. *Science* **128**:714–715.

Cherrett, J. M., and G. R. Sagar, eds. (1977) *Origins of Pest, Parasite, Disease, and Weed Problems*. Blackwell Scientific Publications, Oxford. 413 pp.

Clifford, B. C., and R. B. Clothier. (1974) Physiologic specialization of *Puccinia hordei* on barley hosts with non-hypersensitive resistance. *Trans. Br. Mycol. Soc.* **63**:421–430.

Fasoulas, A. (1979) The honeycomb field designs. Department of Genetics and Plant Breeding, Aristotelian University of Thessaloniki, Greece, Pub. No. 9.

Habgood, R. M. (1972) Resistance to *Rhynchosporium secalis* in the winter barley cultivar Vulcan. *Ann. Appl. Biol.* **72**:265–271.

International Institute of Tropical Agriculture. (1979) *Research Highlights for 1978*. Ibadan, Nigeria.

Johnson, R., and A. J. Taylor. (1976) Spore yield of pathogens in investigations of the race-specificity of host resistance. *Annu. Rev. Phytopathol.* **14**:97–119.

Lozano, J. C., A. Bellotti, A. Van Schoonhoven, R. Howeller, J. Doll, and T. Bates. (1976) Field problems in cassava. CIAT, Series GE-16, Cali, Colombia. 127 pp.

Martin, M. W. (1978) Use of mass selection in early stages of potato breeding. (Abstr.) *Am. Potato J.* **55**:38.

Martin, M. W. (1979) Personal communication. Irrigated Agriculture and Extension Center, USDA, Prosser, WA.

Mussell, H. W. (1980) Tolerance to disease. In *Plant Disease: An Advanced Treatise,* Vol. 5 (J. G. Horsfall and E. B. Cowling, eds.). Academic Press, New York. Pp. 39–52.

Notteghem, J. L. (1977) Mesure au champ de la resistance horizontale du riz a *Pyricularia oryzae. Agron. Trop. (Paris)* **22**:400–413.

Posnette, A. F. (1969) Tolerance of virus infection in crop plants. *Rev. Appl. Mycol.* **48**:113–118.

Robinson, R. A. (1969) Disease resistance terminology. *Rev. Appl. Mycol.* **48**:593–606.

Robinson, R. A. (1973) Horizontal resistance. *Rev. Plant Pathol.* **52**:483–501.

Robinson, R. A. (1976) *Plant Pathosystems. Adv. Ser. Agric. Sci.* 3. Springer-Verlag, Berlin. 184 pp.

Russell, G. E. (1978) *Plant Breeding for Pest and Disease Resistance.* Butterworths, London. 485 pp.

Schafer, J. F. (1971) Tolerance to plant disease. *Annu. Rev. Phytopathol.* **9**:235–252.

Simons, M. D. (1966) Relative tolerance of oat varieties to the crown rust fungus. *Phytopathology* **56**:36–40.

Simons, M. D. (1969) Heritability of crown rust tolerance in oats. *Phytopathology* **59**:1329–1333.

Soto, P. E., and I. W. Buddenhagen. (1981) Development of maize streak virus resistant populations through improved challenge and selection methods. *Plant Dis.* (Submitted for publication)

Thresh, J. M. (1976) Gradients of plant virus diseases. *Ann. Appl. Biol.* **82**:381–406.

Van der Plank, J. E. (1963) *Plant Diseases: Epidemics and Control.* Academic Press, New York. 349 pp.

Van der Plank, J. E. (1968) *Disease Resistance in Plants.* Academic Press, New York. 206 pp.

Zadoks, J. C. (1971) Modern concepts of disease resistance in cereals. *Proc. Sixth Eucarpia Congress*, Cambridge. Pp. 89–98.

Zadoks, J. C. (1972) Reflections on disease resistance in annual crops. In *Biology of Rust Resistance in Forest Trees* (R. T. Bingham, R. J. Hoff, and G. J. McDonald, eds.). *Proc. NATO-IUFRO Adv. Study Inst., USDA Misc. Publ. 1221.*

Zadoks, J. C., and R. D. Schein. (1979) *Epidemiology and Plant Disease Management.* Oxford University Press, New York. 428 pp.

ECOLOGICAL ASPECTS OF DISEASE RESISTANCE

Raoul A. Robinson

This chapter is about the application of the systems concept to the subject of resistance to crop parasites. It is concerned with the ecosystem and those subsystems of an ecosystem which are called pathosystems.

THE ECOSYSTEM

The use of the systems concept in biological systems emphasizes two important distinctions. First is the recognition of systems levels, and this chapter is concerned exclusively with the higher systems levels. The highest level of all is the evolutionary system in its entirety over the whole of geological time. For all practical purposes, the highest level normally considered is the ecosystem, with its complex trophic web of producers, reducers, and consumers. In the study of crop parasites, the highest level is normally the pathosystem, which is described below.

Inherent in the systems concept is the Gestalt principle, which states that systems analysis, synthesis, and management should always proceed from the higher to the lower systems levels. This is the equivalent of the old adage about "arguing from the general to the particular." The reverse procedure leads to suboptimization in which the system as a whole is damaged by attempts to handle it in terms of only one subsystem. Suboptimization in systems analysis and synthesis leads to false conclusions; and, in systems management, it leads to material damage to the system. It is argued below that many of the current problems due to crop parasites are the result of suboptimization and that, once this is recognized, these problems can be solved.

The second advantage of the systems concept is that it makes a clear distinction between structure and behavior; and there is a close analogy with computers. In computer terminology, structure is called hardware, and behavior is called software; and fundamentally different hardware can produce identical software. Behavior should obviously be studied in terms of computer software.

A behavior pattern is a response to a stimulus and it may be exhibited at any systems level such as interspecific populations, the intraspecific population, the individual, the single organ, the cell, and so on down to the single gene. The behavioral response is controlled, and there are two basic kinds of control in biological systems. In an inherited behavior pattern, the response is controlled by the genetic code; and in an acquired behavior pattern, the response is controlled by learned experience. It is safe to conclude that all behavior patterns in plant pathosystems are controlled by the genetic code.

A strategy is a coherent system of behavior patterns and, again, the control of a strategy may be inherited, acquired or both. Of particular relevance is the concept of the evolutionarily stable strategy (ESS) of Maynard Smith and Price (1973), which is here defined as the best possible strategy of inherited behavior patterns at a specified systems level. The strategy, after all, would not have become genetically encoded if it lacked survival value; and it has become the best possible strategy by the competitive elimination of all earlier, inferior strategies. Robinson (1979, 1980a) has argued that any strategy in a wild plant pathosystem must be an EES if only because of the evolutionary survival of that system. It follows that the discernment of an ESS is the best test of validity for any conceptual model of a wild pathosystem. Discovery of a superior ESS will invalidate a postulated ESS and, if no ESS can be discerned, the model is suspect.

Biological systems exhibit two basic strategies, which MacArthur and Wilson (1967) called *r* selection and *K* selection. An *r* strategist has a very high intrinsic rate of increase (*r*) to exploit favorable conditions, which are strictly limited in time and space. *K* strategists have a constant population size that is governed by the carrying capacity (*K*) of the ecosystem. In plants these strategies are most easily discerned in terms of the total biomass; annuals are then typical *r* strategists, while evergreen trees are typical *K* strategists. However, some *K* strategist plants exhibit an *r* strategy with respect to ephemeral tissue such as the leaves of a deciduous tree. Among plant parasites, aphids and rust fungi are typical *r* strategists, while systemic fungi or viruses that continuously parasitize perennial hosts are *K* strategists. The distinction is important because it provides a precise definition of the term epidemic and endemic. An epidemic involves *r* strategists, while an endemic involves *K* strategists (Bradley, 1976). Consequently, among plant hosts, annuals and the leaves of deciduous trees suffer epidemics, while the long-lived parts of perennials suffer endemics.

The most important factor limiting the growth rate of an *r* strategist parasite is the resistance of its host. Equally, many *K* strategist parasites would behave as *r* strategists but cannot because of the resistance of their *K* strategist hosts. When taken by man to a new area where a new *K* strategist host lacks resistance, a *K* strategist parasite is likely to behave as an *r* strategist and cause an epidemic, which, however, may be on an expanded time scale (e.g., chestnut blight in North America).

THE PATHOSYSTEM

A pathosystem is a subsystem of an ecosystem and is defined by the phenomenon of parasitism (Robinson, 1976). As with an ecosystem, the geographical, biological, conceptual, and other boundaries of a pathosystem may be specified as convenient. A pathosystem is normally defined at the systems level of the population, and it concerns the interaction of two populations involving one host species and one parasite species. In a plant pathosystem the host population is a plant (the producer), and the parasite population (the consumer) can be any species in which the individual spends a major proportion of its lifespan inhabiting and obtaining nutrients from one host individual. The term "parasite" includes insects, mites, nematodes, parasitic angiosperms, fungi, bacteria, mycoplasmas, viruses, and viroids; the term is thus used in a wide biological sense. However, other consumers such as herbivores (which graze a population of plants) and carnivores (which may be hyperparasites or parasite predators) are usually considered to be outside the conceptual boundaries of a pathosystem.

There are two quite distinct categories of plant pathosystem. The wild pathosystem is an autonomous (self-regulating) system, and its ESS ensures that it is a stable system; this is testified by the fact of its evolutionary survival to the present. In the crop pathosystem, on the other hand, there is a major component of artificial control exerted by man; the ESS of the wild pathosystem has often been destroyed and, as a result, the crop pathosystem is a notoriously unstable system.

Because the pathosystem concept is primarily concerned with populations, it is necessary to define subpopulations of a pathosystem. A pathodeme is a population of the host in which all individuals have a stated character of resistance in common, although those individuals may differ in other respects. Similarly, a pathotype is a population of the parasite in which all individuals have a stated character of parasitic ability in common, although they also may differ in other respects.

Perhaps the most important aspect of the pathosystem concept is that the behavior of the host population and the behavior of the parasite population are both treated as *one system*. It follows that any subsystem of a pathosystem must be defined in terms of both the host and the parasite. There are three pathosystem components which can only be defined in this way and which consequently have a cardinal role in the definition of pathosystem subsystems. These are the gene-for-gene relationship, the differential interaction and the phenomenon of infection.

The gene-for-gene relationship (Flor, 1942) concerns the concept of pairs of matching or complementary genes. One of each pair is in the host and confers resistance, while the other is in the parasite and confers parasitic ability. A host individual may possess several such genes, and it is then resistant to any parasite individual which lacks one or more of the complementary genes; it is susceptible to any parasite individual which possesses all or more than all of them. When the resistance does not operate, parasitism occurs and the pathodeme and pathotype, or the host individual and the parasite individual, or their genes, are said to match; when the resistance does operate, parasitism does not occur and they are said to be nonmatching. Neither gene of a pair can be demonstrated without the other and this concept is totally dependent on the presence of both the host and the parasite. Gene-for-gene relationships occur in all categories of plant parasite but have been demonstrated most commonly in the fungi and nematodes; among insect parasites, they have been demonstrated in aphids, Hessian fly of wheat and the brown plant hopper of rice.

The concept of the differential interaction is essentially mathematical and it means that a series of parasite differentials is necessary to identify a particular resistance; and a series of host differentials is necessary to identify a particular parasitic ability. (If there is no differential interaction, there is a constant ranking. That is, one host individual, or pathodeme, is either more resistant or less resistant than another, irrespective of which parasite individual, or pathotype, is involved; and there is a similar ranking of parasites according to their parasitic abilities.) The differential interaction is the basis of the old plant pathological concept of physiological specialization; it means that the resistance operates against some populations of the parasite but not others; and this kind of resistance is temporary. When a matching parasite appears, the resistance does not operate and, in common usage, it is said to "break down." All gene-for-gene relationships exhibit a differential interaction; but not all differential interactions are due to a gene-for-gene relationship (see below). Once again, this concept is totally dependent on the presence of both the host and the parasite.

Infection is the contact made by a parasite individual with a host individual for the purposes of parasitism and, yet again, the term is meaningless in the absence of either the host or the parasite. There are two kinds of infection. With autoinfection a host is infected by a new parasite individual derived from a parent parasite which is already parasitizing the same host individual. With alloinfection the parasite individual originates away from the host individual that it infects. The part of the epidemic that involves autoinfection only is called the esodemic, and the part that involves alloinfection only is called the exodemic (Robinson, 1976).

On this basis, it is possible to define the subsystems of a pathosystem. The vertical subsystem is named after Vanderplank's (1963) concept of vertical resistance; it involves a gene-for-gene relationship, and it exhibits a special category of differential interaction called the Person/Habgood differential interaction, which has been described elsewhere (Robinson, 1976, 1979). The horizontal subsystem is named after Vanderplank's (1963) concept of horizontal resistance; it does not involve a gene-for-gene relationship, and it does not exhibit a differential interaction (i.e., there is a constant ranking). The terms vertical and horizontal can be used at any systems level; thus vertical and horizontal subsystems, pathotypes and pathodemes, resistance and parasitic ability (and their mechanisms), genotypes, genes, and so on.

There are a number of other subsystems that do not involve a gene-for-gene relationship but do exhibit a differential interaction (Table 1). These have been described elsewhere (Robinson, 1979) and are beyond the scope of the present chapter.

Perhaps the most important discovery to emerge from the pathosystem concept is that the vertical subsystem can only control alloinfection (the exodemic) and that autoinfection (the esodemic) can only be controlled by the horizontal subsystem. However, while every pathosystem has a horizontal subsystem, some pathosystems lack a vertical subsystem; it follows that the exodemic can also be controlled by the horizontal subsystem. These conclusions are justified below.

MODEL OF A WILD PATHOSYSTEM

On the basis of these various ecosystem and pathosystem principles, it is possible to develop conceptual models of wild pathosystems; and the best test for the validity of such models is the demonstration of an ESS. The present model is hypothetical and concerns two pathosystems involving a species of heteroecious aphid and a species of heteroecious rust fungus, which both share the same species of winter and summer hosts. This choice has been made for two reasons. First, although the aphid and the rust have fundamentally different structure (hardware), their inherited behavior patterns, strategies and ESSs (software) are almost identical at the higher systems levels. Second, the combined model illustrates the essential unity of crop entomology, crop pathology, and breeding plants for resistance to parasites. The pathosystem concept is multidisciplinary. It need hardly be added that the model is a deliberate oversimplification, which will inevitably re-

Table 1

Subsystem	Gene-for-Gene Relationship	Differential Interaction
Vertical	Present	Present
Horizontal	Absent	Absent
Others	Absent	Present

quire modification or elaboration when compared with an actual wild pathosystem.

The winter host is a deciduous tree species, and the summer host is an annual herb. All population numbers are conjectured to the nearest order of magnitude and, obviously, are intended for purposes of illustration only. The aphid and the rust are each referred to as such, but references to "the parasite" indicate that, in this context, the behavior of the aphid and the rust are indistinguishable. All four species are *r* strategists, even if the winter host is only functionally so with respect to its ephemeral leaf tissue. Both pathosystems are thus epidemic rather than endemic in form. Both the pathosystems are also discontinuous in time and space. They have sequential discontinuity due to winter and the alternation of hosts. And they have spatial discontinuity due to genetic heterogeneity; in particular, there is heterogeneity of vertical genes in both of the parasite populations and in the summer host population. Both species of parasite are polycyclic; that is, they have many reproduction cycles in the course of one epidemic cycle. The epidemic cycle has two subcycles referred to as the summer and the winter epidemics, depending on which species of host is involved.

The Primary Exodemic

Because this is a wild ecosystem, it may be assumed that the winter and summer host populations are freely intermingled. The summer epidemic begins with the migration of the parasite (virginoparae or aeciospores) from the winter host to the summer host. Because each summer host seedling is parasite-free, the only possible infection is alloinfection. This is the primary exodemic. Each parasite individual has its own combination of vertical parasitic ability genes; and each summer host individual has its own combination of vertical resistance genes. Each alloinfection is either a matching or a nonmatching infection. If it matches, parasitism begins; and if it does not match, the parasite individual dies and that host individual remains parasite-free. Some matching must obviously occur if the pathosystem is to exist at all.

The primary exodemic is controlled by the vertical subsystem. It is assumed that there are 12 pairs of matching genes in this subsystem, and that every individual in both the host and the parasite populations possesses 6 of these genes (i.e., $n/2$, where n is the number of pairs of vertical genes). There are thus 924 different, 6-gene, vertical pathodemes and pathotypes (this figure is obtained from the binomial mode, m, of Pascal's triangle, shown in Table 2). It is also assumed that all the $n/2$ vertical pathodemes and pathotypes occur with equal frequencies. The probability of any alloinfection being a matching infection is then constant at $1/m$, which, in this model, is 1/924 or 10^{-3} (Robinson, 1980a). This matter is discussed further below.

It is assumed that the summer host population consists of 10^6 individuals

Table 2

Number of Vertical Genes	Binomial Coefficients
0	1
1	1 1
2	1 2 1
3	1 3 3 1
4	1 4 6 4 1
5	1 5 10 10 5 1
6	1 6 15 20 15 6 1
7	1 7 21 35 35 21 7 1
8	1 8 28 56 70 56 28 8 1
9	1 9 36 84 126 126 84 36 9 1
10	1 10 45 120 210 252 210 120 45 10 1
11	1 11 55 165 330 462 462 330 165 55 11 1
12	1 12 66 220 495 792 924 792 495 220 66 12 1

and that there are 10^3 six-gene vertical pathodemes which are randomly distributed in equal proportions; each vertical pathodeme thus consists of 10^3 individuals. Similarly, the summer population of the parasite consists of 10^6 *effective* individuals (i.e., individuals which succeed in making contact with a host individual) with the same equal proportions of 10^3 six-gene vertical pathotypes. On average, therefore, each host is alloinfected only once. Because the probability of that alloinfection being a matching infection is 10^{-3}, only 10^3 host individuals (0.1%) are matched. On average also, each vertical pathotype becomes established on one host individual. Each matching parasite individual then commences asexual reproduction which permits autoinfection; this is the esodemic.

The Esodemic

Both the aphid and the rust are r strategists. Any individual which has made a matching alloinfection must now reproduce as rapidly as possible (i.e., maximum r) to exploit the abundant food resource of that one host individual. This rapid reproduction is greatly assisted by asexual reproduction. Indeed, sexual reproduction is not only unnecessary, it is positively undesirable because, when the parasite is a clone, every autoinfection is a matching infection. Every part of the host individual has the same vertical resistance and every individual in the parasite clone has the same, matching vertical parasitic ability. It follows that the vertical subsystem has no function whatever in the esodemic and autoinfection can only be controlled by the horizontal subsystem. Robinson (1976) used this argument to conclude that horizontal resistance is universal. Because every epidemic (or endemic) must involve matching infection, every host plant must have at least some horizontal resistance to all of its parasites. To deny the existence of horizontal resistance is to postulate an absolute susceptibility.

In addition to abandoning sexual reproduction, the aphid clone also abandons flight as it has no immediate need to migrate. It is only under the stimulus of crowding that winged individuals are produced several asexual generations later. The rust does not possess this structural difference and all its asexual propagules (uredospores) can either autoinfect or be dispersed. In both parasites, however, migration occurs and this permits alloinfection of other host individuals. This is the secondary exodemic.

The Secondary (and Other) Exodemics

It may be assumed that the effective migrating population of the parasite has increased tenfold as a result of the esodemic on 10^3 host individuals. The secondary exodemic thus involves a parasite population of 10^7 effective individuals. Because each vertical pathotype became established on one host individual, all 10^3 vertical pathotypes are again equally represented in the migrating parasite population, and the probability of matching alloinfection is still 10^{-3}. With 10^7 parasite individuals, 10^4 host individuals (1%) are matched and these new esodemics lead to a further tenfold increase in the effective parasite population, which then migrates, causing the tertiary exodemic. On average also, the 10^3 vertical pathotypes are again equally represented. In the tertiary exodemic, therefore, there are 10^8 effective parasite individuals and 10^5 host individuals (10%) are matched. Finally, in the quaternary exodemic, there are 10^9 effective parasite individuals and 10^6 host individuals (100%) are matched. This progression is summarized in Table 3. For simplicity it is assumed that there are only four of these exodemics in the summer epidemic. And all 10^3 vertical pathotypes are still equally represented in the parasite population at the conclusion of the quaternary exodemic; indeed, this is an exact reflection of the equal frequency of vertical pathodemes in the summer host population.

Table 3

Exodemic	Size of Effective Parasite Population	Number of Host Individuals Matched
Primary	10^6	10^3
Secondary	10^7	10^4
Tertiary	10^8	10^5
Quaternary	10^9	10^6

The Effects of the Vertical Subsystem

The vertical subsystem has three major effects on the summer epidemic. In the primary exodemic, only 0.1% of hosts are matched; in the secondary exodemic, 1% are matched; in the tertiary exodemic, 10% are matched; and 100% matching occurs only in the quaternary exodemic, which is close to the end of the summer epidemic.

There is a corresponding effect on the amount of parasite damage to the summer host population. Only 0.1% of host individuals suffer the maximum (100%) damage; 1% of hosts suffer 75% damage; 10% of hosts suffer 50% damage; and 90% suffer only 25% damage. However, the last figure is false in that it assumes a linear growth rate in the parasite. In fact the growth rate is sigmoid, and the lag phase means that the level of parasite damage in the last esodemic will be considerably lower than 25%; and these hosts constitute 90% of the population. It should be noted also that all vertical pathodemes suffer equal damage and none has a survival advantage over the others. Equally, no vertical pathotype has a survival advantage over the others.

Finally, the vertical subsystem prevents a devastating population explosion of the parasite. The probability of matching alloinfection is only 10^{-3} in *each* exodemic. Without the vertical subsystem, the matching parasite population would be 1000 times larger in each successive exodemic. Other factors being equal, there would then be 10^{21} parasite individuals at the end of the summer epidemic, in place of the 10^{9} which actually appear as a result of the control exerted by the vertical subsystem.

The Effects of the Horizontal Subsystem

In this model, it is assumed (quite arbitrarily) that the effective parasite population increases tenfold as a result of each esodemic. If the summer host had no horizontal resistance, this increase would be considerably greater. The effect of the horizontal subsystem is thus the same as that of the vertical subsystem. Both subsystems reduce the intrinsic rate of increase, r, of an r-strategist parasite.

The ESS of the Vertical Subsystem

In a paper written in 1978, Robinson (1980b) argued that there are 2^n vertical genotypes and that, if they all occur with equal frequency in the host and parasite populations, the proportion of matching alloinfection would be $(3/4)^n$. These figures are derived from the Person/Habgood differential interaction. It subsequently became apparent that such an arrangement cannot constitute an ESS. One of those vertical genotypes is the one with no vertical genes at all; in the host such a vertical resistance has the minimum survival advantage because every alloinfection would match; and in the parasite, such a vertical parasitic ability also has the minimum survival advantage because that pathotype will have the minimum rate of matching which is $1/2^n$. Conversely, the vertical genotype with all the available genes will have the maximum survival advantage. Such a host can only be matched by $1/2^n$ alloinfections; and such a parasite can match every host. This system is clearly unstable. The host population would tend to homogeneity of the vertical pathodeme which has all the available resistance genes; and the parasite population would tend to homogeneity of the vertical pathotype which has all the available parasitic ability genes. At this point, the vertical subsystem would cease to function because every alloinfection would be a matching infection.

If the vertical subsystem is to be stable, every vertical genotype must have an equal survival advantage. This is only possible if all vertical genotypes belong to one binomial class; that is, every genotype has the same number of vertical genes, regardless of the identity of those genes. For example, if (as in the present model), there are 12 pairs of matching genes, there are $(n + 1)$ binomial classes ranging from 0 to 12 genes. The number of genotypes within a class is the binomial coefficient, e, and is revealed by Pascal's triangle (see Table 2).

The ESS of the vertical subsystem depends on five factors. First, all the vertical genotypes (pathotypes and pathodemes) must belong to one binomial class; and, second, they must all be equally represented in the two populations. This will ensure equal survival advantage, equal frequencies and a constant probability of matching alloinfection in both space and time which is $1/e$. The third factor concerns the maximum economy of vertical genes which is obtained when the binomial class is also the binomial mode, m, in which there are $n/2$ genes in each genotype. For example, if $n = 12$, and all pathotypes and pathodemes have only one gene, and they occur with equal frequencies, the probability of matching alloinfection will be 1/12 (Table 2). But if all pathotypes and pathodemes have six genes (i.e., $n/2$), and they occur with equal frequencies, the probability of matching alloinfection will be $1/m$, which is 1/924 or 10^{-3}, as already indicated.

The fourth factor contributing to the ESS of the vertical subsystem concerns the value of n. Table 4 shows the constant probabilities for various values of n; there is an optimum n for any pathosystem. It is most unlikely, for example, that both the aphid and the rust pathosystems in the present model would have the same value of n. On the one hand, the rust produces far greater numbers of individuals than the aphid. It is therefore likely to alloinfect with greater frequency and, for this reason, its pathosystem is likely to require a high number of vertical genes. On the other hand, the aphid has tactical mobility and controlled flight; the proportion of its relatively small population which is effective in contacting a host individual is

Table 4

Number of Pairs of Vertical Genes (n)	Binomial Mode (m)	Probability of Matching Alloinfection ($1/m$)
10	252	0.004
12	924	0.001
14	3432	0.0003
16	12870	0.00007
18	48620	0.00002
20	184756	0.000005

thus very high, and this factor will also increase the requirement of n. The nature of the host also affects the requirement of n. In the present model the host is a small annual; the total number of alloinfections per host individual is low, and the requirement of n is also low. But if the summer host were a large tree species, the total number of alloinfections per host individual would be much higher; and the requirement of n would increase accordingly. The *total* alloinfections per host individual are thus governed by the reproductive and dissemination efficiencies of the parasite and the size of the host; the *proportion* of these alloinfections which are matching infections is governed by n.

The mechanisms of the ESS of the vertical subsystem are discussed below. In general terms, they are an example of genetic homeostasis which Lerner (1954) defined as "the tendency for a Mendelian population as a whole to retain its genetic composition arrived at by previous evolutionary history." It seems that this genetic composition requires (1) a value of n appropriate to the size of the host and the reproduction and dissemination efficiency of the parasite; (2) that all pathodemes and pathotypes occur with equal frequency; (3) that all pathodemes and pathotypes belong to the same binomial class; and (4) that this class is the binomial mode. The probability of matching is then constant in both space and time; and it is the optimum probability for the pathosystem in question. Finally, the fifth factor is the "recovery" of vertical resistance.

The Recovery of Vertical Resistance

Because it is an annual, all parts of the summer host (except the seed) die in the fall. At this point, the esodemic ends because there is no further host tissue available to the parasite. At this point also, the vertical resistance "recovers" in the converse sense of its "breakdown." In the following spring, every summer host seedling will be parasite-free, and it can only be alloinfected. The vertical resistance of the entire summer host population will again be as effective as it was in the previous epidemic cycle. This phenomenon is the fifth factor contributing to the ESS of the vertical subsystem, and it is dependent on sequential discontinuity in the pathosystem. There is an annual cycle of breakdown and recovery of vertical resistance; indeed, the esodemic may be said to begin with the breakdown of vertical resistance and to end with its recovery.

The ESS of the Horizontal Subsystem

Horizontal resistance and horizontal parasitic ability are both quantitative variables which are polygenically inherited. Both respond to positive and negative selection pressures and, indeed, each exerts selection pressure on the other.

If the resistance in the host is too low, it increases during the course of a few host generations because of the positive selection pressure exerted by excessive parasite damage. Conversely, an excess of resistance is an unnecessary survival value which tends to decline under negative selection pressure. Similar pressures operate on the parasite population. The point of balance is where the positive and negative selection pressures are equal, both for horizontal resistance in the host and for horizontal parasitic ability in the parasite. This is an ESS which appears to be closely comparable to the parasitic systems studied by Pimental (1968).

The Evolution of the Vertical Subsystem

The cycle of breakdown and recovery of vertical resistance explains why temporary resistance can evolve in a long-lived perennial host such as a deciduous tree in which the individual lifespan is measured in centuries. It also explains the deciduous habit of trees in the wet tropics where there is otherwise no sequential discontinuity. Equally it can be argued that a vertical subsystem cannot evolve in a pathosystem which lacks sequential discontinuity, such as an evergreen tree. In such a pathosystem the vertical resistance would break down but it could never recover. Such a strategy cannot be an ESS. It follows that a vertical subsystem will only occur in a pathosystem with both spatial and sequential discontinuity; that is, in epidemics, which, by definition, involve *r* strategists. And a vertical subsystem cannot occur in a pathosystem with spatial and sequential continuity; that is, in endemics which involve *K* strategists. However, the converse is not necessarily true; a vertical subsystem can evolve in a discontinuous pathosystem, but it need not do so.

The annual and deciduous habits appear to be a relatively recent evolutionary innovation. It is probably safe to conclude that the horizontal subsystem represents the original and fundamental control of the pathosystem, while the vertical subsystem is relatively new, in evolutionary terms, and occurs only in some species of host and against only some species of parasite of those hosts.

Finally, some comment is necessary as to why the vertical subsystem should evolve at all when the more ancient horizontal subsystem had apparently provided pathosystem stability over long periods of geological time. There can be no question that it *has* evolved, possibly in as many as 10^6 different host–parasite associations, and that it has become genetically encoded. Interestingly, the one behavior strategy becomes genetically encoded twice, in the two species of host and parasite; the host genes and the parasite genes clearly constitute one system. Even if no explanation can be found, there can be no doubt that the vertical subsystem must have an ESS and an exceptionally valuable one at that. The most plausible explanation seems to be that the vertical subsystem provides a buffer. A pathosystem with a horizontal subsystem only is relatively unbuffered against the extremes of seasonal variation in the external environment. Abnormal weather may so favor the parasite that the host population is decimated. The pathosystem will then exhibit fairly wide fluctuations in the population size of both the host and the parasite. Such fluctuations do not represent the greatest possible stability over periods of evolutionary time. With a vertical subsystem, however, an abnormally high growth rate in the parasite would be controlled by the gene-for-gene relationship. There is then a far greater stability over periods of evolutionary time; and this is what the ESS is all about.

The Winter Exodemic

At the end of the summer epidemic, the parasite must migrate in order to alloinfect the winter host. This is the winter exodemic. However, it is most unlikely that this exodemic is controlled by a vertical subsystem. The winter host does not need the protection of vertical resistance because there is only the one exodemic and the parasite is essentially a temporary resident, which soon departs for the summer host. Similarly, the parasite needs an equal frequency of all $n/2$ vertical pathotypes in its spring migrant population and a vertical subsystem in the winter epidemic would greatly impede the sexual recombination necessary to achieve this equal frequency. It is probable, therefore, that the winter exodemic is controlled by the horizontal subsystem only.

The Winter Epidemic

The parasite can only undergo sexual recombination on the winter host. Possibly the only behaviorial difference between the aphid and the rust is that the aphid overwinters after sexual recombination (as fertilized eggs), while the rust overwinters (as teliospores) before sexual recombination and, indeed, before alloinfecting the winter host. This difference is obviously essential in view of the deciduous habit of the winter host and the differences in hardware of the two species of parasite. However, it does not appear to affect the ESS of either pathosystem.

Sexual Recombination

The summer host must maintain equal frequencies of $n/2$ vertical pathodemes; if it does not do so, its rate of parasite damage will increase. The parasite must also maintain equal frequencies of $n/2$ vertical pathotypes; if it does not do so, its rate of reproduction will decline. In general terms, these equal frequencies can be maintained by genetic homeostasis and random mating. For example, individuals with too few vertical genes suffer a directly reduced fitness and there is evidence (Vanderplank, 1968) that individuals with too many vertical genes suffer an indirectly reduced fitness. The entire population (of either host or parasite) would then tend to $n/2$ vertical genotypes only. But other mechanisms may also operate. Robinson (1980a) pointed out that many of the more recently evolved plant hosts are obligately allogamous, and that a pollen grain is closely similar to a fungal spore. Self-fertilization thus corresponds to autoinfection; and cross-fertilization corresponds to alloinfection. The self-incompatibility genes have a function that is closely similar to the function of the vertical resistance genes, with the obvious difference that they control the equivalent of autoinfection (self-fertilization). It is perhaps possible that the vertical resistance genes are linked to the self-incompatibility genes in such a way that only equal frequencies of $n/2$ vertical pathodemes occur. A similar system might also operate in the parasite, as many fungi are heterothallic. The four homeostatic systems would then be mutually reinforcing and would constitute a supersystem. However, this is not directly relevant to the present model, as there is no evidence for such mechanisms in either the rusts or aphids.

The Relation Between the Vertical and the Horizontal Subsystems

These two models concern wild pathosystems which have a high degree of discontinuity. Robinson (1979, 1980a) suggested that plant pathosystems differ in their degrees of discontinuity and that a classification of wild pathosystems is possible on this basis, ranging from virtual continuity in time and space, to maximum discontinuity in time and space. The discontinuity is measured by a number of variables and it seems that, in any one pathosystem, the degrees of discontinuity are approximately equal, both in all these variables and in the host and the parasite. It seems also that the survival value of a vertical subsystem is directly proportional to the degree of discontinuity, both in relation to the horizontal subsystem and to the number of parasite species against which it occurs in that one host species. The survival value of the horizontal subsystem, on the other hand, is inversely proportional to the degree of discontinuity and is at its greatest in the continuous pathosystem.

Because the present models have a high degree of discontinuity, the survival value of the vertical subsystem is obvious, while that of the horizontal subsystem is much less obvious. However, if the summer host had been a large deciduous tree, instead of a small annual herb, the value of the horizontal subsystem would become more apparent. Each esodemic would then involve a very large population of genetically identical leaves and, without horizontal resistance, this population would be totally destroyed by uncontrolled autoinfection. (It is worth digressing at this point to comment that, in the crop pathosystem, one cultivar is normally a single vertical pathodeme. Each plant in the crop is then the epidemiological equivalent of one leaf in a single, abnormally large tree.)

This argument has relevance to plant breeding because the degree of discontinuity in the wild pathosystem of the progenitors of a crop species will determine the ease of breeding. If that wild pathosystem was continuous, vertical resistance will not occur and breeding for horizontal resistance is easy. Examples of this situation include sugarcane and cassava. But, as the degree of discontinuity of the wild progenitor pathosystem increases, so the importance of the horizontal subsystem decreases and the frequency and prominence of vertical subsystems increase. Breeding for resistance to crop parasites is thus more likely to encounter difficulties in crop species derived from a pathosystem with a high degree of discontinuity, such as many of the cereals and grain legumes.

THE CROP PATHOSYSTEM

The crop pathosystem must be regarded as a distortion of the wild pathosystem. Various components of the wild pathosystem have been changed in the general process of domestication and cultivation. This is due to an element of artificial control exerted by man. If that control were to cease abruptly, all evidence of cultivars and cultivation would disappear very quickly. It follows that any attempt to explain the wild pathosystem in terms of the crop pathosystem is unrealistic. Equally, if we are to be realistic, we should only attempt to explain the crop pathosystem in terms of the wild pathosystem. It is, perhaps, one of the weaknesses of crop science that almost no studies whatever have been made of wild pathosystems. Indeed, any discussion of the wild pathosystem is currently restricted to conceptual models, such as the one presented above.

It is immediately obvious that the crop pathosystem is very different from this conceptual model of the wild pathosystem and, if the model is even approximately accurate, these differences must be due to various artifacts of agriculture. The following discussion is primarily concerned with some of the more important of those artifacts. For simplicity, it is assumed that the hypothetical wild pathosystem (discussed above) has been domesticated; the summer host is now a cultivar and it may be thought of as one of the cereals or grain legumes.

Domestication

As a result of some thousands of years of more or less unconscious selection and perhaps three quarters of a century of scientific selection, the summer host is so changed that it can no longer survive unaided in a wild ecosystem. On the other hand, the yield and quality of the crop product have been raised far beyond such capacities in a wild plant. But the domestication of resistance to crop parasites has been considerably less successful. Cultivars are notoriously more prone to parasite damage than are wild plants. This failure has not been for want of trying; indeed, the development of resistance to parasites seems to dominate most breeding programmes. The failure appears to be due to an almost complete lack of understanding of the original evolutionary functions of the various components of the crop pathosystem (see below).

Cultivation

The level of resistance in a wild plant pathosystem is enough, but only enough, to ensure the evolutionary survival of both the host and the parasite in a stable, dynamic equilibrium. The requirement of the crop pathosystem is that neither the yield nor the quality of the crop product is damaged by parasites. And this means that the levels of resistance necessary in the crop pathosystem are likely to be considerably higher than those of the wild pathosystem. But there is one factor in cultivation that dwarfs all other considerations. The cultivar is a pure line. The spatial discontinuity of the wild summer host population, due to its genetic heterogeneity, has gone. The ESS of the vertical subsystem has been destroyed. When the summer host population is homogeneous, a similar homogeneity is imposed on the parasite population. From the primary exodemic onward, every alloinfection is a matching infection. Clearly, no great weight can be given to the conjectured orders of magnitude of a conceptual model. But they are an indication of an overall effect. It will be recalled that, in the model, heterogeneity of vertical genotypes reduced the final parasite population from 10^{21} to 10^{9}. There can be little doubt that an effect of this kind has been lost in many crop pathosystems.

However, there is one major effect of homogeneity which, in evolutionary terms, is completely new and quite different from the wild pathosystem. If a new cultivar, with a different vertical resistance, is introduced quite suddenly and on a large scale, the parasite is thrown off balance. Every alloinfection is then a nonmatching infection and the crop remains entirely free from the parasite. This effect may last for some years until the parasite is able to adjust; there is then a population explosion of the matching vertical pathotype and the vertical resistance is said to have "broken down." This disabled cultivar is then replaced with another which has a different vertical resistance; and the cycle is repeated. This has been dubbed the "boom-and-bust cycle of cultivar production." As such, it is an artificial strategy which is fundamentally different from the ESS of the wild pathosystem. This is not to say that it is necessarily a bad strategy or that it cannot be improved (see below).

The Vertifolia Effect

Vanderplank (1963) first recognized this effect, which he named after a potato cultivar in which it was particularly prominent. When breeding for vertical resistance there is negative selection pressure for horizontal resistance which tends to decline. As a result, most modern cultivars have a very low level of horizontal resistance and this is another reason why the "bust" of the boom-and-bust cycle is so severe. It was suggested above that the function of both the vertical and the horizontal subsystems was to reduce the intrinsic rate of increase, *r*, of an *r* strategist parasite. In the crop pathosystem homogeneity has destroyed the ESS of the vertical subsystem; and the vertifolia effect has destroyed the ESS of the horizontal subsystem.

The Conventional Approach to Breeding for Resistance

It was probably inevitable that the beauty and fascination of Mendel's laws should dominate plant breeding procedures. The conventional approach to breeding for resistance is firmly based on gene-transfer methodologies and, for this reason, it is axiomatic that breeding for resistance requires a "good source" of resistance, which is then transferred genetically to a suitable cultivar by backcrossing. The whole procedure is obviously much easier if it involves a single prominent resistance mechanism, whose inheritance is controlled by a single Mendelian gene. If there is a vertical subsystem in the wild pathosystem, this procedure cannot fail to find it; indeed, it cannot find anything else. This is suboptimization; the system as a whole is damaged by attempts to improve it in terms of the one vertical subsystem. Further sub-optimization has occurred within the vertical subsystem both by employing only one (or a very few) vertical genes at a time and by utilizing them on a basis of homogeneity.

The problem of "physiologic specialization" has been known since the earliest days of scientific plant breeding; it even predates the term "pathology." As already discussed, physiologic specialization means that there is a differential interaction, the most common being the Person/Habgood differential interaction due to the gene-for-gene relationship, although there are others (see above) which are usually of minor importance.

Physiologic specialization is due to the fact that all screening is conducted during the exodemic; that is, the host population is screened against nonparasitic (i.e., nonmatching) pathotypes of the parasite. And if a matching pathotype happens to appear during the screening process, the host individual (or pathodeme) in question is discarded as susceptible. But, in this sense, *all* the host individuals in the screening population are "susceptible"; it is just that their "susceptibility" has not yet been manifested. This is essentially a problem of terminology. It is often argued that there is no real difference between a high resistance and a low parasitic ability, or vice versa. But there is a difference; the effects may be the same, but the causes are quite different. The level of resistance can only be measured in terms of the level of parasite damage. Given a constant, medium level of resistance, high and low parasitic abilities will produce high and low levels of parasitism, respectively. And similarly with a constant, medium level of parasitic ability. The confusion arose because the terminology is based on the host; thus vertical "resistance." The terminology could equally be based on the parasite and then vertical "resistance" is not resistance at all; it is an absence of parasitic ability. This is suboptimization in systems analysis and it can only be resolved by regarding the host and the parasite as one system; the terminology must be based on the system rather than one or another of the components of that system. It is consequently important to refer only to matching and nonmatching. Indeed, even though they have been freely employed in this chapter, the terms "vertical resistance" and "vertical parasitic ability" are misnomers.

Any differential interaction involves matching and nonmatching. So long as the screening is conducted with nonmatching pathotypes, the resulting 'resistance' is liable to fail. It is not even resistance. Indeed, the only genuine resistance occurs when there is no differential interaction; with a high level of resistance there is then a low level of parasite damage in spite of a high level of parasitic ability in the parasite. Such resistance is horizontal resistance and, for all practical purposes, it is permanent resistance.

Quantitative Differential Interactions

So far, the discussion has concerned differential interactions which are qualitative in their effects; the host is protected either completely or not at all. In the crop pathosystem, vertical subsystems (as well as other differential interactions) often confer incomplete protection against nonmatching pathotypes and the effects are then quantitative. This is the cause of much scientific dispute. Robinson (1979, 1980a) has argued that this phenomenon is an artifact of agriculture while others (see Parlevliet and Zadoks, 1977, for a comprehensive review) regard quantitative differential interactions as grounds for denying the existence of a horizontal subsystem. The latter viewpoint is essentially pessimistic, as its clear implication is that all resistance involves a differential interaction and that, consequently, all resistance is temporary. Resolution of this dispute is obviously of the utmost importance. However, it cannot be finally resolved until there are some exhaustive studies of wild pathosystems; and these studies are both complex and difficult. In the meantime, the battle will no doubt continue unabated. But, so long as, of necessity, it is confined to conceptual models, the demonstration of an ESS remains the test of validity. And the "pessimistic school" have yet to produce an ESS for either the quantitative differential interaction or the absence of a horizontal subsystem. Indeed, it can be argued that any tendency to incompleteness of protection would detract from the best possible strategy of the vertical subsystem just as decisively as a tendency to homogeneity; and the absence of a horizontal subsystem can only mean an absolute susceptibility.

The "Strength" of Vertical Resistance

Vanderplank (1968) produced the new concept of "strength" of vertical resistance which must be briefly explained. Strength refers to the rarity of the matching vertical pathotype; if it is rare, the resistance endures for a long period and is described as strong; but if it is common, the resistance breaks down quickly and is described as weak. Equally, when a strong resistance is abandoned, the matching pathotype becomes rare again quickly, after having been common following the breakdown of a widely used vertical resistance; with a broken down, weak resistance, the matching pathotype remains common. By analogy, the insecticide DDT confers a weak protection against houseflies and a relatively strong protection against malarial mosquitoes. Obviously, strong vertical resistances are more valuable in agriculture; they endure for a longer period and can be reemployed after a shorter interval following their breakdown.

If the ESS of the wild vertical subsystem, described above, is correct, it is clear that the vertical resistances which are actually employed in agriculture must have a considerably greater strength than they did in the wild pathosystem. This increase in strength is an artifact of agriculture. In some crop pathosystems, such as *Puccinia polysora* of maize in tropical Africa, vertical resistance breaks down in the course of one screening season and it is agriculturally useless (Robinson, 1976). This crop pathosystem is close to the postulated ESS of the wild vertical subsystem in which all the vertical resistances must be both weak and *equally* weak if genetic heterogeneity is to be maintained.

The mechanism of an increased strength in agriculture is apparently a loss of epidemiological competence in the matching vertical pathotype. This occurs when a vertical resistance gene is transferred from a wild host (which may even be a different species) to a cultivar which is cultivated in a very different environment. Some vertical pathotypes apparently have a reduced epidemiological competence on the new host and in the new environment. When breeding for vertical resistance, it is inevitable that weak resistances will be lost; they may even escape observation. And only relatively strong vertical resistances, which endure until at least the completion of the breeding process, will be retained. Indeed, it seems reasonable to postulate that vertical resistance would never have been employed at all in agriculture but for the phenomenon of increased strength. Other artifacts of agriculture may contribute to an increased strength. For example, eradication of the winter host in the present model would greatly increase the rarity of vertical pathotypes.

Vanderplank (1968) recognized yet another aspect of strength. Various combinations of vertical genes may have an increased strength such that the matching vertical pathotype lacks epidemiological competence entirely. Such a vertical resistance is effectively permanent and it apparently occurs in the spring wheats of North America which possess the genes Sr_6Sr_{9d} for vertical resistance to *Puccinia graminis tritici*.

Interplot Interference

Vanderplank (1963) was also the first to recognize the importance of interplot interference; he called it the "cryptic error" in field trials. Subsequently, James et al. (1973, 1976) have undertaken the most detailed studies of the phenomenon. Positive interference occurs when the influx of propagules into a field plot is greater than the exodus, thus providing a false impression of susceptibility. Negative interference occurs when the exodus is greater than the influx and there is a false impression of resistance. Positive interference can cause errors of more than a hundredfold in the levels of parasitism.

Perhaps the most important aspect of interplot interference is that it involves alloinfection (Robinson, 1979). The alloinfection may be either matching or nonmatching infection. If it is nonmatching infection, vertical resistance appears perfect and neither its impermanence nor a probable low level of horizontal resistance (due to the vertifolia effect) are apparent. Conversely, if the alloinfection is matching infection, a high level of horizontal resistance can be completely obscured and neither its permanence nor its value under interference-free conditions is apparent.

Interplot interference is at its greatest when the plots are small, the distance between plots is small, and the difference in parasite levels is great. These are exactly the conditions which prevail in the pedigree breeding processes of conventional plant breeding. In these circumstances, the false value of vertical resistance is so enhanced, and the real value of horizontal resistance is so obscured, that it was inevitable that the old fashioned plant breeders should screen for vertical resistance. In fact, many of them were entirely convinced that this was the only kind of resistance that existed. There is no suggestion of blame in this criticism as these scientists had no way of knowing about very recent developments in epidemiology and plant pathosystem concepts. But a continuing failure to recognize this source of gross error will become increasingly difficult to excuse.

THE CROP STABLE STRATEGY

The crop stable strategy (CSS) may be defined as the best possible agriculturally stable strategy; it must necessarily be an artificial strategy in the sense that the inherited behavior patterns of the wild pathosystem have to be externally controlled by man. The CSS must consequently be a coherent system of both inherited and acquired behavior patterns. The acquired behavior patterns are ours; they are controlled by learned experience; and we are the ones who must do the learning. So far, many of our crop strategies have been unstable and inappropriate. This is no different from many other artificial strategies, such as military, political, and economic strategies, which have not infrequently been major disasters.

It was argued above that the crop pathosystem is a distortion of the wild pathosystem and that we should only attempt to explain the crop pathosystem in terms of the wild pathosystem, difficult though this may be with the current state of our knowledge. While obviously accepting the possibility that the model of a wild pathosystem presented here may be incorrect, it seems indisputable that the vertical and horizontal subsystems do represent inherited behavior patterns, which have not become genetically encoded for nothing; they must each have an ESS. And the essence of the ESS is *stability*, over periods of geological time. If the crop pathosystem is unstable, the reason can only lie with the artificial control exerted by man. Our learned experience is not yet good enough.

It was also mentioned above that the requirement of resistance in the crop pathosystem is greater than in the wild pathosystem. This is mainly because the wild pathosystem can tolerate considerable losses from parasites while the acceptable level of loss in the crop pathosystem is negligible. Other factors increase this requirement even further. These include various artificial aspects of cultivation such as an unnatural crowding of the host population, cultivation beyond the natural limits of the wild ecosystem of the crop progenitors and so on. This means that resistance must be domesticated; it must be raised by artificial selection to levels above those of the wild pathosystem. If the resistance is still inadequate when the limits of domestication have been reached, the resulting resistance should then be employed with the best possible crop-stable strategy.

Domestication of Horizontal Resistance

There can be little doubt that horizontal resistance is the most promising candidate for domestication. For all practical purposes, it is permanent resistance. It is also a polygenically inherited variable which can be increased above its natural optimum to levels approaching its maximum. There is increasing evidence that these levels may be sufficient to achieve a control of crop parasites which is permanent, complete and comprehensive (Robinson, 1979, 1980b), at least in some crops in some areas and, possibly, in many crops in many areas. It is only when the maximum levels of horizontal resistance have been shown to be both maximal and inadequate that resort should be made to vertical resistance.

Because it is a quantitative variable, horizontal resistance can be domesticated by artificial selection in the same way that the sucrose content of sugarcane or sugar beet was raised far above its natural level in the wild progenitors. This will require population breeding methods and changes in gene frequencies, rather than pedigree breeding and gene transfers. And this is a considerable departure from tradition in autogamous crops such as most cereals and grain legumes. But there is an even greater departure from tradition. Horizontal resistance is only manifested during the esodemic; that is, after vertical resistance has broken down. It follows that all screening for horizontal resistance must commence with susceptibility; there must be no "good source" of resistance. Indeed, the process must begin with screening potential parents for susceptibility to one vertical pathotype of each parasite species in which a vertical subsystem occurs. These pathotypes are then used in all screening to ensure that the vertical resistance, although present, is inoperative in all parents and all their progenies in all subsequent generations. All infection must be matching infection.

Because polygenically inherited characters are not amenable to gene transfer, screening should be conducted for all desirable variables simultaneously, with a small increase in all the acceptable minima in each generation. These many variables include all the components of crop yield and quality as well as horizontal resistance to all locally important parasites. At the highest systems level, these variables collectively constitute a single survival value, which may be called "balanced domestication," and which is the agricultural equivalent of Darwinian fitness in a wild ecosystem. Any cultivar which is deficient in even one component of domestication is an inferior cultivar, and the agricultural competition should be so effective that such a cultivar cannot survive. The screening process should proceed from the higher to the lower systems levels; the Gestalt principle is essential if suboptimization is to be avoided. More detailed descriptions of these breeding procedures have been published elsewhere (Robinson, 1973, 1976, 1979, 1980a,b).

Domestication of Vertical Resistance

The domestication of vertical resistance must clearly aim at increased strength. This may not always be easy and may necessitate deliberate hybridization in the parasite in order to determine the epidemiological competence of the matching vertical pathotype. In general, however, vertical resistance should only be employed when the maximum obtainable levels of horizontal resistance prove inadequate. Given extensive breeding for horizontal resistance, this may well prove to be a rather uncommon situation.

If vertical resistance must be employed, the breeding target should be the maximum strength. Such resistance, if it can be achieved, will be effectively permanent. If it cannot be achieved, reasonably strong vertical resistances should be employed in some form of pattern in either time or space. A sequential pattern would provide a succession of "booms" while avoiding the "busts." A spatial pattern may be regional or local. A regional pattern would be designed against a parasite migrating over long distances and would ensure that each successive migration (exodemic) would challenge the parasite with a different vertical pathodeme. A local pattern involves a mixture of different vertical pathodemes within one crop; that is, a multiline. The stronger the vertical resistance of each pathodeme within the pattern, the more effective the CSS will be.

REFERENCES

Bradley, D. J. (1976) *The origins of pest, parasite, disease and weed problems;* Symp. Brit. Ecol. Soc., Bangor, abstr. P.A.N.S. 22:422.

Flor, H. H. (1942) Inheritance of pathogenicity in *Melampsora lini*. *Phytopathology* **32**:653–659.

James, W. C., C. S. Shih, L. C. Callbeck and W. A. Hodgson. (1973) Interplot interference in field experiments with late blight of potato (*Phytophthora infestans*). *Phytopathology* **63**:1269–1275.

James, W. C., W. A. Hodgson, and L. C. Callbeck. (1976) Representational errors due to interplot interference in field experiments with late blight of potato. *Phytopathology* **66**:695–700.

Lerner, I. M. (1954) *Genetic Homeostasis*. Oliver & Boyd, London. 134 pp.

MacArthur, R. H., and E. O. Wilson. (1967) *The Theory of Island Biogeography*. Princeton University Press, Princeton, N.J. 203 pp.

Maynard-Smith, J. and G. R. Price (1973) The logic of animal conflict. *Nature* **246**:15–18.

Parlevliet, J. E., and J. C. Zadoks. (1977) The integrated concept of disease resistance; a new view including horizontal and vertical resistance in plants. *Euphytica* **26**:5–21.

Pimental, D. (1968) Population regulation and genetic feedback. *Science* **159**:1432–1437.

Robinson, R. A. (1973) Horizontal resistance. *Rev. Plant Pathol.* **52**:483–501.

Robinson, R. A. (1976) *Plant Pathosystems.* Springer-Verlag, Berlin. 184 pp.

Robinson, R. A. (1979) Permanent and impermanent resistance to crop parasites. *Z. Pflanzenzuecht.* **83**:1–39.

Robinson, R. A. (1980a) New concepts in breeding for disease resistance. *Annu. Rev. Phytopathol.* **18**:189–210.

Robinson, R. A. (1980b) The pathosystem concept. In *Breeding Plants Resistant to Insects* (F. G. Maxwell and P. R. Jennings, eds.). Wiley, New York.

Vanderplank, J. E. (1963) *Plant Diseases: Epidemics and Control.* Academic Press, New York. 349 pp.

Vanderplank, J. E. (1968) *Disease Resistance in Plants.* Academic Press, New York. 206 pp.

MULTIPLE MECHANISMS, REACTION RATES, AND INDUCED RESISTANCE IN PLANTS

Joseph Kuć

The envelopes carrying genetic information, whether they are animal or plant bodies, are well suited for their ultimate purpose—the survival and propagation of the genetic information they carry (Dawkins, 1976). Though diversity, order, and regulation are characteristic of life, change is a necessary ingredient for survival. Life does not have a "status quo." In order to survive, a life form must have evolved effective mechanisms to minimize damage, caused by other life forms and the environment, to its envelope and content of genetic information. It follows, therefore, that all plants have highly effective mechanisms for resistance to disease caused by infectious agents. If not, they would not have survived. It also follows that these mechanisms are subject to change and modification. The changes and modification, though perhaps slight, are sufficient for survival.

It seems a paradox, therefore, that plants growing in the wild are seldom, if ever, free of disease. The goal of producing disease-free plants originates with man and is not the rule in our ecosystem. Man has also introduced other concepts into our ecosystem which have direct bearing on plant disease and survival—the emphasis on uniformity and departure from diversity, and the emphasis on maximum yield under optimal conditions for a crop's growth. Clearly, man has been interested in the survival of plants only as their survival effects the survival of man, and mechanisms effective for the survival of plants may not be adequate to the intensive high-yield, high-quality, and monocultural demands placed on modern agriculture by man. In the process of maximizing his own survival, man, however, has not ignored the inherent mechanisms for plant disease resistance. Resistant plants are our number one defense against plant disease, and they are our only effective defense against diseases caused by bacteria and viruses. Until relatively recently, however, plant breeders and the manufacturers of pesticides seem to have established as goals the production of disease or insect-free plants. Pesticides generally have been developed with a single purpose, to directly kill the disease-causing organism. Are these goals realistic? Can we make better use of disease resistance mechanisms present in all plants as a practical means for disease control?

RESISTANCE AS A FACTOR OF DIFFERENTIATION, AGE, OR TEMPERATURE

Disease resistance and susceptibility are not absolute, but, rather, are subjective evaluations within certain parameters of growth and environment. The first three or four expanding leaves of apple shoots are susceptible to apple scab, caused by *Venturia inaequalis,* and all others are resistant (Biehn et al. 1966; Grijseels et al. 1964; Nusbaum and Keitt, 1938). From a practical viewpoint, the reaction of leaves at the growing point is determining resistance or susceptibility. Nevertheless, all apple trees have resistance to apple scab caused by all races of *V. inaequalis* if we consider the older leaves. The resistance is even evident in apple varieties the grower would call completely susceptible or lacking genes for resistance. Many similar situations exist. Cells of ȩtiolated green bean hypocotyls close to the growing point are susceptible to anthracnose caused by *Colletotrichum lindemuthianum.* Resistance increases as the root zone is approached, and tissue near and in the root zone, as well as the roots themselves, are highly resistant (Kuć and Caruso, 1977). This is even true of bean cultivars, which breeders or growers might consider lack genes for resistance to *C. lindemuthianum.*

Cucumber seedlings of cultivars susceptible to scab, caused by *Cladosporium cucumerinum,* are highly susceptible at the growing point and for a distance of approximately 0.5 cm below the cotyledon. Tissue beneath this zone is highly resistant (Kuć, 1962). Tissue sections taken from the susceptible zone become macerated within 1 hr when placed in culture filtrates of the fungus or solutions containing a combination of proteolytic, pectinolytic, and cellulolytic enzymes, whereas sections taken from the resistant zone remain intact 24 hr after treatment. Temperature also has a profound effect on resistance. All cucumber cultivars are highly resistant to scab at temperatures of 22–24°C, and some are at 20–22°C. Cultivar resistance is determined at 17–20°C, which is the optimum temperature for scab. Cultivar resistance at 17–20°C, resistance of all tissues at 20–24°C, and resistance of older tissues are associated with rapid lignification around sites of penetration, and the zone of lignification extends for many cells beyond the fungus (Hammerschmidt, 1979). The pathogen grows and sporulates well *in vitro* at 22–24°C, and spores germinate and the pathogen penetrates the foliage at these temperatures.

Physalospora obtusa, Botryosphaeria ribis, and *Glomerella cingulata* cause disease on older rather than younger tissue. The three fungi cause disease on ripening apple fruit late in the season (Kuć et al. 1967; Sitterly and Shay 1960; Wallace et al., 1962a,b). The transition from immunity to susceptibility occurs within 1 to 2 weeks in the field and the date of change varies from year to year depending on climatic conditions. When the fungi are introduced into wounds made in green apple fruit in June, the fungus is rapidly contained in the area of the wound. However, fungus introduced into ripening fruit late in July, or perhaps early August, rapidly spreads through the tissue and a rapidly expanding area of rot is evident within 2–3 days. Fungitoxic compounds were not detected in resistant apples and resistance was apparently due to the inability of the fungi to degrade cell walls, possibly because of the presence of pectin–protein–polyvalent cation complexes.

Leaves are generally immune to root pathogens and roots are generally immune to leaf pathogens even when the pathogens are introduced directly into tissues. The picture is made even more complex by the observation that some pathogens penetrate a host directly through the cuticle, and others penetrate through stomata or wounds.

Many additional cases can be cited in which resistance and susceptibility are a function of tissue differentiation, age, or temperature. Genetic information for resistance is present in all tissues of the plants discussed, but differentiation, age or temperature determine its expression. Is this true for cultivar resistance under optimal conditions for disease?

RESISTANCE AS A FACTOR OF PHYTOALEXIN ACCUMULATION

It is unlikely that resistance to a pathogen is determined solely by the presence of a single constitutive chemical substance. The competition of life forms and requirement of change for survival suggest that a plant would not survive with such a static form of defense. The existence of distinct races of some pathogens, which cause disease in some and not other cultivars of a plant, also argues against the role of a single constitutive chemical compound in resistance. The ability of pathogens to develop resistance to commercial pesticides supports the importance of multiple mechanisms and active or response-directed mechanisms for disease resistance.

Early experimentation suggested the concept of response-directed mechanisms for disease resistance (Chester, 1933). The classical work by Muller and Borger (1940) clearly established the concept. Based on simple but extremely perceptive and scientifically sound experimentation, Muller and his colleagues developed the phytoalexin theory (Muller, 1956, 1961). The theory has gained wide acceptance. It states that an important mechanism for disease resistance in plants is their ability to accumulate rapidly relatively simple chemical substances (phytoalexins) to levels inhibitory to the growth and development of pathogens. Cruickshank and his group established the mechanism's role in resistance of nonhosts to fungi (Cruickshank, 1963; Cruickshank and Perrin, 1964, 1971; Cruickshank et al., 1971), and recent investigations support the role of phytoalexins in non-host and cultivar resistance (Keen, 1971; Kuć, 1976a,b; Kuć and Lisker, 1978; Sato and Tomiyama, 1976; Van Etten and Pueppke, 1976; Yoshikawa et al., 1978).

The role of phytoalexins in resistance does not, however, infer that plants have a unique metabolic mechanism for disease resistance which is activated by unique constituents in the pathogen. Many abiotic and biotic factors cause the accumulation of phytoalexins. Unlike the protein antibodies synthesized by animals, phytoalexins are relatively simple organic molecules with little biological specificity. Like antibodies, however, phytoalexins accumulate in both susceptible and resistant individuals, and resistance and susceptibility, as influenced by phytoalexins, is determined by the speed and magnitude of phytoalexin accumulation. Multiple phytoalexins are produced in response to infection (Deverall, 1977; Ingham, 1972; Kuć, 1966, 1968, 1972, 1976a,b; Kuć and Lisker, 1978; Stoessl et al., 1976; Van Etten and Pueppke, 1976). Some arise via distinct synthetic pathways in the host, others because of modifications arising from metabolic activity of the pathogen and/or host (Fuchs and Hijwegen, 1979; Heuvel and Glazener, 1975; Ishiguri et al., 1978, Yoshikawa et al, 1979).

The accumulation of phytoalexins does not depend upon the presence or absence of genetic information for their synthesis. It appears to depend upon whether the information is expressed soon enough and with sufficient magnitude to contain an infectious agent. Clearly the rate at which an event occurs is critical, and the accumulation of phytoalexins appears to be controlled by the rate of synthesis and degradation.

It is also apparent that the characterized phytoalexins reported in the literature have little direct influence on the resistance of plants to bacterial pathogens and no influence on diseases caused by viruses. Expression of genetic information, multiple mechanisms, and reaction rates apparently are central in disease resistance. It also appears unrealistic that resistance mechanisms should be active at only one stage of the infection process. The effectiveness of multiple mechanisms for survival would be enhanced if the mechanisms were active at different stages of the infection process. Can multiple mechanisms for resistance be induced in plants which are susceptible under normal conditions of growth? Does the induction of disease resistance have practical significance for disease control?

INDUCED RESISTANCE

If resistance to disease is dependent on multiple mechanisms and the expression of genetic information, it is reasonable to speculate that resistance can be induced in susceptible plants and that it will be effective against a broad spectrum of pathogens. The induction of resistance, or susceptibility, may or may not be dependent upon the molecular recognition of a unique component in host or pathogen directly, for example, rejection of a foreign substance in animals. It may rely on the establishment of a physiological change. This change can arise from infection by many unrelated infectious agents or by synthetic chemical substances. The physiological change may be due to a change brought about by what the pathogen does rather than by a unique component in the pathogen.

Many examples of induced resistance in plants exist, but few examples of induced susceptibility are available. Since resistance is the rule and susceptibility the exception in nature, this is expected. Susceptibility is dependent upon a fine adjustment of abiotic and biotic conditions, and minor modifications in physiology might be expected to shift the balance to resistance.

Many cases of induced resistance to fungal, bacterial, and viral diseases are documented in the literature. Cultivar nonpathogenic races of pathogens, nonpathogens of a host, pathogens, and metabolic products of hosts or infectious agents have been used to to induce resistance (Bell, 1969; Cruickshank and Mandryk, 1960; Hargreaves and Bailey, 1978; Heale and Sharman, 1977; Kuć, 1968; Kuć and Caruso, 1977; Lovrekovich and Farkas, 1965; Main, 1968; Matta, 1971; Matta and Garibaldi, 1977; McIntyre et al., 1975; Randall and Helton, 1976; Ross, 1966; Skipp and Deverall, 1973; Yarwood, 1954, 1956). The metabolic mechanisms controlling the phenomenon and its practical implications have received relatively little attention.

INDUCED RESISTANCE IN GREEN BEAN AND CUCURBITS

Studies of the interaction of green bean with *C. lindemuthianum* indicated that bean hypocotyls were protected from anthracnose if inoculated with a cultivar nonpathogenic race of the fungus or some nonpathogens of bean before inoculation with a cultivar pathogenic race (Rahe et al., 1969a). Subsequently the protection was demonstrated to be systemic (Elliston et al., 1971, 1976a) and bean cultivars susceptible to all races of the fungus were protected by prior inoculation with *Colletotrichum lagenarium*, the incitant of anthracnose on cucurbits (Elliston et al., 1976a) or with pathogenic races attenuated by heat in the host (Rahe and Kuć, 1970; Rahe, 1973). Isoflavonoid phytoalexins did not accumulate in systemically protected tissue until after the tissue was challenged with the pathogen (Rahe et al., 1969b; Elliston et al., 1977). Thus it was possible to differentiate two facets of the induced resistance phenomenon: (1) the conditioning of host cells to respond to the presence of the pathogen as if it was a nonpathogen, and (2) the response itself, that is, the accumulation of phytoalexins. Clearly the signal conditioning cells was systemic though the accumulation of phytoalexins was localized at the site of fungal penetration. The evidence supported our hypothesis that all plants have disease resistance mechanisms which can be effective if expressed soon enough and with sufficient magnitude. Other aspects of the work with the green bean–*C. lindemuthianum* interaction are reported elsewhere (Kuć and Caruso, 1977; Elliston et al., 1976a,b,c).

Recent studies of cucurbits, *C. lagenarium,* the incitant of cucurbit anthracnose, and *C. cucumerinum,* the incitant of scab, supported our work with induced resistance in green bean. Cucumber cultivars resistant to scab, when inoculated with *C. cucumerinum,* were protected against disease caused by *C. lagenarium* (Hammerschmidt et al., 1976). Cultivars susceptible to scab and anthracnose, when inoculated with *C. lindemuthianum,* a pathogen of green bean, were protected against both diseases. Once again, it was apparent that effective mechanisms for disease resistance exist in susceptible cultivars. The data also suggested that defense reactions could be effective against unrelated organisms. Subsequent studies indicated that infection of cotyledon or first true leaf (leaf 1) of eight cucumber cultivars with *C. lagenarium* systemically protected foliar tissue above (developed or not yet developed) against disease caused by the pathogen (Kuć et al., 1975). Protection was evident as a reduction in the number and size of lesions. Physical damage or chemical injury, even when repeatedly applied, did not elicit protection. Susceptibility in this interaction is characterized by the formation of large but defined lesions. Since many such lesions form on leaves, stems, and fruits of cucurbits, the productivity of plants is adversely affected. The restriction of lesion development suggests the presence of a mechanism(s) for resistance in susceptible plants. Whether the mechanism(s) that restricts lesion development is identical to that which is systemically induced by the pathogen is unknown, but similarities are evident.

At this stage it was decided to more closely characterize the biology of induced systemic resistance in cucurbits. Inoculation of leaf 1, when the second true leaf (leaf 2) was one fourth to one third expanded, systemically protected plants for four to five weeks (Kuć and Richmond, 1977). A second or booster inoculation, three weeks after the first inoculation, extended protection into the fruiting period. Protection was elicited by and effective against two races and four isolates of the fungus and with a total of 20 susceptible and 5 resistant cultivars. A single lesion produced significant protection and protection was maximal with 5–10 lesions on leaf 1. Protection of leaf 2 was evident 72–96 hr after leaf 1 was inoculated and corresponded to the time necrosis became evident on leaf 1. Excising leaf 1 96 hr after inoculation did not reduce protection of leaf 2. Leaf 2 was protected if excised 120 hr after leaf 1 was inoculated. The inducer leaf, therefore, is not necessary for protection once protection has been elicited, and once elicited in a leaf, the leaf can be excised without a loss of protection.

In similar experiments to those conducted with cucumber, four cultivars of muskmelon and watermelon were systemically protected against anthracnose by prior infection with *C. lagenarium* (Caruso and Kuć, 1977a). In three separate field trials, cucumber plants were protected against anthracnose by a prior inoculation with the pathogen (Caruso and Kuć, 1977b).

Protection of watermelons was evident in two trials and indications were that muskmelons could also be protected.

Protection, however, was not restricted to fungi. Infection of cotyledons or leaf 1 with tobacco necrosis virus (TNV) protected leaves above against anthracnose (Jenns and Kuć, 1977). Subsequently, Caruso and Kuć (1979) reported that infection with *Pseudomonas lachrymans,* the incitant of angular leaf spot of cucurbits, protected cucumbers against angular leaf spot and anthracnose. As with *C. lagenarium,* protection was evident when symptoms appeared on the inducer leaf, that is, 48 hr and 72–96 hr for TNV and *P. lachrymans,* respectively. Recent studies indicate that infection by TNV, *P. lachrymans* or *C. lagenarium* protected cucumber against disease caused by the inducer or either of the two other infectious agents (Caruso and Kuć, 1979; Jenns and Kuć, 1980a,b). The three organisms also protected cucumber against scab, a nonlocal lesion disease (Staub and Kuć, 1980; Jenns and Kuć, 1980b), and *C. cucumerinum* protected cucumber against disease caused by *C. lagenarium* (Staub and Kuć, 1980).

The complexity of the induced protection phenomenon was emphasized by the work of Henfling (1979). He reported that cucumber plants infected with *C. lagenarium* were systemically protected against the hypersensitive reaction elicited by *Phytophthora infestans,* a pathogen of potato. Infection of cucumber with *P. infestans,* which produced numerous small necrotic lesions, did not protect cucumber against disease caused by *C. lagenarium* or the hypersensitive reaction caused by *P. infestans.* This indicates that necrosis produced by a microorganism does not always result in induced systemic protection.

If a systemic signal for protection is produced by the infected leaf, it might be graft transmissible. Kuć and Richmond (1977) had demonstrated that once protection was induced, the inducer leaf could be removed from intact plants without reducing protection in developed leaves or leaves still in the apical bud. Jenns and Kuć (1979) reported that anthracnose resistance, induced by infection of leaf 1 with *C. lagenarium* or TNV, was transmitted to a scion of a susceptible cucumber cultivar grafted onto the infected plant above leaf 1. Resistance was also transmitted if grafting preceded the inoculation. Resistance to anthracnose in susceptible cultivars was not transmitted by grafting onto uninoculated resistant cultivars, but it was transmitted if the resistant rootstocks were inoculated with *C. lagenarium.* Watermelon and muskmelon scions grafted onto susceptible cucumber rootstocks were protected against anthracnose by inoculating the rootstocks with *C. lagenarium.* The "signal" for induced protection is graft transmissible but not cultivar-, genus-, or species-specific.

What metabolic mechanism(s) can explain the effectiveness, systemic nature, and broad-spectrum effect of induced resistance? It seems unlikely that a component common to viruses, bacteria, and fungi acts as an elicitor of resistance, and that a single mechanism in the host is responsible for restricting development of viral, bacterial, and fungal pathogens.

Richmond et al. (1979) reported that penetration of *C. lagenarium* into leaves was reduced in both resistant and susceptible cultivars when plants were systemically protected by previous infection with the pathogen. The formation of appressoria was not reduced on protected plants. Protection, as determined by the number of lesions, was markedly reduced at sites where the epidermis was damaged or removed. The noninduced resistance of three cultivars tested was not due to a reduction in the formation of appressoria or penetration, but it was associated with rapid necrotization of penetrated cells. Jenns and Kuć (1980a) also reported reduced fungal penetration into plants infected with TNV. They indicated that protection was reduced but not lost when spores of *C. lagenarium* were infiltrated into leaves, which bypassed the epidermis. Thus, either a single mechanism reduces penetration and restricts development within the host or different mechanisms are involved. It is also evident that systemic resistance can be induced in resistant cultivars, but the mechanism for induced resistance is in addition to the resistance normally evident in these cultivars. Multiple mechanisms for resistance could have their effect on different stages in a pathogen's development, for example, germination of condia, formation of appressoria, penetration of host from appressoria, development within the host, and sporulation of a fungus. For survival, multiple mechanisms acting on different stages of a pathogen's development would be highly advantageous to a plant. Bacteria and viruses enter through wounds; nevertheless, systemic resistance can be induced against these pathogens once they have entered the host. This resistance is apparent as a marked reduction of bacterial population (Caruso and Kuć, 1979) or infectivity of TNV (Jenns and Kuć, 1980b). The evidence to date suggests that induced resistance in cucurbits is not elicited by unique constituents common to fungi, bacteria and viruses but rather that it is due to a common function of the three infectious agents. The common function may be the ability to cause a low level, but persistent, metabolic stress. If this is the case, induced systemic resistance would be a manifestation of a wound response. The importance of persistent low level stress in induced resistance is supported by the work of Hammerschmidt (1979). He found markedly enhanced lignification at the sites of challenge in cucumber plants systemically protected with *C. lagenarium, C. cucumerinum,* or TNV. Rapid lignification was also apparent around sites of penetration in non-induced resistant cucumbers challenged with *C. cucumerinum*. Lignification is a common response in cucumber around sites of injury. Further evidence is provided by Bergstrom (1979), who found that aphid infestation of leaf 1 protected leaf 2 against disease caused by *C. lagenarium*.

PRACTICAL IMPLICATIONS OF INDUCED RESISTANCE

Induced resistance makes use of the plant's natural defense mechanism(s). It is systemic, can protect annuals throughout the growing season, and may involve the activation of multiple mechanisms which are effective against fungi, bacteria and viruses. Economically feasible chemical pesticides are not available to control bacterial diseases, and there are no effective chemicals for the control of viral diseases. In both instances we completely rely on vector control, sanitation, and the development of disease resistant varieties.

The induction of resistance to a pathogen in plants considered completely susceptible implies that resistance is determined by the rate and magnitude of reactions rather than the presence or absence of genetic information for the reactions. Many of the highest-yielding and quality crops become susceptible to disease. Breeders search throughout the world for new lines of resistance and often find them in agronomically inferior plants. Induced resistance may enable agriculture to make greater use of existing agronomically superior plants.

At this time, induced resistance is most applicable to crops that are transplanted rather than directly grow in the field, for example, melons, tomato, and tobacco. Inoculation with TNV offers promise of controlling disease on transplanted as well as field-grown plants. The virus appears inocuous, can be applied to foliage in the form of a high pressure spray, and it doesn't require the exacting conditions of moisture and temperature necessary for infection by fungal and bacterial inducers. When the signal for induced resistance is isolated, it may be possible to utilize it as a spray or seed treatment.

Plant immunization would not eliminate the need for chemical pesticides. It could reduce our dependence on this form of disease control and thereby reduce the presence of pesticide residues on food crops and danger to the ecosystem. Of course, chemicals responsible for induced resistance, either as signals or agents which restrict development of pathogens, may be no safer than chemical pesticides. It seems reasonable to speculate, however, that induced resistance is as safe as our first line of defense against disease—the disease resistant plants.

The ability to transmit resistance via grafts offers an added dimension to protection. Many food crops and ornamentals are propagated by grafting, and the use of immunized rootstocks may present another vehicle to get the signal into susceptible plants.

Many questions remain unanswered. A major emphasis is being placed on the elucidation of the molecular mechanisms responsible for induced resistance. At the same time it is important to test the validity of the concept in the field under conditions of natural infection and with different pathogens, for example, those that cause wilts, foliar leaf spots, fruit, and seedling rots.

The implications of long range cell-to-cell communication, the possible relation of induced resistance to a wound response, and the role of such a response in evolutionary development, are in themselves ample reason for emphasis in this area of research. The practical control of disease on a preventive level, utilizing and perhaps improving on a mechanism that has stood the evolutionary test for survival, is a distinct possibility.

ACKNOWLEDGMENTS

The author's work reported in this manuscript has been supported in part by Cooperative State Research Service Grant 316-15-51, P.L. 89-106 of the United States Department of Agriculture and a grant from the Ciba-Geigy Corporation.

Journal Paper No. 79-11-254 of the Kentucky Agricultural Experiment Station, Lexington, Kentucky.

REFERENCES

Bell, A. (1969) Heat-inhibited or heat-killed conidia of *Verticillum albo-atrum* induce disease resistance and phytoalexin synthesis in cotton. *Phytopathology* **59**:1147–1151.

Bergstrom, G. (1979) Personal communication. Department of Plant Pathology, University of Kentucky, Lexington, Ky.

Biehn, W. L., E. B. Williams, and J. Kuć. (1966) Resistance of mature leaves of *Malus atrosanguinea* 804 to *Venturia inaequalis* and *Helminthosporium carbonum*. *Phytopathology* **56**:588–589.

Caruso, F. L., and J. Kuć. (1977a) Protection of watermelon and muskmelon against *Colletotrichum lagenarium* by *Colletotrichum lagenarium*. *Phytopathology* **67**:1285–1289.

Caruso, F., and J. Kuć. (1977b) Field protection of cucumber against *Colletotrichum lagenarium* by *Colletotrichum lagenarium*. *Phytopathology* **67**:1290–1292.

Caruso, F., and J. Kuć. (1979) Induced resistance of cucumber to anthracnose and angular leaf spot by *Pseudomonas lachrymans* and *Colletotrichum lagenarium*. *Physiol. Plant Pathol.* **14**:191–201.

Chester, K. S. (1933) The problem of acquired physiological immunity in plants. *Q. Rev. Biol.* **8**:129–154, 275–324.

Cruickshank, I. (1963) Phytoalexins. *Annu. Rev. Phytopathol.* **1**:351–374.

Cruickshank, I., and M. Mandryk. (1960) The effect of stem infestation of tobacco with *Peronospora tabacina* on foliage infection to blue mould. *J. Aust. Inst. Agric. Sci.* **26**:369–372.

Cruickshank, I., and D. Perrin. (1964) Pathological function of phenolic compounds in plants. In *Biochemistry of Phenolic Compounds*, (J. Harborne, ed.), Academic Press, London. Pp. 511–544.

Cruickshank, I., and D. Perrin. (1971) Studies on phytoalexins. XI. The induction, antimicrobial spectrum and chemical assay of phaseollin. *Phytopathol. Z.* **70**:209–229.

Cruickshank, I., D. Biggs, and D. Perrin. (1971) Phytoalexins as determinants of disease reaction in plants. *J. Indian Bot. Soc.* **50A**:1–11.

Dawkins, R. (1976) *The Selfish Gene*. Oxford University Press, New York. 224 pp.

Deverall, B. (1977) *Defense Mechanisms in Plants*. Cambridge University Press, London. 110 pp.

Elliston, J., J. Kuć, and E. Williams. (1971) Induced resistance to anthracnose at a distance from the site of the inducing interaction. *Phytopathology* **61**:1110–1112.

Elliston, J., J. Kuć, and E. Williams. (1976a) Protection of bean against anthracnose by *Colletotrichum* species nonpathogenic on bean. *Phytopathol. Z.* **86**:117–126.

Elliston, J., J. Kuć, and E. Williams. (1976b) A comparative study of the development of compatible, incompatible and induced incompatible interactions between *Colletotrichum* species and *Phaseolus vulgaris*. *Phytopathol. Z.* **87**:289–303.

Elliston, J., J. Kuć, and E. Williams. (1976c) Effect of heat treatment on the resistance of *Phaseolus vulgaris* to *Colletotrichum lindemuthianum* and *Colletotrichum lagenarium*. *Phytopathol. Z.* **88**:43–52.

Elliston, J., J. Kuć, E. Williams, and J. Rahe. (1977) Relation of phytoalexin accumulation to local and systemic protection of bean against anthracnose. *Phytopathol. Z.* **88**:114–130.

Fuchs, A., and T. Hijwegen. (1979) Specificity in degradation of isoflavonoid phytoalexins. *Acta Bot. Neerl.* **28**:227–229.

Grijseels, A. J., E. B. Williams, and J. Kuć. (1964) Hypersensitive response in selections of *Malus* to fungi nonpathogenic to apple. *Phytopathology* **54**:1152–1154.

Hammerschmidt, R. (1979) Personal communication. Department of Plant Pathology, University of Kentucky, Lexington, Ky.

Hammerschmidt, R., S. Acres, and J. Kuć. (1976) Protection of cucumber against *Colletotrichum lagenarium* and *Cladosporium cucumerinum*. *Phytopathology* **66**:790–793.

Hargreaves, J. A., and J. A. Bailey. (1978) Phytoalexin production by hypocotyls of *Phaseolus vulgaris* in response to constitutive metabolites released by damaged bean cells. *Physiol. Plant Pathol.* **13**:89–100.

Heale, J., and S. Sharman. (1977) Induced resistance to *Botrytis cinerea* in root slices and tissue cultures of carrot (*Daucus carota* L.). *Physiol. Plant Pathol.* **10**:51–61.

Heuvel, J. van Den, and J. A. Glazener. (1975) Comparative abilities of fungi pathogenic and nonpathogenic to bean (*Phaseolus vulgaris*) to metabolize phaseollin. *Neth. J. Plant Pathol.* **81**:125–137.

Henfling, J. W. D. M. (1979) *Colletotrichum lagenarium* systemically protects cucumber against the hypersensitive reaction elicited by *Phytophthora infestans*, but *P. infestans* does not protect cucumber against disease caused by *C. lagenarium*. In *Aspects of the Elicitation and Accumulation of Terpene Phytoalexins in the Potato–Phytophthora Infestans Interaction*. Ph.D. Thesis, University of Kentucky, Lexington, Ky. Pp. 189–194.

Ingham, J. (1972) Phytoalexins and other plant natural products as factors in plant disease resistance. *Bot. Rev.* **38**:343–424.

Ishiguri, Y., K. Tomiyama, N. Doke, A. Murai, N. Katsui, F. Yogihashi, and T. Masamune. (1978) Induction of rishitin-metabolizing activity in potato tuber tissue disks by wounding and identification of vishitin metabolites. *Phytopathology* **68**:720–725.

Jenns, A., and J. Kuć. (1977) Localized infection with tobacco necrosis virus protects cucumber against *Colletotrichum lagenarium*. *Physiol. Plant Pathol.* **11**:207–212.

Jenns, A. E., and J. Kuć. (1979) Graft transmission of systemic resistance of cucumber to anthracnose induced by *Colletotrichum lagenarium* and tobacco necrosis virus. *Phytopathology* **69**:753–756.

Jenns, A. E., and J. Kuć. (1980a) Characteristics of anthracnose resistance induced by localized infection with tobacco necrosis virus. *Physiol. Plant Pathol.* **17**:81–91.

Jenns, A. E., and J. Kuć. (1980b) Non-specific resistance to pathogens induced systemically by local infection of cucumber with tobacco necrosis virus, *Colletotrichum lagenarium* Pass (Ell et Halst) or *Pseudomonas lachrymans* (Sm et Bryan) Carsner. *Phytopathol. Mediterr.*, in press.

Keen, N. T. (1971) Hydroxyphaseollin production by soybean resistant and susceptible to *Phytophthora megasperma* var. *sojae*. *Physiol. Plant Pathol.* **1**:265–275.

Kuć, J. (1962) Production of extracellular enzymes by *Cladosporium cucumerinum*. *Phytopathology* **52**:961–963.

Kuć, J. (1966) Resistance of plants to infectious agents. *Annu. Rev. Microbiol.* **20**:337–364.

Kuć, J. (1968) Biochemical control of disease resistance in plants. *World Rev. Pest Control* **7**:42–55.

Kuć, J. (1972) Phytoalexins. *Annu. Rev. Phytopathol.* **10**:207–232.

Kuć, J. (1976a) Phytoalexins in the specificity of plant-parasite interaction. *In* Specificity in Plant Disease. (R. K. S. Wood and A. Graniti, eds.) Plenum Press, New York. Pp. 253–268.

Kuć, J. (1976b) Phytoalexins. *In* Encyclopedia of Plant Physiology, Vol. 4. Physiological Plant Pathology (R. Heitefuss and P. Williams, eds.) Springer-Verlag, New York. Pp. 632–652.

Kuć, J., and F. L. Caruso. (1977) Activated coordinated chemical defense against disease in plants. In *Host–Plant Resistance to Pests* (P. A. Hedin, ed.) American Chemical Society, Washington, D.C. Pp. 78–89.

Kuć, J., and N. Lisker. (1978) Terpenoids and their role in wounded and infected plant storage tissue. *In* Biochemistry of Wound Plant Tissues. (G. Kahl, ed.) Walter de Gruyter, Berlin, New York. Pp. 203–242.

Kuć, J., and S. Richmond. (1977) Aspects of the protection of cucumber against *Colletotrichum lagenarium* by *Colletotrichum lagenarium*. *Phytopathology* **67**:533–536.

Kuć, J., E. B. Williams, M. A. Maconkin, A. F. Ross, and Lesly J. Freedman. (1967) Factors in the resistance of apple to *Botryosphaeria ribis*. *Phytopathology* **57**:38–42.

Kuć, J., G. Shockley and K. Kearney. (1975) Protection of cucumber against *Colletotrichum lagenarium* by *Colletotrichum lagenarium*. *Physiol. Plant Pathol.* **7**:195–199.

Lovrekovich, L., and G. Farkas. (1965) Induced protection against wildfire disease on tobacco leaves treated with heat-killed bacteria. *Nature* **205**:823–824.

Main, C. E. (1968) Induced resistance to bacterial wilt in susceptible tobacco cuttings pretreated with avirulent mutants of *Pseudomonas solanacearum*. *Phytopathology* **58**:1058–1059.

Matta, A. (1971) Microbial penetration and immunization of uncongenial host plants. *Annu. Rev. Phytopathol.* **9**:387–410.

Matta, A., and A. Garibaldi. (1977) Control of verticillium wilt of tomato by preinoculation with avirulent fungi. *Neth. J. Plant Pathol.* **83**(Suppl. 1):457–462.

McIntyre, J., J. Kuć, and E. Williams. (1975) Protection of Bartlett pear against fireblight with deoxyribonucleic acid from virulent and avirulent *Erwinia amylovora*. *Physiol. Plant Pathol.* **7**:153–170.

Muller, K. O. (1956) Einige Einfache Versuche zum Nachweis von Phytoalexinen. *Phytopathol. Z.* **27**:237–254.

Muller, K. O. (1961) The phytoalexin concept and its methodological significance. *Recent Adv. Bot.* **1**:396–400.

Muller, K. O., Borger, H. (1950) Experimentelle Untersuchungen über die Phytophthoraresistenze der Kartoffel. *Arb. Biol. Reichsanstalt. Land Forstwirtsch., Berlin-Dahlem* **23**:189–231.

Nusbaum, C. J., and G. W. Keitt. (1938) A cytological study of host–parasite relations with *Venturia inaequalis* on apple leaves. *J. Agric. Res.* **56**:595–618.

Rahe, J. (1973) Phytoalexin in nature of heat-induced protection against bean anthracnose. *Phytopathology* **63**:572–577.

Rahe, J. E., and J. Kuć. (1970) Metabolic nature of the infection-limiting effect of heat on bean anthracnose. *Phytopathology* **60**:1005–1009.

Rahe, J. E., J. Kuć, Chien-Mei-Chuang, and E. B. Williams. (1969a) Induced resistance in Phaseolus vulgaris to bean anthracnose. *Phytopathology* **59**:1641–1645.

Rahe, J., J. Kuć, C. Chuang, and E. Williams. (1969b) Correlation of phenolic metabolism with histological changes in *Phaseolus vulgaris* inoculated with fungi. *Neth. J. Plant Pathol.* **75**:58–71.

Randall, R., and A. Helton. (1976) Effect of inoculation date on induction of resistance to *Cytospora* in Italian prune trees by *Cytospora cincta*. *Phytopathology* **66**:206–207.

Richmond, S., J. Kuć, and J. E. Elliston. (1979) Penetration of cucumber leaves by *Colletotrichum lagenarium* is reduced in plants protected by previous infection with the pathogen. *Physiol. Plant Pathol.* **14**:329–338.

Ross, A. F. (1966) Systemic effects of local lesion formation. *Proc. Int. Conf. Plant Viruses, Wageningen, 1965*. Pp. 127–150.

Sato, N., and K. Tomiyama. (1976) Relation between rishitin accumulation and degree of resistance of potato tuber tissue to infection by an incompatible race of *Phytophthora infestans*. *Ann. Phytopathol. Soc. Jpn.* **42**:431–435.

Sitterly, W. R., and J. E. Shay. (1960) Physiological factors affecting the onset of susceptibility of apple fruit to rotting by fungal pathogens. *Phytopathology* **50**:91–93.

Skip, R., and B. Deverall. (1973) Studies on cross-protection in the anthracnose disease of bean. *Physiol. Plant Pathol.* **3**:299–313.

Staub, T., and J. Kuć. (1980) Systemic protection of cucumber plants against disease caused by *Cladosporium cucumerinum* and *Colletotrichum lagenarium* by prior localized infections with either fungus. Submitted for publication in *Phytopathology*.

Stoessl, A., J. B. Stothers, and W. B. Ward. (1976) Sesquiterpenoid stress metabolites of the Solanacea. *Phytochemistry* **15**:855–872.

Van Etten, H. D., and S. G. Pueppke. (1976) Isoflavonoid phytoalexins. In *Biochemical Aspects of Plant–Parasite Relationships*. (J. Friend and D. R. Threlfall, eds.) Academic Press, London. Pp. 239–289.

Wallace, J., J. Kuć, and H. N. Draudt. (1962a) Biochemical changes in the water-insoluble material of maturing apple fruit and their possible relationship to disease resistance. *Phytopathology* **52**:1023–1027.

Wallace, J., J. Kuć, and E. Williams. (1962b) Production of extracellular enzymes by four pathogens of apple fruit. *Phytopathology* **52**:1004–1009.

Yarwood, C. E. (1954) Mechanism of acquired immunity to a plant rust. *Proc. Natl. Acad. Sci. USA* **40**:374–377.

Yarwood, C. E. (1956) Cross protection with two rust fungi. *Phytopathology* **46**:540–544.

Yoshikawa, M., K. Yamauchi, and H. Masago. (1978) Glyceollin: its role in restricting fungal growth in resistant soybean hypocotyls infected with *Phytophthora megasperma* var. *sojae*. *Physiol. Plant Pathol.* **12**:73–82.

Yoshikawa, M., K. Yomauchi, and H. Masago. (1979) Biosynthesis and biodegradation of glyceollin by soybean hypocotyls infected with *Phytophthora megasperma* var. *sojae*. *Physiol. Plant Pathol.* **14**:157–169.

EXPLOITING DISEASE TOLERANCE BY MODIFYING VULNERABILITY

Harry Mussell

Tolerance is usually considered to be an incomplete form of resistance, and, as such, less desirable than "true resistance." Cobb (1894), writing about wheat reactions to stem rust, noted a tolerant response, which he labeled "rust-enduring." He characterized tolerance as of limited and less certain application than other forms of resistance, but nevertheless, probably useful. This characterization of tolerance as a phenomenon of limited usefulness has persisted to the present time. In this chapter I make a case for the importance and potential of disease tolerance, to discuss some of the strategies and tactics that might be employed to exploit this method of disease management to its fullest, and to consider its implications and limitations.

WHAT IS TOLERANCE?

With respect to infectious plant pathogens, disease tolerance is the ability of a cell, plant, or field to produce an acceptable yield while providing at least a limited habitat for the growth and reproduction of pathogens of that cell, plant, or field. This may be achieved through host neutralization of the irritants generated by a pathogen, by the absence of receptor sites for the pathogenic irritants in the host, or because a host can compensate for the stresses inflicted by pathogens due to the natural resiliency and redundancy present in green plants. The provision of a habitat for pathogens is an important aspect of tolerance, since it relieves some of the selection pressure on pathogen populations for the generation of "new" pathogenic races of the pests.

I make a distinction between host–pathogen interactions which affect the health and growth of the pathogen (resistance or susceptibility) and those aspects of the host–pathogen interactions which determine the health and performance of the host (tolerance and vulnerability). The multiple component hypothesis of disease dynamics (Bateman, 1978) represents an excellent conceptual basis for making these distinctions. Within the framework provided by this hypothesis, we can categorize the interactions between host and pathogen into four possible results (Mussell, 1980). These interactions provide either a favorable or an unfavorable environment for the pathogen and concurrently, a favorable or an unfavorable environment for the host.

Any event or interaction which contributes to the establishment of an environment favorable to pathogen development contributes to the susceptibility of the host. Host determinants which contribute to this environment include both recognition and compatibility phenomena, and the nature and magnitude of the host response once the pathogen is established. The pathogen contributes to the generation of its own favorable environment by its ability *not* to alert the host of its presence e.g., the absence of specific elicitors, (Keen, 1975) and/or through its capacity to deal with any host response that is triggered e.g., ability to metabolize phytoalexins, (Van Etten, 1973). Resistance results from any shift in the balance of the above interactions that generates an unfavorable environment for the pathogen. Most of the research in disease physiology and breeding for resistance is concentrated on shifting the above balance to the detriment of the pathogen.

The balance of the interactions between host and pathogen which impinge on host performance determine the vulnerability or tolerance of the host. The host contributes to this determination through the presence or absence of receptor sites for pathogen-generated irritants, the host's ability to neutralize or inactivate these irritants, and the intrinsic ability of the host to compensate for the stresses imposed by the pathogen. Pathogen irritants include toxins, enzymes and hormones which disrupt normal host physiology (Horsfall and Cowling, 1979). Vulnerability of a host will be determined by its sensitivity to these irritants. Although many of the irritants have been well characterized, and their effects on host physiology well documented, little effort has been directed toward characterizing the potential range of qualitative and quantitative host responses to these irritants. This area of plant disease physiology should receive a higher research priority because it is the net impact of these interactions on host function which determine the effects of infection of plant performance, and thus determine the magnitude of the disease problem.

The schematic model of potential host–parasite interactions presented in Figure 1 emphasizes the fact that the sequence of events that leads to disease manifestation may be separate from those events that determine the success or failure of an infection. The diagram does not imply that all four possible interactions will exist as separable events in all diseases, nor should it be interpreted to imply fixed or constant relationships between any two of these categories. In some diseases (e.g., eyespot disease of sugarcane) susceptibility and vulnerability are inextricably linked because the fungus can-

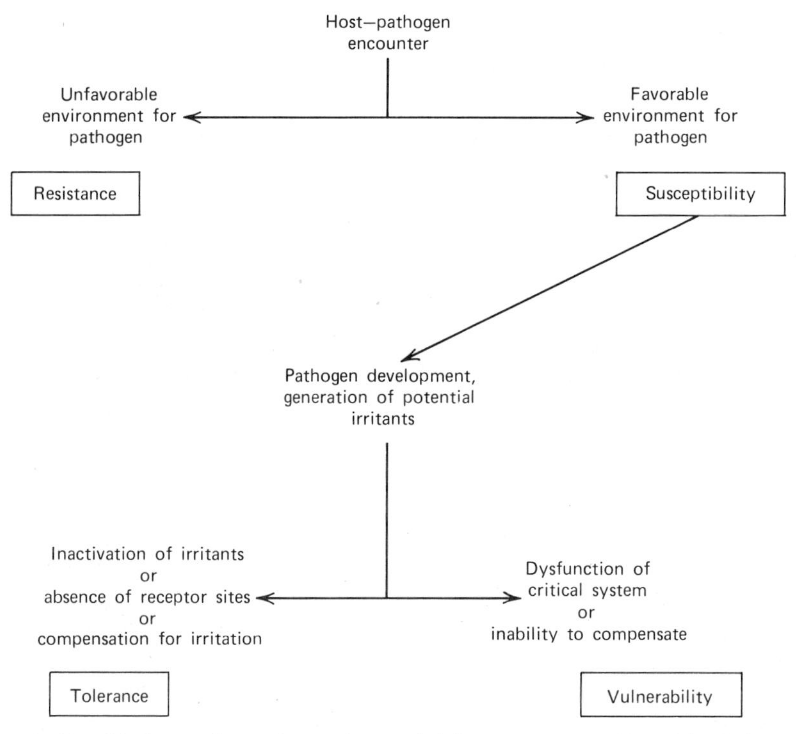

Figure 1 Possible host-pathogen interactions and their consequences. Copyright © Academic Press, 1980.

not colonize host tissues until they have been killed or weakened considerably by toxins elaborated by the pathogen. In other diseases (such as shot-hole disease of maple), extreme cellular vulnerability results in whole plant tolerance, a small portion of host tissue is lost, but the plant as a whole does not suffer. The diagram is meant only to emphasize that disease vulnerability can result from a highly specific sequence of stimulus–response events, and that this vulnerability is not necessarily directly linked to host susceptibility to infection. The mechanisms of tolerance will vary greatly from one host plant to another, and from disease to disease; however, the one common characteristic of all tolerant situations is the generation of an adequate yield coupled with the provision of the habitat necessary for pathogen growth and development.

FUNCTIONAL LEVELS OF TOLERANCE

Many of the mechanisms of disease tolerance function at the cellular level; however, tolerance can also be achieved at the whole plant and field levels as long as the two important criteria of maintaining plant performance and providing a habitat for the pathogen are both present.

Cellular Tolerance

At the cellular level tolerance can be achieved through any mechanism which reduces the level of stress imposed on the host as the result of infection. A good example of cellular tolerance, involving deletion of a toxin receptor site is southern corn leaf blight, caused by *Helminthosporium maydis*. This disease was a relatively minor disorder of corn in the United States for many decades (Ulstrup, 1970). In the early 1970s a new race of the pathogen (race T) appeared in the United States, which produced a toxin specific for corn containing the cytoplasmic Texas male sterility (TMS) factor. Almost all commercial corn hybrids in the United States contained this TMS factor, and were thus extremely vulnerable to the disease, resulting in a devastating blight in 1970, and considerable losses in 1971.

The genetics of this pathogen are very complex, and not totally understood; however, it is possible to examine the interactions between toxin-producing and nontoxin-producing races of the pathogen and all combinations of inbred corn lines which are susceptible or resistant and possessing or lacking the cytoplasmic factor for vulnerability. When the results of the eight possible interactions are examined, host vulnerability is only apparent if the TMS vulnerability factor is present to interact with toxin, and this vulnerability is expressed as greater tissue damage regardless of the presence or absence of genes for resistance to the pathogen (Yoder, 1976, Fig. 1B). This particular host pathogen combination is unique in that the unusual combination of vulnerability to disease and resistance to the causal agent can be observed in one plant.

In most plant diseases, the host traits which determine susceptibility and those which determine vulnerability are not as easily separated as they are in southern corn leaf blight. In eyespot disease of sugarcane, for example, vulnerability to the toxin generated by the pathogen appears to be a prerequisite for susceptibility to the pathogen. Isolates of the pathogen which are incapable of generating the toxin are incapable of infecting sugarcane, and lines of sugarcane that do not contain the receptor site for the toxin are not diseased by any isolates of the pathogen (Strobel, 1975).

The contrasting associations between susceptibility and vulnerability in the two diseases discussed above highlight the point that infection and pathogenesis are not necessarily closely linked. The amount of stress imposed on a host plant during infection can vary according to both the amount of irritation generated by the pathogen and the relative sensitivity of the host to that irritation. By reducing the irritability of the host, we can achieve reasonable levels of yield without imposing the selection pressure on pathogen populations that is the inevitable result of incorporating specific genes for pathogen resistance into a host plant (Jennings, 1976).

Reducing irritability by deleting receptor sites from plants is the most obvious potential mechanism for obtaining cellular tolerance, but it is by no means the only way. Many plants have at least some ability to metabolically neutralize the irritants generated by a pathogen (Agrios, 1978; Patil, 1980). A fuller understanding of the metabolism and physiology responsible for host neutralization of pathogen irritants should lead to a greater understanding of one of the cellular mechanisms of tolerance, and would lead to a more rapid development of tolerant plants.

It is important to emphasize that a systematic exploitation of the benefits of cellular tolerance will require a more generous understanding of the mechanisms of pathogenesis and the target sites for pathogen-irritants in the host. It is not sufficient to understand only the nature of the pathogenic irritants nor just the nature of the host response. For example, the rapid cellular death observable in a hypersensitive response represents an extreme case of cellular vulnerability; yet, depending upon the nature of the specific host–parasite combination in which this response occurs, this cellular vulnerability can result in whole plant vulnerability (e.g., wildfire disease of tobacco), whole-plant resistance (e.g., stem rust of wheat), or tolerance (e.g., shothole disease of maple). Even knowledge of a specific interaction within a host–parasite combination is not sufficient to predict the ultimate performance of the host. Binding of cell wall degrading enzymes to host cell walls or tissues appears to be a characteristic of vulnerability in *Rhizoctonia* rot of strawberries (Cervone et al., 1978), while a similar binding phenomenon appears to contribute to the tolerance of cotton to *Verticillium* wilt (Mussell and Strand, 1977).

Whole Plant Tolerance

Tolerance at the whole plant level can be achieved through several diverse mechanisms. In wheat, tolerance to stem rust occurs through several phenomena. The rust endurance observed by Cobb (1894), and later by many other authors, is apparently due to the ability of certain wheat varieties to allocate resources to seed production in spite of the presence of high levels of infection. Other varieties of wheat are tolerant because they cause a lengthening of the period between infection and sporulation ("slow rusting"), while a third mechanism of tolerance is manifest by varieties which stimulate the fungus toward early teliospore formation, preventing the exponential buildup of inoculum capable of reinfecting the wheat. These three separate responses of wheat plants to rust all meet the criteria of tolerance in that they result in acceptable yields while providing the habitat necessary for growth and reproduction of the fungus.

A novel phenomenon which may be useful in developing whole plant tolerance is the induced immunity exhibited by susceptible plants after an early and heavy infection by a pathogen (Caruso and Kuć, 1979). If the ability of crop plants to respond to infection by induced immunity can be amplified, and especially if this trait can be incorporated into older foliage, the result would be tolerance as defined in this chapter. These plants would provide an extensive habitat for the pathogen early in the infection cycle. Later in the growing season, these "immunized" plants would not become reinfected, and they should produce a satisfactory yield.

Another aspect of host performance, which could be amplified to provide whole-plant tolerance, involves exploiting the resiliency and redundancy inherent in most plant species. Here we can take a lesson from the entomologists who have successfully developed tolerance to rootworm in corn. This tolerance is based on the ability of certain corn lines to outgrow the damage inflicted by the borer (Owens et al., 1974). With the proper emphasis, it should certainly be possible to produce crop lines that are capable of outgrowing the damage inflicted by many of the localized, lesion-type plant diseases.

Field or Crop Tolerance

The multiline approach to disease control (Jensen, 1952; Browning and Frey, 1969) represents a mechanism for the generation of a tolerant field. This strategy reduces the stress on an entire field, rather than on individual plants by providing the pathogen population with a genetic mosaic of plants vulnerable to different pathogenic races of the pathogen. Because each host gene is present at a low frequency, no one race of the pathogen can reproduce to epiphytotic levels; however, the presence of the genetic mosaic of host cytoplasm insures that most races of the pathogen will encounter a suitable habitat for at least limited growth and reproduction. The end result of this tactic is that all races of the pathogen are allowed to procreate (albeit at reduced levels), while the field as a whole generates an acceptable yield of grain. This multiline strategy has, in fact, been described as "synthetic tolerance" (Browning et al., 1962). In a recent discussion of the potentials of this type of gene deployment, Browning et al. (1977) have suggested a strategy that, if implemented, would result in the entire North American continent becoming tolerant to certain cereal rust diseases!

VULNERABILITY AND TOLERANCE IN VASCULAR WILTS

The systemic vascular wilt diseases caused by *Verticillium* spp. and *Fusarium* spp. are complex diseases involving several levels of interaction between host and pathogen. This complexity has been an impediment to a total understanding of these diseases; however, the general events which determine resistance or susceptibility to these pathogens are spatially and temporally distinct from the determinants of host tolerance or vulnerability. This distinction makes these diseases excellent systems for examining and evaluating the unique interactions which lead to disease tolerance. The events which determine the ultimate resistance or susceptibility of the host plants generally occur early in the infection sequence, and are apparently confined to the root and crown system of the host (Tjamos and Smith, 1975). The relative tolerance or vulnerability of the host, on the other hand, appears to be related to the abilities of the above ground portions of the hosts to cope with the postinfection irritants generated by the fungi (Cronshaw and Pegg, 1976; Mussell and Strand, 1977). Several lines of evidence for tolerance, as a unique physiological attribute of host tissues, have been developed from observations of these diseases. These lines of evidence can be divided into three general categories: chemical amplification of tolerance, chemical enhancement of cellular vulnerability, and evidence on the mechanisms of pathogenesis operative in these diseases.

Chemical Enhancement of Tolerance

The synthetic growth retardants tributyl [(5-chloro-2-thienyl)methyl] phosphonium chloride (TTMP), chlormequat (CCC), and *N,N*-dimethylpiperidinium chloride (DCP) reduced the severity of *Verticillium* wilt of cotton when applied as foliar sprays (Erwin et al., 1979). The materials used were neither fungistatic nor fungicidal at the concentrations tested, and did not eliminate the fungus from the host plants. The results of Erwin et al. are similar to those reported by Davis and Dimond (1953) in which 2,4-D reduced symptom expression in *Fusarium* wilt of tomato but was not toxic to the fungus. The authors of both of the above papers speculated about the possibility of the active materials inducing the production of antifungal compounds by the host plants; however, their evidence is equally compatible with the interpretation that these active growth regulators modified the level of irritability of the host, generating a more tolerant situation. Using CCC, Sinha and Wood (1967) observed the reduction of symptoms in *Verticillium* wilt of tomato, and Buchenauer (1971) reported the delay and reduction in symptom severity of both *Fusarium* and *Verticillium* wilt of tomato. In all of the above examples, the chemical treatment did not eliminate the pathogen from the host tissues. Also, all of the above observations would argue that the inherent sensitivity of the hosts plants to the irritants generated by the pathogens was reduced by treatment with synthetic plant growth regulators. Retig and Chet (1974) reported that treatment of tomato plants with low levels of catechol prior to inoculation with *Fusarium* reduced or eliminated the appearance of disease symptoms without materially lowering the levels of pathogen population in the vascular system of treated plants. In a followup of this work, Retig and Lisker (1975) reported that catechol treated tomato plants were heavily colonized by *Fusarium,* and that the fungus produced cell wall degrading enzymes in these plants; however, the plants remained symptomless. Because of the low levels of catechol used in these studies, the long term effectiveness of the treatment, and the absence of toxin effects of these levels of catechol on the pathogen *in vitro*, it seems obvious that the mechanism of action here involved desensitization of the host plants to the irritants generated by the pathogen.

All of the above information suggests that the inherent level of sensitivity of host tissues to the irritants produced by vascular wilt fungi can be modified by chemical treatment. This would suggest that the same desensitization can be achieved by genetic means, resulting in the production of more tolerant plants.

Chemical Enhancement of Vulnerability

The miticide Kelthane, when applied to cotton seedlings, increased the vulnerability of these plants to *Verticillium* wilt (Mussell and Ramftl, 1972), possibly through inhibition of a cell wall phosphatase in the host (Daley et al., 1979). This miticide also increased the sensitivity of cotton leaves to a purified endopolygalacturonase (endoPG) produced by *Verticillium* (Mussell and Ramftl, 1972). This is of interest, since ethylene, which naturally occurs during *Verticillium* pathogenesis in tomatoes (Pegg and Cronshaw, 1976), also enhanced the vulnerability of tomato stem sections to a protein fraction containing endoPG activity (Cronshaw and Pegg, 1976). These authors also demonstrated an intrinsic difference between the responses of susceptible and resistant tomato cuttings to both symptom expression and generation of pathological ethylene. These observations, coupled with their work demonstrating ethylene-enhanced sensitivity to a carbohydrate toxin produced by the fungus (Cronshaw and Pegg, 1976), and their demonstration that the pectolytic enzyme fraction from *Verticillium* cultures would induce the generation of ethylene in treated plants, contribute evidence to the multicomponent nature of the complex interactions between *Verticillium* and its hosts.

MECHANISMS OF PATHOGENESIS

Our investigations have also been aimed at elucidating the role of endoPG in *Verticillium* wilts. We have developed evidence that purified endoPG from *Verticillium* will induce symptoms identical to those of the disease (Mussell, 1973; Mussell and Strand, 1977), and that this purified endoPG will catalyze the release of cell-wall proteins from the host, including several of the enzymes shown to be involved in the generation of ethylene (Strand and Mussell, 1975; Strand et al., 1976; Lund and Mapson, 1970). We have also observed that this purified endoPG binds to cell walls of cotton (Mussell and Strand, 1977) and that walls from tolerant cultivars bind more endoPG than to walls from vulnerable cotton plants (Mussell and Strand, 1977). We have theorized that this binding ability is an important characteristic of tolerant plants. From the above information, we have constructed a model of the sequence of *initial* events which might lead to tolerance of vulnerability in *Verticillium* wilts (Figure 2).

Recently, we have observed that virulent, but not avirulent isolates of *Verticillium*, elaborate a material which prevents the binding of endoPG to host cell walls (Figure 3). This observation argues strongly for the participation of this "blocking factor" in pathogenesis by *Verticillium*, and substantiates our contention that the ability of host cell walls to immobilize endoPG through binding is an important determinant of disease tolerance in the vascular wilt diseases. Further characterization of the specific interactions between host cell walls, pathogen endoPG and blocking factor should substantially increase our understanding of at least one unique mechanism for disease tolerance in plants.

SUMMARY

I have presented evidence that one of the prime goals of disease physiology research should be to characterize the range and nature of host responses to irritants generated by pathogenic microorganisms. This emphasis would focus attention on the principal goal of agricultural pest management, the elimination of disease-induced crop losses. The development of disease tol-

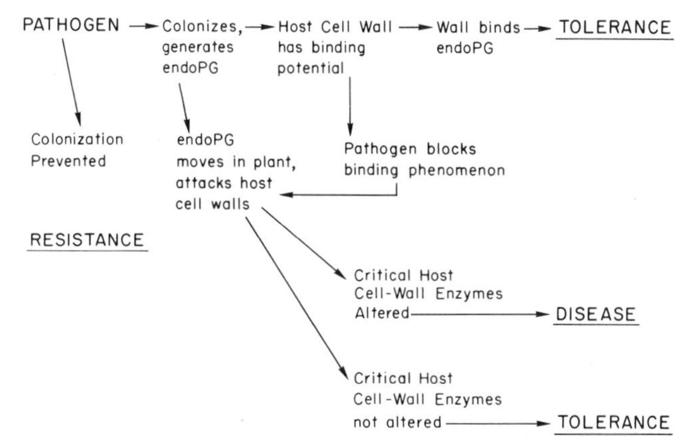

Figure 2 Hypothetical sequence of possible interactions between the endopolygalacturonase of *Verticillium* and host cell walls leading to disease or disease tolerance.

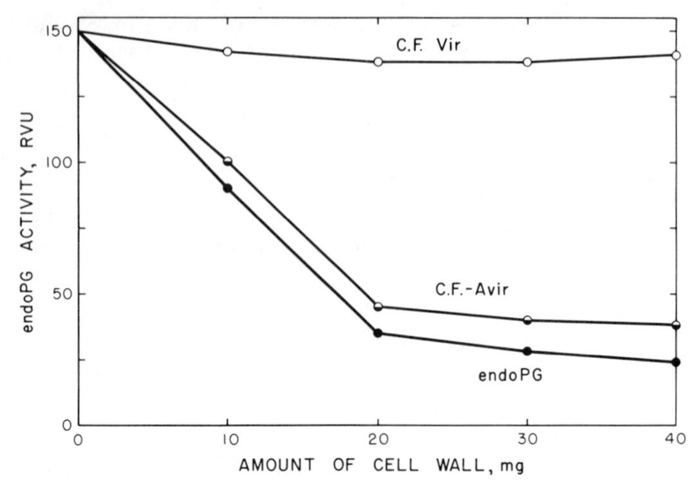

Figure 3 Binding of purified endopolygalacturonase from *Verticillium* (endoPG) and endopolygalacturonase in culture fluids (C.F.) to cell walls prepared from Bonny Best Tomato foliage. Binding was measured as a loss of activity from solution.

erant varieties as characterized in this chapter should lend some much needed long term stability to disease management because it would eliminate one of the most important factors responsible for change and adaptation in pathogen populations—the selection pressure generated by using management tactics aimed at pathogen exclusion. Infectious microorganisms are a normal part of any natural plant environment and it is doubtful that we can eliminate them. What we can eliminate, however, is the crop loss attributable to disease-induced stress (Mussell and Malone, 1979). It is apparent that we have to adopt management strategies that will allow for the presence of infections in our crops. Yet this does not mean that we must also accept disease losses from these infections. Disease tolerance should play an ever increasing role in our crop management strategies.

REFERENCES

Agrios, G. N. (1978) Plant Pathology, 2nd Ed. Academic Press, New York. Pp. 79–84.

Bateman, D. F. (1978) The dynamic nature of disease. In *Plant Disease—An Advanced Treatise,* Vol. III (J. G. Horsfall and E. B. Cowling, eds.) Academic Press, New York. Pp. 53–83.

Browning, J. A., and K. J. Frey. (1969) Multiline cultivars as a means of disease control. *Annu. Rev. Phytopathol.* **7**:355–382.

Browning, J. A., M. D. Simons, and K. J. Frey. (1962) The potential value of synthetic tolerant or multiline varieties for control of cereal rusts in North America. *Phytopathology* **52**:726.

Browning, J. A., M. D. Simons, and E. Torres. (1977) Managing host genes: epidemiologic and genetic concepts. In *Plant Disease—An Advanced Treatise*, Vol. I (J. G. Horsfall and E. B. Cowling, eds.) Academic Press, New York. Pp. 191–211.

Buchenauer, H. (1971) Einfluss einiger Wuchschemmstoffe (CCC, CMH, Amo-1618, Phosfon-D, B-995) auf die Fusarium- und *Verticillium* Welke der Tomate. *Phytopathol. Z.* **72**:53–66.

Caruso, F. L., and J. Kuć. (1979) Induced resistance of cucumber to anthracnose and angular leaf spot by *Pseudomonas lachrymans* and *Colletotrichum lagenarium*. *Physiol. Plant Pathol.* **14**:191–202.

Cervone, F., A. Scala, and F. Scala. (1978) Polygalacturonase from *Rhizoctonia fragariae:* Further characterization of two isozymes and their action towards strawberry tissue. *Physiol. Plant Pathol.* **12**:19–26.

Cobb, N. A. (1894) Contributions to an economic knowledge of Australian rusts (Uridineae). *Agric. Gaz. N.S.W.* **5**:239–250.

Cronshaw, D. K., and G. F. Pegg. (1976) Ethylene as a toxin synergist in *Verticillium* wilt of tomato. *Physiol. Plant Pathol.* **9**:33–44.

Daley, L. S., P. Carroll, and H. Mussell. (1979) The cell wall phosphatase of cotton (*Gossypium*) is inhibited by kelthane. *Biochem. J.* **179**:719–721.

Davis, D., and A. E. Dimond. (1953) Inducing disease resistance with plant growth regulators. *Phytopathology* **43**:137–140.

Erwin, D. C., S. D. Tsai, and R. A. Kahn. (1979) Growth retardants mitigate *Verticillium* wilt and influence cotton yields. *Phytopathology* **69**:283–287.

Horsfall, J. G., and E. B. Cowling (eds.). (1969) *Plant Disease—An Advanced Treatise*, Vol. IV. *How Pathogens Induce Disease*. Academic Press, New York. 466 pp.

Jennings, P. R. (1976) The amplification of agricultural production. *Sci. Am.* **235**:180–194.

Jensen, N. F. (1952) Intra-varietal diversification in oat breeding. *Agron. J.* **44**:30–44.

Keen, N. T. (1975) Specific elicitors of plant phytoalexin production: Determinants of race specificity in pathogens? *Science* **187**:74–75.

Lund, B. M., and L. W. Mapson. (1970) Stimulation by *Erwinia carotovora* of synthesis of ethylene in cauliflower tissue. *Biochem. J.* **119**:251–263.

Mussell, H. W. (1973) Endopolygalacturonase: Evidence for involvement in *Verticillium* wilt of cotton. *Phytopathology* **63**:62–70.

Mussell, H. (1980) Tolerance to disease. In *Plant Disease—An Advanced Treatise*, Vol. 5 (J. G. Horsfall and E. B. Cowling, eds.) Academic Press, New York. Pp. 39–54.

Mussell, H. W., and M. J. Malone. (1979) Disease tolerance: Reducing the impact of disease-induced stress on crop yields. In *Stress Physiology in Crop Plants* (H. Mussell and R. C. Staples, eds.). Wiley, New York. Pp. 15–23.

Mussell, H., and A. Ramftl. (1972) Increased susceptibility of dicofol-treated cotton to *Verticillium albo-atrum*. *Phytopathology* **62**:780.

Mussell, H. W., and L. L. Strand. (1977) Pectic enzymes: Involvement in pathogenesis and possible relevance to tolerance and specificity. In *Cell Wall Biochemistry*

Related to Specificity in Host–Plant Pathogen Interactions (J. Raa and B. Solheim, eds.) Columbia University Press, New York. Pp. 31–70.

Owens, J. C., D. C. Peters, and A. R. Hallauer. (1974) Corn rootworm tolerance in maize. *Environ. Entomol.* **3**:767–772.

Patil, S. S. (1980) Defenses triggered by the invader: Detoxifying the toxins, *in* Plant Disease–An Advanced Treatise Vol V. How Plants Defend Themselves. (J. G. Horsfall and E. B. Cowling, eds.) Academic Press, New York. Pp 269–277.

Pegg, G. F., and D. K. Cronshaw. (1976) Ethylene production in tomato plants infected with *Verticillium albo-atrum. Physiol. Plant Pathol.* **8**:279–295.

Retig, N., and I. Chet. (1974) Catechol-induced resistance of tomato plants to *Fusarium* wilt. *Physiol. Plant Pathol.* **4**:469–475.

Retig, N., and N. Lisker. (1975) Changes in polygalacturonase, pectin lyase and cellulase in tomato plants, associated with *Fusarium* inoculation. *Phytoparasitica* **3**:113–119.

Sinha, A. K., and R. K. S. Wood. (1967) The effect of growth substances on *Verticillium* wilt of tomato plants. *Ann. Appl. Biol.* **60**:117–128.

Strand, L. L., and H. Mussell. (1975) Solubilization of peroxidase activity from cotton cell walls by endopolygalacturonases. *Phytopathology* **65**:830–831.

Strand, L. L., C. Rechtoris, and H. Mussell. (1976) Polygalacturonases release cell wall proteins. *Plant Physiol.* **58**:722–725.

Strobel, G. A. (1975) A mechanism of disease resistance in plants. *Sci. Am.* **232**:80–88.

Tjamos, E. C. and I. M. Smith. (1975) The expression of resistance to *Verticillium albo-atrum* in monogenically resistant tomato varieties. *Physiol. Plant Pathol.* **6**:215–225.

Ulstrup, A. J. (1970) History of southern corn leaf blight. *Plant Dis. Rep.* **54**:1100–1102.

Van Etten, H. D. (1973) Differential sensitivity to pisatin and phaseollin. *Phytopathology* **63**:1477–82.

Yoder, O. C. (1976) Evaluation of the role of *Helminthosporium maydis* Race T toxin in southern corn leaf blight. In *Biochemistry and Cytology of Plant–Parasite Interactions* (K. Tomiyama, J. M. Daly, I. Uritani, H. Oku and S. Ouchi, eds.). Elsevier, New York. Pp.16–24.

THE ROLE OF PROTOPLASTS AND CELL CULTURES IN PLANT DISEASE RESEARCH

Elizabeth D. Earle and Vernon E. Gracen

Cell and tissue culture techniques are opening new vistas for research with higher plants. A remarkably wide range of manipulations with cultured plant materials is now possible (see Thorpe, 1978 and Sharp et al., 1979 for recent summaries).

Cultures can be started from almost any part of a plant, from the petals to the roots. Very small explants such as epidermal peels, isolated apical domes, and even single cells may yield successful cultures. Materials may be grown either on stationary agar medium or in agitated liquid medium. Some liquid-grown cultures are so finely dispersed that they can be transferred by pipetting or plated like microorganisms. By providing appropriate selective conditions, variant cell types can be obtained.

The morphology of many cultures is dramatically affected by the composition of the nutrient medium used. Changes in the levels of hormones and other nutrients can shift growth patterns (often reversibly) from relatively unorganized callus tissue to a more differentiated state in which roots or shoots appear. Whole plants can be recovered from many cultures. Some of these plants simply represent hormone-induced outgrowths of axillary buds from original explants, but others have been obtained from material maintained in undifferentiated form for months or years. Such shifts from unorganized material back to whole plants permit conventional genetic studies and possible agricultural use of material subjected to manipulations *in vitro*. Some variants selected at the cellular level have already been regenerated into plants with altered heritable characteristics. Attempts to produce plants with improved nutritional composition or with resistance to biological and environmental stress are underway in many laboratories.

Haploid as well as diploid material can be cultured and used to recover plants. Both anther cultures and isolated microspores can give rise to haploid tissue and/or plants. Haploids are particularly useful for selection experiments because recessive alleles are expressed. Moreover, haploid material can undergo chromosome doubling to give rise to completely homozygous diploid plants. The repeated inbreeding usually required to achieve homozygosity can thus be bypassed.

Some of the most exciting new culture techniques use protoplasts, plant cells whose walls have been removed with wall-degrading enzymes. The absence of walls allows manipulations with protoplasts not possible with intact plant cells. Protoplasts can be induced to fuse with other protoplasts, including ones from taxonomically remote plants, and even with animal cells (Jones et al., 1976). Whole populations of protoplasts can be enucleated and separated into nucleated "miniprotoplasts" and enucleated cytoplasmic portions, each of which can be used in fusion experiments (Wallin et al., 1979). Protoplasts can also take up various microorganisms such as viruses, bacteria and algae, as well as isolated organelles and even isolated DNA (Kleinhofs and Behki, 1977). Many protoplasts can regenerate walls after they are removed from enzyme solution, washed, and cultured in nutrient media. Some divide, form multicellular clumps and even grow back into whole plants. The success of such manipulations with protoplasts opens the possibility of transferring genetic material in novel ways and recovering new types of plants from the treated protoplasts.

It is essential to point out that most of these techniques are far from routine. With a few plants, notably tobacco, the procedures usually work well. With many other species, regeneration of plants or even callus tissue from cells, microspores, or protoplasts has not yet succeeded. Some important crop plants such as soybean or cereals seem to pose special difficulties. On the other hand, systems for regenerating plants from protoplasts of potato (Shepard and Totten, 1977), pearl millet (Vasil and Vasil, 1980) and alfalfa (Kao and Michayluk, 1980) have recently been developed. Several intergeneric hybrid plants have been produced by protoplast fusion (e.g., Melchers et al., 1978; Dudits et al., 1979). There is thus some reason for optimism about applying existing techniques to additional plant species and about developing more powerful new techniques in well-characterized systems.

The techniques now available are already being used in many areas of plant biology. They are ready to be exploited in efforts to understand and limit plant diseases. Current and potential contributions of cell and protoplast techniques to such work can conveniently be grouped into three general categories:

1. Elucidation of the mechanisms of some plant diseases.
2. Selection of disease resistant material.
3. Transfer of genes for disease resistance.

THE MAIZE–*H. MAYDIS,* RACE T EXPERIMENTAL SYSTEM

All three approaches listed above can be illustrated with work using a single plant species and a single pathogen, namely *Zea mays* and *Helminthosporium (Bipolaris) maydis,* race T. The interaction between these two organisms has been studied intensively at many levels (Gregory et al., 1977); southern corn leaf blight produced by *H. maydis,* race T on sensitive maize (corn) is one of the plant diseases closest to a molecular explication. Background information about the system can be briefly summarized as follows:

1. Both *H. maydis* race T and *H. maydis* race O produce leaf blights on corn.
2. Resistance to race O is conditioned by nuclear genes, while resistance to race T involves cytoplasmic as well as nuclear genes.
3. Plants with Texas (T) male sterile cytoplasm are much more sensitive to race T than ones with normal male fertile (N) or other male-sterile cytoplasms. Male sterile T and fertility restored (TRf) plants are both sensitive. This specificity for the widely used T cytoplasm caused severe crop losses in 1970 and led to the abandonment of T cytoplasm for hybrid seed production.
4. *H. maydis,* race T produces a toxin (HmT toxin) which is responsible for specificity for T-cytoplasm material. This toxin can substitute for the fungus in laboratory studies, thereby eliminating complications associated with introduction of a living organism into the experimental system.
5. Mitochondria isolated from T-cytoplasm tissues are rapidly damaged by HmT toxin; mitochondria isolated from resistant cytoplasms are unaffected.

Cytoplasm-Specific Effects of HmT Toxin on Corn Callus and Protoplasts

In vitro studies have followed the interaction of HmT toxin with corn callus and corn protoplasts. HmT toxin can completely inhibit growth of T-cytoplasm callus but has no effect on growth of N callus (Gengenbach and Green, 1975). This cytoplasm-specific inhibition is the basis of the selection studies to be discussed later. HmT toxin also damages T but not N protoplasts (Pelcher et al., 1975). This result could not be taken for granted, since not all host-specific toxins reveal their specificity at the protoplast level (Earle, 1978).

We have studied the effect of HmT toxin on corn mesophyll protoplasts in considerable detail (Earle et al., 1978). These protoplasts can readily be obtained by rubbing young corn leaves with carborundum and incubating them for several hours in wall-degrading enzymes. The protoplasts do not divide, but they can survive for a week or more. N-cytoplasm protoplasts are unaffected even after a week of culture in medium containing very high toxin concentrations. Exposure of T-cytoplasm protoplasts to HmT toxin has no immediate effects at the light microscope level, but within 18–24 hours treated protoplasts show bulging, distortion of chloroplasts, and failure to enlarge normally. Within a few days, all toxin-treated T-cytoplasm protoplasts collapse. The exact timing of the effects is influenced by several factors including toxin concentration, light, nuclear genotype, and age of the protoplasts. T-cytoplasm protoplasts are very sensitive to HmT toxin; the same low concentrations of toxin active against isolated T mitochondria eventually collapse T protoplasts, although the mitochondria are affected more rapidly (Gregory et al., 1980). Protoplasts therefore can serve as a convenient and sensitive assay system for identification and purification of HmT toxin.

The specificity of HmT toxin for cultured T-cytoplasm cells and protoplasts has been clearly demonstrated. The next question is how this information can advance our understanding or control of a plant disease. What can we now do that was not possible before?

The Use of Protoplasts for Studies of Disease Mechanisms

The first answers come from studies of disease mechanisms. Although it is well known that isolated T mitochondria are rapidly affected by HmT toxin, there has been controversy over the primary site of toxin action in living cells (Gregory et al., 1977). Suspensions of protoplasts have many attractive features for studies of the cellular effects of HmT toxin. They allow work with millions of individual cells, aliquots of which can be handled by pipetting. Each protoplast is directly exposed to toxin or other reagents. Tissue interactions are eliminated, and toxin penetration problems are minimized. Protoplasts are visibly affected by very low concentrations of toxin. Only small amounts of toxin or plant material are required. For these reasons, protoplasts have made possible sensitive and convenient ultrastructural and biochemical studies of toxin action.

These ultrastructural studies showed that HmT toxin rapidly damages both mitochondria in living T-cytoplasm cells and isolated T mitochondria (Aldrich et al., 1977; York et al., 1980). Isolated T mitochondria swell and lose their characteristic internal structure after toxin treatment. So do mitochondria in toxin-treated T protoplasts. Some mitochondria in T protoplasts show such damage 5 minutes after exposure to toxin, and within 30 minutes, all mitochondria are disrupted. No ultrastructural changes in other cellular components are seen until several hours later when chloroplast damage is apparent in dark-grown protoplasts (Earle and Gracen, 1977). Changes in mitochondrial ultrastructure have also been seen in toxin-treated T-cytoplasm roots and leaves (Aldrich et al., 1977; Malone et al., 1978), but problems with the penetration of toxin and fixatives make interpretation of the effects less clearcut.

Biochemical studies also have been facilitated by the protoplast system. If HmT toxin uncouples mitochondria in cells as it does isolated mitochondria, one would expect to see a rapid drop in the ATP levels. Using a luciferin–luciferase assay, Walton et al. (1979) found that the ATP content of dark-grown protoplasts drops sharply within 1.5 minutes after toxin treatment. This is the fastest *in vivo* response to HmT toxin reported to date. The results support the idea that mitochondria are indeed the primary site of action *in vivo* and that other effects are secondary to the drop in ATP level.

The experiments with protoplasts confirm the biological importance of studies of the interaction of HmT toxin with isolated T mitochondria and submitochondrial particles. Further biochemical and ultrastructural studies will help clarify the effects of this interaction on the structure and function of living T-cytoplasm cells. With the *H. maydis* race T–maize experimental system, it may soon be possible to trace a plant disease from a molecular interaction to organelle damage to cellular disruption to visible symptoms on a whole plant.

Selection of Disease Resistant Material

Tissue culture techniques also can facilitate attempts to identify, distinguish, and select genes for disease resistance. In the HmT system such attempts have used two different approaches. One involves selection of resistant material from T cytoplasm callus cultures followed by regeneration of plants. The other combines *in vitro* screening of many T-cytoplasm corn lines with field screening and conventional breeding.

The first approach to selection of material resistant to HmT starts with culture of T callus on medium containing HmT toxin. Although growth of most of the callus is inhibited by the toxin, a few sectors sometimes grow, especially after gradual transfers from lower to higher toxin concentrations (Gengenbach and Green, 1975). Gengenbach et al. (1977) selected such resistant sectors of callus and regenerated plants from them. A leaf bioassay showed that these plants were resistant to HmT toxin. The plants were as resistant to *H. maydis* race T as N-cytoplasm plants, indicating that selection with HmT toxin *in vitro* had resulted in increased pathogen resistance at the field level. This resistance was cytoplasmically inherited. Mitochondria from the resistant plants were resistant to HmT toxin.

The plants regenerated from toxin-resistant callus no longer showed the Texas type of male sterility and so were not of agricultural value. Apparently toxin sensitivity and T male sterility are very closely linked traits, probably involving the same mechanisms. It is noteworthy, however, that the mitochondrial DNA of the selected resistant plants was not like that of N plants (Pring, 1979). Instead it was a variant of the T type of mitochondrial DNA. In other words, Gengenbach et al. had selected a new type of resistant material, not just recovered the same type of resistance already available. In cases not complicated by the linkage of pathogen sensitivity to a desirable trait, such a selection could be a valuable contribution to a breeding program.

The second approach, still in preliminary stages, uses protoplasts rather than callus. It starts with the fact that some corn inbreds and hybrids in T cytoplasm are more resistant to *H. maydis* race T than others; that is, nuclear genes can affect the severity of the leaf blight. Comparisons of protoplasts from the same inbreds and hybrids showed that nuclear genes also influence the response of protoplasts to HmT toxin; some genotypes were markedly more resistant than others (Earle et al., 1978; Pham, 1980). The correlation between the field response to the fungus and the protoplast response to toxin was quite good, suggesting that some of the genes for disease resistance act at the cellular level. Screening for resistance at the protoplast level may therefore be a useful way to predict field resistance.

One T-cytoplasm hybrid appeared sensitive at the protoplast level but quite resistant to the fungus in the field (Pham, 1980). In this hybrid, resistance seems to involve genes active only at a supracellular level. Comparison of the behavior of T-cytoplasm material at the plant and protoplast levels may thus offer a new way to distinguish between two types of resistance. Comparison of the effect of HmT toxin on protoplasts and isolated mitochondria from the same genotypes might further distinguish different genes for resistance. The relation of toxin resistance to pathogen resistance requires further clarification, but the combination of rapid protoplast assays with field screening and breeding may help in production of plants with increased and more stable disease resistance.

Gene Transfer Experiments

Fusion of protoplasts can bring together genes in combinations not possible by conventional sexual crosses. Protoplasts from plants that are sexually incompatible (including hosts and nonhosts of a pathogen) or that have different cytoplasms can be fused. In most sexual crosses, only maternal cytoplasm is transmitted to the zygote, but in asexual protoplast fusion, two cytoplasms as well as two nuclei are brought together. It is now important to learn which genes in such novel combinations are active and how long they remain present and/or active.

The HmT toxin–corn protoplast system, which involves a cytoplasmic trait expressed *in vitro,* is an excellent one for studying the early interaction of cytoplasmic genes for disease resistance. What happens if a T protoplast is fused with a toxin-resistant protoplast and then treated with toxin? We first approached this question using conveniently obtained soybean callus protoplasts as the toxin-resistant protoplasts. Green T-cytoplasm leaf protoplasts and colorless soybean callus protoplasts can be distinguished both visually and by their response to HmT toxin. The boybean protoplasts survive well even when cocultured with toxin-damaged T-corn protoplasts. Treatment with polyethylene glycol makes protoplasts adhere, and, after a high calcium-high pH rinse, some corn and soybean protoplasts fuse. Fusions can be recognized at the light microscope level by the presence of green chloroplasts and the characteristic granular cytoplasm of the soybean protoplasts. Electron microscopy confirms that these apparent fusions do indeed contain nuclei and cytoplasmic components of both corn and soybean. The fusion products can survive several weeks and sometimes even divide, unlike unfused corn mesophyll protoplasts, which never divide.

Corn–soybean fusions are resistant to HmT toxin; they can survive several weeks and occasionally divide in toxin concentrations that collapse all unfused T protoplasts in a few days. Thus toxin resistance appears to be "dominant" in a hybrid protoplast. This result means that HmT toxin can be used to identify and select fusions of green T protoplasts with toxin-resistant, nongreen protoplasts. All surviving green protoplasts must be hybrids.

It is now possible to follow the two mitochondrial populations in the hybrid protoplasts, using HmT toxin as a marker. Both normal (presumably soybean) and damaged (presumably corn) mitochondria can be seen in fusion products treated briefly with toxin immediately after fusion. What if the fusions are cultured for various lengths of time before toxin treatment? How long are the two different types of mitochondria maintained in a hybrid cytoplasm? Studies designed to answer these questions are underway. They should help us determine whether the creation of stable hybrid cytoplasms is feasible or whether elimination or alteration of one mitochondrial type occurs after fusion.

Now that it is clear that resistance is dominant in the fused corn–soybean protoplasts, we are trying to determine whether resistance can similarly be transferred by N cytoplasm corn protoplasts, by enucleated corn and soybean protoplasts, or by isolated mitochondria. It may someday even be possible to influence the response to HmT toxin by supplying T protoplasts with mitochondrial DNA from resistant material.

If techniques for regenerating plants from corn protoplasts are developed, hybrid corn cytoplasms could be examined at the plant level as well as at the cellular level. "Cybrid" tobacco and petunia plants derived from fused protoplasts are already being studied (e.g., Belliard et al., 1978; Zelcer et al., 1978; Izhar and Power, 1979).

IN VITRO STUDIES WITH OTHER DISEASE SYSTEMS

The work described above illustrates the wide range of *in vitro* plant disease studies possible when a host-specific toxin active at the cellular level is available. Valuable culture studies have been done using other toxins besides HmT toxin, notably toxins produced by *Helminthosporium victoriae, H. sacchari, Alternaria kikuchiana, A. alternata,* and *Phytophthora infestans* (Earle, 1978). Methionine sulfoximine, a putative analog to tabtoxin from *Pseudomonas tabaci*, was used by Carlson (1973) to select for resistance to some aspects of wildfire disease of tobacco.

The number of host-specific toxins now available is small, however, and the number shown to have *in vitro* effects parallel to those seen *in vivo* is even smaller (Earle, 1978). Do cell and protoplast techniques therefore have only very limited application to disease studies? This is an important question. For a number of reasons, the answer should be no. This conclusion is based on the feasibility of several additional lines of work described in the next two sections.

Isolation and Study of Additional Toxins

Many of the toxins already available have not yet been tested against cells and protoplasts or have received attention in only a few laboratories. Even nonspecific toxins might fruitfully be studied *in vitro,* since resistance or tolerance to these toxins may improve plant yield or vigor. Moreover, the list of toxins involved in plant diseases is steadily growing (Yoder, 1980).

The extreme sensitivity of T-cytoplasm protoplasts to HmT toxin suggests that protoplast assays may actually aid in the isolation of some new toxins. Filtrates from cultures of pathogens suspected of acting via toxins could be tested against protoplasts from plants known to be relatively sensitive and relatively resistant to the pathogen. If a host-specific toxin active at the cellular level is present in the filtered culture medium, survival or appearance of protoplasts from the two types of plants may differ. This kind of assay can be reliable and sensitive even if nothing is known about the putative toxin's mode of action. Many plant lines and many "toxin" samples can readily be tested. We are now using this approach to see whether toxins are produced by *Kabatiella zeae* and by *Helminthosporium carbonum* race 3 (Fox et al., 1979). Both fungi cause leaf-blight diseases of corn which have been appearing with increasing frequency in the Northeast and in other locations.

The use of protoplasts in the search for new toxins can easily encounter difficulties since many different interpretations of negative results are possible.

1. Failure of a filtrate to affect any of the treated protoplasts may mean that the pathogen does not act via a toxin. On the other hand, the culture conditions may not have been optimal for toxin production. Possibly the filtrate contains a toxin whose action requires cell walls or metabolic processes or cell types not present in the protoplast preparation.
2. What if the filtrate damages protoplasts from both resistant and sensitive

plants? This could mean that a nonspecific substance with no relevance to disease is present. Alternatively, the filtrate may contain a substance involved in disease development, but which shows no specificity at the cellular level; specificity may occur at the tissue, organ or plant level, perhaps in the way the fungus behaves on the leaf or in the way multicellular tissues respond to the toxin.

While the problems are obvious, so are the possible rewards. New toxins could have great value both in field screening for disease resistant lines and in a variety of *in vitro* studies.

In Vitro Disease Studies Without Toxins

Most of the preceding comments assume a need for some type of disease-related screening *in vitro,* but this is not always necessary to take advantage of tissue culture techniques. Screening at the cellular level may save time and effort, but it is not essential.

Consider first the identification and selection of genes for resistance. Material can be generated *in vitro* but screened in conventional ways. For example, haploid plants from anther cultures can facilitate conventional disease screening by exposing recessive alleles normally present in heterozygous combinations. This approach could be particularly useful with polyploid crop plants. Populations of plants regenerated from cultures of somatic cells or from leaf protoplasts sometimes show surprising variation in morphology, growth habit and response to stress. Such variation might arise if regenerated plants come from cells that had already diverged cytologically or genetically in the original explant. Alternatively, variation may be induced *in vitro* by culture procedures, such as the exposure to high levels of 2,4-D.

Shepard and co-workers recently regenerated large numbers of Russet Burbank potato plants from leaf protoplasts, screened these plants for resistance to *Phytophthora infestans* (Shepard et al., 1980) and to an *Alternaria solani* toxin (Matern et al., 1978). They reported recovery of plant lines having resistance that remained stable after several cycles of vegetative propagation. Although the basis of the variability in the regenerated plants is not yet understood, this work clearly illustrates how culture procedures and disease screening can be separated.

Culture techniques can also facilitate *transfer* of genes for disease resistance even in the absence of *in vitro* screening methods. One simple example is *in vitro* fertilization to overcome certain types of prezygotic incompatibility. Another is the rescue via culture of embryos from wide crosses that would normally abort because of endosperm failure (e.g., James, 1979).

A more dramatic approach is use of protoplast fusion to break down barriers to transfer of disease resistance to crop plants. Fusion of protoplasts from two sexually incompatible species could be followed by screening of regenerated hybrid plants for disease resistance. This is already being done with tobacco (Evans and Flick, 1979). Protoplasts of two disease resistant wild species (*Nicotiana nesophila* and *N. stocktonii*) were fused with protoplasts from cultivated tobacco (*N. tabacum*). Hybrid plants regenerated from these interspecific fusions were identified by biochemical, morphological, and cytological methods similar to those used in analysis of other tobacco somatic hybrids (Evans et al., 1980). The hybrid plants will soon be checked for resistance to brown spot, black shank, and other diseases and pests. The point is that although some method for identifying hybrid plants is needed, the *initial* selection need not involve toxins or disease resistance. Screening of regenerated plants for disease resistance is necessary in any case, since traits expressed at the cellular level might not be expressed in mature plants. This approach has great potential and it should be applied to a variety of crop and disease symptoms.

CONCLUSIONS

Tissue culture technology can be applied to work on plant diseases in many novel and promising ways. Some of these do not require regeneration of plants and are quite feasible now. For the most rapid progress in exploiting the potential of recent cell culture techniques, many types of additional effort will be required. They include the following:

1. More work on regeneration of plants from cells and protoplasts of important crops.
2. Isolation and study of more phytotoxins.
3. Greater understanding of the biochemistry of diseases not involving toxins; this may permit development of new *in vitro* screening methods.
4. Work on identification and isolation of DNA coding for disease resistance. Such work must precede attempts to transfer resistance into protoplasts via plasmids or isolated DNA. In fact, studies of all aspects of the characterization of plant DNA, especially DNA from crop plants, are needed.
5. More work on all areas of crop improvement by unconventional means. The techniques and concepts involved in disease studies and in attempts to improve other agonomic traits are very similar.
6. Continued attention to "model systems" even if they do not use crop plants or address problems of great agricultural importance. Well-characterized systems are often the most suitable for developing new techniques and for answering basic questions about the relations of cells *in vitro* and plants regenerated from them.

Also needed is a balanced view of the importance of cell and protoplast techniques. Enthusiasm about these techniques should not obscure the fact that they are an adjunct or supplement to conventional methods, not a replacement for them. Even if culture manipulations provide us with useful new material, much additional breeding, testing, and selection will be required before new varieties are ready for agricultural use. On the other hand, inflated publicity about unconventional approaches to agriculture tends to make the results achieved so far seem almost disappointing. Unrealistic expectations and premature rejection of new techniques both must be avoided. Increased communication between plant pathologists, plant breeders, agronomists, plant physiologists and cell biologists will speed progress in applying these techniques to agricultural problems.

ACKNOWLEDGMENTS

Some of the work described was supported by grants from the Rockefeller Foundation and the National Science Foundation (PCM-7822572).

REFERENCES

Aldrich, H. C., V. E. Gracen, D. W. York, E. D. Earle, and O. C. Yoder. (1977) Ultrastructural effects of *Helminthosporium maydis* race T toxin on mitochondria of corn roots and protoplasts. *Tissue Cell* **9**:167–177.

Belliard, G., G. Pelletier, F. Vedel, and F. Quetier. (1978) Morphological characteristics and chloroplast DNA distribution in different cytoplasmic parasexual hybrids of *Nicotiana tabacum*. *Mol. Gen. Genet.* **165**:231–237.

Carlson, P. S. (1973) Methionine sulfoximine-resistant mutants of tobacco. *Science* **180**:1366–1368.

Dudits, D., G. Hadlaczky, G. Y. Bajszar, C. Koncz, G. Lazar, and G. Horvath. (1979) Plant regeneration from intergeneric cell hybrids. *Plant Sci. Lett.* **15**:101–112.

Earle, E. D. (1978) Phytotoxin studies with plant cells and protoplasts. In *Frontiers of Plant Tissue Culture* (T. E. Thorpe, ed.). International Association for Plant Tissue Culture, Calgary, Canada. Pp. 363–372.

Earle, E. D., and V. E. Gracen. (1977) Unpublished data. Department of Plant Breeding and Biometry, Cornell University, Ithaca, N.Y.

Earle, E. D., V. E. Gracen, O. C. Yoder, and K. P. Gemmill. (1978) Cytoplasm-specific effects of *Helminthosporium maydis* race T toxin on survival of corn mesophyll protoplasts. *Plant Physiol.* **61**:420–424.

Evans, D. A., and C. Flick. (1979) Personal communication. Department of Biological Sciences, SUNY, Binghamton, N.Y.

Evans, D. A., L. R. Wetter, and O. L. Gamborg. (1980) Somatic hybrid plants of *Nicotiana glauca* and *Nicotiana tabacum* obtained by protoplast fusion. *Physiol. Plant.* **48**:225–230.

Fox, D., E. D. Earle, and V. E. Gracen. (1979) Unpublished data. Department of Plant Breeding and Biometry, Cornell University, Ithaca, N.Y.

Gengenbach, B. G., and C. E. Green. (1975) Selection of T cytoplasm maize callus cultures resistant to *H. maydis* race T pathotoxin. *Crop Sci.* **16**:465–470.

Gengenbach, B. G., C. E. Green, and C. M. Donovan. (1977) Inheritance of selected pathotoxin resistance in maize plants regenerated from cell cultures. *Proc. Natl. Acad. Sci. USA* **74**:5113–5117.

Gregory, P., E. D. Earle, and V. E. Gracen. (1977) Biochemical and ultrastructural aspects of Southern corn leaf blight disease. *In* Host Plant Resistance to Pests (P. A. Hedin, ed.). American Chemical Society, Washington, D.C. Pp. 90–114.

Gregory, P., E. D. Earle, and V. E. Gracen. (1980) Effects of purified *Helminthosporium maydis* race T toxin on the structure and function of corn mitochondria and protoplasts. Plant Physiol. (in press).

Izhar, S., and J. B. Power. (1979) Somatic hybridization in *Petunia:* A male sterile cytoplasmic hybrid. *Plant Sci. Lett.* **14**:49–55.

James, J. (1979) New maize X tripsacum hybrids for maize improvement. *Euphytica* **28**:239–247.

Jones, C. W., I. A. Mastrangelo, H. H. Smith, H. Z. Liu and R. A. Meek. (1976) Interkingdom fusion between human (HeLa) cells and tobacco hybrid (GGLL) protoplasts. *Science* **193**:401–403.

Kao, K. N., and M. R. Michayluk. (1980) Plant regeneration from mesophyll protoplasts of alfalfa. *Z. Pflanzenphysiol.* **96**:135–141.

Kleinhofs, A., and R. Behki. (1977) Prospects for plant genome modification by nonconventional methods. *Annu. Rev. Genet.* **11**:79–101.

Malone, C. P., R. J. Miller, and D. J. Koeppe. (1978) The *in vivo* response of corn mitochondria to *Bipolaris* (*Helminthosporium*) *maydis* (race T) toxin. *Physiol. Plant.* **44**:21–25.

Matern, U., G. Strobel, and J. Shepard. (1978) Reactions to phytotoxins in a potato population derived from mesophyll protoplasts. *Proc. Natl. Acad. Sci. USA* **75**:4935–4937.

Melchers, G., M. D. Sacristan, and H. A. Holder. (1978) Somatic hybrid plants of potato and tomato regenerated from fused protoplasts. *Carlsberg Res. Commun.* **43**:205–218.

Pelcher, L. E., K. N. Kao, O. L. Gamborg, O. C. Yoder, and V. E. Gracen. (1975) Effect of *Helminthosporium maydis* race T toxin on protoplasts of resistant and susceptible corn (*Zea mays*). *Can. J. Bot.* **53**:427–431.

Pham, H. N. (1980) Nuclear and cytoplasmic control of resistance to *Helminthosporium maydis* race T and its toxin in corn plants, protoplasts, and mitochondria. Ph.D. Thesis, Cornell University, Ithaca, N.Y.

Pring, D. R. (1979) Personal communication. Department of Plant Pathology, University of Florida, Gainesville, Fla.

Sharp, W. R., P. O. Larsen, E. F. Paddock, and V. Raghavan (eds.). (1979) Plant Cell and Tissue Culture. Ohio State University Press, Columbus, Ohio.

Shepard, J. F., and R. E. Totten. (1977) Mesophyll cell protoplasts of potato: Isolation, proliferation, and plant regeneration. *Plant Physiol.* **60**:313–316.

Shepard, J. F., D. Bidney, and E. Shahin. (1980) Potato protoplasts in crop improvement. *Science* **208**:17–24.

Thorpe, T. E. (ed.). (1978) Frontiers of Plant Tissue Culture. Proc. 4th International Congress of Plant Tissue and Cell Culture, Calgary, Canada.

Vasil, V., and I. K. Vasil. (1980) Isolation and culture of cereal protoplasts. II. Embryogenesis and plantlet formation from protoplasts of *Pennisetum americanum*. *Theor. Appl. Genet.* **56**:97–99.

Wallin, A., K. Glimelius, and T. Eriksson. (1979) Formation of hybrid cells by transfer of nuclei via fusion of miniprotoplasts from cell lines of nitrate reductase deficient tobacco. *Z. Pflanzenphysiol.* **91**:89–94.

Walton, J. D., E. D. Earle, O. C. Yoder, and R. M. Spanswick. (1979) Reduction of adenosine triphosphate levels in susceptible maize mesophyll protoplasts by *Helminthosporium maydis* race T toxin. *Plant Physiol.* **63**:806–810.

Yoder, O. C. (1980) Toxins in pathogenesis. *Annu. Rev. Phytopathol.* **18**:103–129.

York, D. W., E. D. Earle, and V. E. Gracen. (1980) Ultrastructural effects of *Helminthosporium maydis* race T toxin on isolated corn mitochondria and mitochondria within corn protoplasts. Can. J. Bot. (in press).

Zelcer, A., D. Aviv, and E. Galun. (1978) Interspecific transfer of cytoplasmic male sterility by fusion between protoplasts of normal *Nicotiana sylvestris* and X-ray irradiated protoplasts of male-sterile *N. tabacum*. *Z. Pflanzenphysiol.* **90**:397–407.

BASIC RESEARCH FOR PLANT DISEASE MANAGEMENT

Richard C. Staples and Gary H. Toenniessen

Research directed to practical control of plant disease differs markedly from that carried out by physiologists and biochemists, who seek to find the ways that plants defend themselves from colonization by pathogens. Plant breeders have been remarkably successful in developing new varieties resistant to disease often without even knowing the identity of the pathogen (Robinson, 1976). So long as genetic variation exists within a crop, and pools of germplasm remain unexploited (Coulter, 1980), plant breeding in field plots will continue to be the procedure of choice for developing resistant crop lines.

However, basic research may prove useful to the plant breeder; for example, where barriers to breeding exist, such as intergeneric sterility, alternative procedures could provide high-performance crop lines not presently available. The contributions to this book have demonstrated that potential opportunities for exploiting new ways of managing plant disease, derived from fundamental knowledge of host and pathogen, are legion. In this chapter the authors cite a few especially promising ones which depend upon application of techniques from molecular biology, studies on resistance mechanisms, induced susceptibility, integration of plant growth, and bacteriocins.

APPLICATION OF TECHNIQUES FROM MOLECULAR BIOLOGY

Throughout the world, a range of new technology is emerging from advances made in molecular biology (Nathans, 1979; Earle and Gracen, this volume; Yoder, this volume). These techniques may provide opportunities to combine genes from a much wider range of organisms than normally possible through sexual mechanisms. Synthetic genes constructed by recombination or chemical synthesis might even be introduced. The utility of these techniques is obvious, but their application seems years away. For example, protoplasts from higher plants have yet to be transformed using plant DNA, and for many species techniques for regenerating protoplasts to complete plants have not been developed. However, where acquisition of resistance may require very rare gene combinations, protoplast techniques offer the possibility for screening large numbers of individuals without the huge burden of labor and expense required in making selections from field plots. A research effort to develop and exploit sophisticated new tools provided by molecular biology in plant pathology should be enormously useful.

STUDIES ON RESISTANCE MECHANISMS

The interaction between host and pathogen implies a recognition one of the other. Scientists who study this relationship include those who investigate the resistance response of the host when invaded by the pathogen, and those who conduct research on recognition of the correct host by the pathogen. Merger of these differing views would allow the concerns of the plant breeder to be dealt with more adequately. These breeding objectives should include the development of tolerant plants capable of yielding well, even though colonized by the pathogen (Robinson, 1976; Mussell, 1980).

The progress made in basic studies on plant resistance confirms that plants may have several defense mechanisms for use as the situation warrants. These defense mechanisms include the accumulation of fungitoxic phytoalexins (Keen, this volume), while Heath (1977, 1980) has demonstrated that nonhost plants actively block development of haustoria normally induced by hyphal contact with a host cell wall. Both phytoalexins and the "blockers" of haustorial formation require elicitors which have not been suitably identified (Albersheim and Valent, 1978; Keen, this volume; Heath, this volume). This lack of knowledge has prevented even a start on the identification of receptors which the cell is likely to employ in order to recognize and defend itself against the pathogen.

Beyond the active defense mechanisms employed by the plant, there are cell-surface phenomena which guide the development of pathogens that are closely adapted to their host plants, that is, the rusts, powdery mildews, anthracnoses, and a wide range of other fungi (Ellingboe, 1972; Wynn, 1976; Dey, 1933; Wynn and Staples, this volume). These organisms develop infection structures when their germ tubes elongate from germinating conidia and encounter features of host topography such as venations or stomatal guard cells. Pathogen recognition of the host apparently results in chemical bonding of the germ tube to the surface of the host, possibly involving the surface waxes (Ellingboe, 1972; Dickinson, 1949). Little is known about the genetic controls over development of surface features such as waxes and configuration of stomatal guard cells. Once this information becomes available, it should be possible to introduce field resistance in plants by selecting surface characteristics that inhibit pathogen infection and disease development.

A wide range of opportunities exist for directly interfering with development of fungi, much more so than for the viruses and bacteria. Among fungi, for example, one might prevent development of asexual spores in order to limit growth of the pathogen (Trione, this volume). This would reduce expression of symptoms and make the plant apparently resistant. Long-cycle rusts, which require an alternate host to complete their life cycles, are poorly understood with respect to the basis for the specificity involved in the requirement for alternate hosts. Better knowledge of alternate host resistance would provide usable technology for disease control. In a related way, smuts (*Ustilago* spp.) undergo a dikaryon to monokaryon transition. The dikaryons are pathogens, while after conversion to monokaryons they become saprophytes. The smuts are especially unstable in this sense, and an effort should be made to exploit this conversion in order to prevent dikaryons from infecting plants.

INDUCED SUSCEPTIBILITY

Certain toxins secreted by parasites are known to be required for all or part of the molecular events that lead to symptom expression in their respective hosts, yet for many of these toxins, the chemical structures and primary sites of action are unknown (Yoder, 1980). Despite this lack of knowledge, it has been possible to use some of these toxins to select for disease resistance. Toxins are particularly useful in this regard because they are effective at the whole plant level, at the level of cultured callus tissue, or at the level of isolated protoplasts.

Uncertainty concerning toxin structure has prevented toxin synthesis as well as identification of the receptor mechanisms which appear to reside in membranes. The biosynthetic pathways for the toxins likewise remain obscure as does a knowledge of pathway control points (Yoder, this volume).

INTEGRATION OF PLANT GROWTH

A group of rather simple chemical hormones controls the sequence of growth in higher plants. They might be exploited for pathogen control if we knew more about how they worked (Pegg, this volume). Even the mode of action of the conceptually uncomplicated indoleacetic acid, which has been studied for more than 50 years, remains a mystery. The growth hormone, abscissic acid, is involved in the adaptive responses of various plant organs to environmental stress such as drought. This suggests that continued research on the chemical control of plant growth will provide fruitful procedures for the control of other stresses including plant diseases. We will need detailed information about the processes which control plant growth before we will have reached a point where such research can be applied.

BACTERIOCINS

Bacteriocins, a group of chemical substances produced by bacteria, are quite toxic to other bacteria that are susceptible (Cuppels and Kelman, unpublished information; Vidaver, 1976). Although the host range of known bacteriocins is very specific, research may lead to broader range bacteriocins useful for control of bacterial plant pathogens. For example, the bacteriocin "Agrocin 84" provides a highly effective form of control of crown gall caused by *Agrobacterium tumefaciens* (Moore and Warren, 1979).

SUMMARY

In looking over the field of host–parasite relations, one is struck by the enormous biological complexity faced by researchers. Elements of this complex biology include biochemical modes of action in pathogen–host recognition, control of plant growth, genetics of toxin production, and reproduction by fungi. All of these research subjects are exceedingly difficult. However, recent advances in understanding animal and microbial cell systems lead one to expect that intensive research into plant cell systems will yield results. The present shift of some molecular biologists from dealing with phages and bacteria toward studying the development of eukaryotes is one hopeful sign that new technology will become available to break present technical barriers. This hungry world needs vastly more efficient crops than are now available. Varieties of food crops are needed that are resistant against pests, that yield adequately with low fertilizer and energy inputs, and that are adapted to land impacted by a broad array of environmental and biotic stresses. The current trend toward a plateau in production of our major crops (Wittwer, 1979) suggests that basic innovations in plant protection must supplement and support the traditional procedures of plant breeding.

REFERENCES

Albersheim, P. and B. S. Valent. (1978) Host–pathogen interactions in plants. Plants when exposed to oligosaccharides of fungal origin defend themselves by accumulating antibiotics. *J. Cell. Biol.* **78**:627–643.

Coulter, J. (1980) Crop Improvement. In *Linking Research to Crop Production.* (R. C. Staples and R. J. Kuhr, eds.) Plenum Press, New York. Pp. 35–49.

Dey, P. K. (1933) Studies in the physiology of the appressorium of *Colletotrichum gloesporioides. Ann. Bot.* **47**:305–312.

Dickinson, S. (1949). Studies in the physiology of obligate parasitism. I. The stimuli determining the direction of growth of the germ tubes of rust and mildew spores. *Ann. Bot. (London) (N.S.)* **13**:89–104.

Ellingboe, A. H. (1972) Genetics and physiology of primary infection by *Erysiphe graminis. Phytopathology* **62**:401–406.

Heath, M. C. (1977) A comparative study of non-host interactions with rust fungi. *Physiol. Plant Pathol.* **10**:73–88.

Heath, M. C. (1980) Effects of infection by compatible species or injection of tissue extracts on the susceptibility of nonhost plants to rust fungi. *Phytopathology* **70**:356–360.

Moore, L. W. and G. Warren. (1979) *Agrobacterium radiobacter* strain 84 and biological control of crown gall. *Annu. Rev. Phytopathol.* **17**:163–179.

Mussell, H. (1980) Tolerance to disease. In *Plant Disease-An Advanced Treatise*, Vol. 5 (J. G. Horsfall and E. B. Cowling, eds.) Academic Press, New York, Pp 39–54.

Nathans, D. (1979) Restriction endonuclease, simian virus 40, and the new genetics. *Science* **206**:903–909.

Robinson, R. A. (1976) *Plant Pathosystems.* Springer-Verlag, New York.

Vidaver, A. K. (1976) Prospects for control of phytopathogenic bacteria by bacteriophages and bacteriocins. *Annu. Rev. Phytopathol.* **14**:451–465.

Wittwer, S. H. (1979) Future technological advances in agriculture and their impact on the regulatory environment. *BioScience* **29**:603–610.

Wynn, W. K. (1976) Appressorium formation over stomates by the bean rust fungus: Response to a surface contact stimulus. *Phytopathology* **66**:136–146.

Yoder, O. C. (1980) Toxins in pathogenesis. *Annu. Rev. Phytopathol.* **18**:103–129.

AUTHOR INDEX

SUBJECT INDEX

Phosphatase, of cell wall, role in vulnerability, 280

CHEMICAL CONCEPTS IN POLLUTANT BEHAVIOR
Ian J. Tinsley

RESOURCE RECOVERY AND RECYCLING
A. F. M. Barton

QUANTITATIVE TOXICOLOGY
V.A. Filov, A.A. Golubev, E.I. Liublina, and N.A. Tolokontsev

ATMOSPHERIC MOTION AND AIR POLLUTION
Richard A. Dobbins

INDUSTRIAL POLLUTION CONTROL—Volume I: Agro-Industries
E. Joe Middlebrooks

BREEDING PLANTS RESISTANT TO INSECTS
Fowden G. Maxwell and Peter Jennings, Editors

NEW TECHNOLOGY OF PEST CONTROL
Carl B. Huffaker, Editor

THE SCIENCE OF 2,4,5-T AND ASSOCIATED PHENOXY HERBICIDES
Rodney W. Bovey and Alvin L. Young

INDUSTRIAL LOCATION AND AIR QUALITY CONTROL: A Planning Approach
Jean-Michel Guldmann and Daniel Shefer

PLANT DISEASE CONTROL: RESISTANCE AND SUSCEPTIBILITY
Richard C. Staples and Gary H. Toenniessen, Editors

AQUATIC POLLUTION
Edward A. Laws

MODELING WASTEWATER RENOVATION: Land Treatment
I. K. Iskandar